W0107651

Gong · Lu · Wang · Yang (Eds.)

International Symposium in Memory of Hua Loo Keng

Volume I Number Theory

Gong Sheng Lu Qi-keng
Wang Yuan Yang Lo (Eds.)

International Symposium in Memory of Hua Loo Keng

Volume I Number Theory

Springer-Verlag Berlin Heidelberg GmbH

Gong Sheng Lu Qi-keng
Wang Yuan Yang Lo (Eds.)

This proceedings volume serves as the 21st Volume in the series in Pure and Applied Mathematics published by Science Press, Beijing, China.

Mathematics Subject Classification (1980): 10BXX 10CXX 10GXX

ISBN 978-3-662-07983-6 ISBN 978-3-662-07981-2 (eBook)
DOI 10.1007/978-3-662-07981-2

This work is subject to copyright. All rights are reserved, whether the whole or part of the material is concemed, specifically the rights of translation, reprinting, re-use of illustrations, recitation, broadcasting, reproduction on microfilms or in other ways, and storage in data banks. Duplication of this publication or parts thereof is only permitted under the provisions of the German Copyright Law of September 9, 1965, in its current version and a copyright fee must always be paid. Violations fall under the prosecution act of the German Copyright Law.

© Springer-Verlag Berlin Heidelberg 1991
Originally published by Springer-Verlag Berlin Heidelberg New York in 1991
Softcover reprint of the hardcover 1st edition 1991

41/3140-543210

Preface

The international symposium on number theory and analysis in memory of the late famous Chinese mathematician Prof. Hua Loo Keng was co-sponsored by the Institute of Mathematics, Academia Sinica and the University of Science and Technology of China. It took place between August 1st and 7th of 1988 on the campus of Tsing Hua University, and some 150 mathematicians were present. The symposium was carried out in two separate sections: number theory and analysis. This is reflected in the publication of a set of two volumes, the first one on Number Theory edited by Professor Wang Yuan and the second on Analysis by Professors Gong Sheng, Lu Qi-keng and Yang Lo.

The distinguished list of main speakers and the contents of these two volumes reflect the high level of the mathematical activity throughout the seven days. We pay special tribute to our main speakers professors Chuang, Conn, Ding, Drasin, Fitzgerald, Gaier, Gong, Grauert, Gu, Hejhal, Iyanaga, Karatsuba, Koranyi, Liao, Lu, Pan, Richert, Satake, Schmidt, Siu, Tatuzawa, Tsang, Vladimirov, Y. Wang, G. Y. Wang, Wüstholz and Yang, who gave the excellent one hour lectures, and also to the participants who gave contributed talks on their own research work. The discussions among the mathematicians were always in a warm atmosphere. Our thanks go to professors Chern, Subbarao and Yau for their contributions to these proceedings.

We wish to express our thanks to the following institutions for their generous financial support: The Division of Mathematics, Physics and Chemistry of Academia Sinica, the University of Science and Technology of China, the Institute of Mathematics, Academia Sinica, and the Chinese National Natural Science Foundation. We also received support from the Third World Academy of Sciences for the international travelling expenses of Hua Loo Keng's students who were in foreign countries, and of the mathematicians of the third-world countries. We are also grateful to Springer-Verlag and Science Press (Beijing) for their help in preparing of this publication, and to the Tsing Hua University for providing lodging, lecture rooms and other facilities.

Gong Sheng, Lu Qi-keng,
Wang Yuan, Yang Lo

Contents

Contents of Volume II

Hua Loo Keng: A Brief Outline of his Life and Works

WANG Yuan
Institute of Mathematics, Academia Sinica, Beijing, China

Hua Loo Keng was born on 12 November, 1910 in Jintan county in the south of Jiangsu Province of China. His father managed a small family grocery store, and they were too poor to allow him to enrol in senior middle school when he graduated from Jintan junior middle school (the equivalent of the first three years of high school). He then attended the Shanghai Chung-Hua Vocational School where he completed one and a half years of its two-year accounting course. He was forced by the poor circumstances of his family to leave school at the age of fifteen and return home to help his father in the family shop. He could only learn mathematics from books in his spare time. Very soon, he became so interested in his mathematics that he could not pay full attention to the shop. His father was unhappy with him and often threatened to burn his books.

In 1927 Hua got a job as a clerk in his Jintan middle school and married Wu Xiao Yuan, a Jintan girl. They had a daughter the following year, and there are also three sons and two more daughters, the youngest being born in 1951. In 1928 Hua was struck by typhoid, followed by arthritis which burdened him for life with a lame left leg.

Hua showed mathematical talent in his early life. His first article appeared in the journal *Science* (Shanghai) in 1929. His second paper "On the incorrectness of Su Jia Ju's paper" appeared in the same journal the following year. This paper was noticed by C. L. Hsiung, the chairman of the Mathematics Department of Tsing Hua University in Beijing, but, of course, Hsiung had never heard of anyone called Hua Loo Keng. Later P. C. Tang, a Jintan born teacher in the Mathematics Department, informed Hsiung that Hua was not even a middle school graduate, but a mere clerk in a small village. Hsiung was very much impressed and he invited Hua to Tsing Hua University. Hua then started work as a clerk in Tsing Hua's Mathematics Department in 1931, and was appointed departmental assistant the following year. He was promoted to the rank of lecturer and made a research fellow of the China Cultural Foundation in 1934. During this period his contemporaries at Tsing Hua who were to become distinguished mathematicians were S. S. Chern, P. L. Hsu and Chao Ko. Hua's initial research interest was in Waring's problem in number theory, in which he was encouraged by Professor W. Z. Yang, who obtained his Ph.D. from the University of Chicago under the supervision of L. E. Dickson.

In 1936 Hua went to Cambridge, England at the invitation of G. H. Hardy, to whom Hua had been recommended by N. Wiener. Although Hardy himself was in the United States when Hua arrived in England, he soon acquired an adequate command of English to allow him to be acquainted with several young mathematicians, among whom were H. Davenport, T. Estermann, R. A. Rankin and E. C. Titchmarsh. With at least fifteen papers written during his Cambridge period it is obvious that he benefited much from these mathematicians who remained his life long friends.

In July 1937 Japan invaded China and Tsing Hua, together with Peking and Nanking universities, had to be evacuated to Kunming in Yunnan Province where they formed the Southwest Associated University of China, and Hua returned from Cambridge to become a professor there from 1938 to 1945. His research interests had now broadened to include the geometry of matrices, automorphic functions, functions of several complex variables and group theory. He initiated a seminar program in the 1940s, and those who participated included Tuan Hsio Fu, Min Szu Hoa, Fan Ky and Shu Shien Siu who all became well known mathematicians.

In 1946, at the invitation of the Soviet Academy of Sciences and the Soviet International Cultural Association, Hua visited the Soviet Union for three months and he met I. M. Vinogradov and Yu. V. Linnik.

From 1947 to 1948 Hua was a visiting member of the Institute for Advanced Studies, Princeton, and he taught number theory at Princeton University. During 1948-1950 Hua was a professor at the University of Illinois at Urbana-Champaign where he supervised the research of several students (including R. Ayoub, J. Mitchell and L. Schoenfeld). Besides number theory he also worked in equations over finite fields, classical groups and field theory.

In 1950 Hua returned to China with his wife and children and took part in the preparatory work for the establishment of the Institute of Mathematics, Academia Sinica, since the former institute had been moved to Taiwan with the Nationalist government of Chiang Kai Shek. He was appointed director of the Institute when it opened in 1952, and he immediately took charge of the reconstruction program for the Institute which was to consist of sections in pure mathematics, applied mathematics and also computing techniques. He paid special attention to the training of young mathematicians, among whom were Chen Jing Run, Pan Cheng Dong and Wang Yuan in number theory, Wan Zhe Xian in algebra and Kung Sheng and Lu Qi Keng in complex analysis. For their benefit, and also for other Chinese mathematicians, he embarked on the writing of a series of books: *Additive Prime Number Theory* (Chinese edition, 1957), *Introduction to Number Theory* (Chinese edition 1957), *Harmonic Analysis of Functions of Several Complex Variables in the Classical Domains* (1959), *The Estimation of Exponential Sums with Applications to Number Theory* (Chinese edition, 1963), *Introduction to Higher Mathematics* (1963) and *Classical Groups* (1963, with Wan Zhe Xian).

Hua was subjected to harassing interrogation during the "Cultural Revolution" in 1966. His house was searched by the "Red Guards" and many of his manuscripts were confiscated and irretrievably lost. Fortunately, due to the personal intervention and protection of Mao Ze Dong and Zhou En Lai, his own situation became better in 1967. He could stay safely at home and even travel inside China to popularise mathematical methods in industrial departments.

In 1958 Hua was appointed Vice-Rector of the Chinese University of Science and Technology, and began work on applied mathematics, particularly on the application of number theory to numerical integration in multi-dimensional space. He also worked on popularising the "Optimum seeking method" (Fibonacci search) and the "Critical path method" in Chinese factories and industry departments. For nearly twenty years he and his assistants Chen De Quen and Ji Lei travelled to over twenty provinces, gave lectures to workers and taught them how to use the above two methods in their work. As a result the output of factories was increased and the quality of goods produced was improved.

The "Cultural Revolution" came to an end in 1976 and China adopted an open policy in 1979. With the aid of his students, Hua published two more books: *Applications of Number Theory to Numerical Analysis* (1978, with Wang Yuan) and *Starting with the Unit Circle* (1979). His *Selected Papers,* edited by H. Halberstam, was published by Springer-Verlag in 1983. Hua was Vice-President of the Chinese Academy of Sciences in 1978 and Director of the Institute of Applied Mathematics in 1980. He was elected President of the Chinese Mathematical Society, holding the position from 1950 to 1983.

Following the adoption of the open policy in China, Hua received honorary doctorates from the University of Nancy (1980), the Chinese University of Hong Kong (1983) and the University of Illinois (1984). He became a foreign associate of the U.S. Academy of Sciences (1983) and the Deutsche Akademie der Naturforscher Leopoldina. He was elected a member of the Third World Academy of Sciences in 1983. Although Hua was in poor health he continued to work in mathematics and its applications. He was permitted to travel abroad as a scholar and he visited Europe, the United States and Japan several times. He died on 12 June, 1985 of a heart attack at the end of his lecture at Tokyo University, Japan.

Mathematical Works

1. NUMBER THEORY

1) ESTIMATION OF TRIGONOMETRIC SUMS

Let q be a positive integer and $f(x)$ be a polynomial with integer coefficients

$$f(x) = a_k x^k + \cdots + a_1 x,$$

where $(a_k, \ldots, a_1, q) = 1$. Consider the complete exponential sum

$$S(q, f(x)) = \sum_{x=1}^{q} e\left(\frac{f(x)}{q}\right), \qquad e(x) = e^{2\pi i x}.$$

If $f(x) = x^2$, then $S(q, x^2)$ is the well known sum of Gauss who proved that

$$|S(q, x^2)| = q^{1/2}.$$

The problem of estimating $S(q, f(x))$ has a long history, and it had only been tackled for special polynomials until Hua came along. More specifically he proved in a characteristically elegant way that

$$|S(q, f(x))| = O(q^{1-1/k+\epsilon}), \tag{1}$$

where ϵ is any pre-assigned positive number, and the implied constant depends only on k and ϵ. This was given in a paper [49A] in 1940, while in an earlier paper [38] he had obtained the result with the implied constant depending also on the coefficients of $f(x)$. It is easy to see that (1) is best possible apart from the improvement of the factor q^ϵ; for example, $|S(p^k, x^k)| = p^{k-1}$ when p is prime not dividing k. Later Hua [100] generalised (1) to any algebraic number field K of degree n.

For incomplete exponential sums, Hua [119] proved that

$$\sum_{x=1}^{P} e\left(\frac{hx^k}{q}\right) = \frac{P}{q} \sum_{x=1}^{q} e\left(\frac{hx^k}{q}\right) + O(q^{1/2+\epsilon}), \tag{2}$$

a result with an important application to Waring's problem.

Hua simplified and improved on Vinogradov's method for the estimation of Weyl's sums by pointing out that the essence of the method is the following mean value theorem [95]: Let $f(x) = \alpha_k x^k + \ldots + \alpha_1 x$ and

$$C_k = C_k(P) = \sum_{x=a+1}^{a+P} e(f(x)).$$

If $t_1 - t_1(k) \geq k(k+1)/4 + \ell k$, then

$$\int_0^1 \cdots \int_0^1 |C_k|^{2t_1} d\alpha_1 \cdots d\alpha_k \leq (Ft_1)^{4t_1\ell} P^{2t_1 - \frac{1}{2}k(k+1)+\delta} (\log P)^{2\ell},$$

where $\delta = \frac{1}{2}k(k+1)(1-1/k)^\ell$. From this one derives immediately the following theorem: Suppose that $k \geq 12$, $2 \leq r \leq k$ and

$$\left|\alpha_r - \frac{h}{q}\right| \leq \frac{1}{q^2}, \qquad (h,q) = 1, \quad 1 \leq q \leq P^r.$$

Then, for $P \leq q \leq P^{r-1}$, we have

$$S = \sum_{x=1}^{P} e(f(x)) \ll P^{1-1/\sigma_k+\epsilon},$$

where $\sigma_k = 2k^2(2\log k + \log\log k + 3)$.

In all current monographs on analytic number theory, Vinogradov's method is stated according to Hua's formulation; see, for example, *The Hardy-Littlewood Method,* R. C. Vaughan, Cambridge tract in Mathematics, **80,** 1981.

Concerning character sums, Hua [61] proved in 1942 that the estimate

$$\frac{1}{A+1}\left|\sum_{a=1}^{A} \sum_{n=-a}^{a} \chi(n)\right| \leq \sqrt{p} - \frac{A+1}{\sqrt{p}}, \qquad 1 \leq A < p$$

holds for all non-principal characters χ modulo p. This led to the improvement of the estimation of the least primitive root modulo p, and of the least solution of Pell's equation [62].

2) WARING'S PROBLEMS AND RELATED PROBLEMS

In 1770 Waring conjectured that, for any integer $k \geq 2$, there exists an integer $s = s(k)$ depending only on k such that every positive integer can be expressed as a sum of s k-th powers of non-negative integers. Waring's conjecture was proved by Hilbert in 1900. In the 1920s Hardy and Littlewood created and developed a very powerful new analytic method in additive number theory, the so-called "circle method". The method allows one to give much more precise results on Waring's problem. Let $G(k)$ denote the least integer s such that every sufficiently large integer N can be represented by

$$N = x_1^k + \cdots + x_s^k, \tag{3}$$

where $x_1, \ldots, x_s$ are non-negative integers. Hardy and Littlewood established an asymptotic formula for the number $r_{s,k}(N)$ of solutions of (3), namely:

$$r_{s,k}(N) \sim \mathfrak{S}(N) \frac{\Gamma^s(1+1/k)}{\Gamma(s/k)} N^{s/k-1} \tag{4}$$

when $s \geq (k-2)2^{k-1}+5$. Here $\mathfrak{S}(N)$ is the so-called singular series which has a positive lower bound independing of N, and therefore $G(k) \leq (k-2)2^{k-1}+5$. This result of Hardy and Littlewood was improved by Hua [36] in 1938 to

$$G(k) \leq 2^k + 1,$$

and he also proved that the asymptotic formula for $r_{s,k}(N)$ holds for $s \geq 2^k+1$. For this he used what has come to be called Hua's inequality:

$$\int_0^1 \left| \sum_{x=1}^{P} e(\alpha x^k) \right|^{2^k} d\alpha \ll P^{2^k-k+\epsilon}. \tag{5}$$

However, Vinogradov (see Vinogradov [5]) gave a great improvement on the results of Hardy and Littlewood on Waring's problem for large values of k. He proved that (4) holds for $s \geq 10k^2 \log k$, $k > 10$, and Hua improved this to $s \geq 2k^2(\log k + \log\log k + 2.5)$, $k > 10$. It may be of interest to remark that, for small k, Hua's condition $s \geq 2^k + 1$ for (4) was improved only in recent years by R. C. Vanghan and D. R. Heath-Brown to $s \geq 2^k$ $(k \geq 3)$ and $s \geq \frac{7}{8}2^k+1$ $(k \geq 6)$ respectively. A new elementary proof of Hilbert's theorem, based on Schnirelman's method on the density of a sequence of natural numbers, was given in 1943 by Yu. V. Linnik. Commenting on this proof H. Davenport [1] wrote "The underlying ideas of this proof were undoubtedly suggested by certain features of the Hardy-Littlewood method, and in particular by Hua's inequality".

In the 1930s many mathematicians studied the generalisation of Waring's problem by replacing x^k with a polynomial of degree k. The main difficulty in the generalisation was removed by the estimate (1) of Hua, who gave a very general formulation of the Hilbert theorem. Let $f_i(x)$ $(1 \leq i \leq s)$ by any given s integral valued polynomials of degree k with positive leading coefficients. Between 1937 and 1940, Hua proved that the number of solutions of the equation

$$N = f_1(x_1) + \cdots + f_s(x_s)$$

has an asymptotic formula for $s \geq 2^k + 1$ when $1 \leq k \leq 10$, and for $s \geq 2k^2(\log k + \log\log k + 2.5)$ when $k > 10$. Note that the positivity of the corresponding singular series has not been considered. Let $f(x)$ be an integral valued polynomial and let $G(f)$ denote the least integer s such that

$$N = f(x_1) + \cdots + f(x_s)$$

is solvable for all sufficiently large integer N. Let $\partial^0 f$ denote the degree of f.

Hua established in 1940 that $G(f|\partial^0 f = 3) \le 8$, $G(f|\partial^0 f = k) \le (k-1)2^{k-1}$ and $\max_f G(f) \ge 2^k - 1$ for $k \ge 5$, with $f(x)$ running over all integral valued polynomials of degree k ([30, 45, 47]).

The main idea of the circle method can be sketched as follows. The number of solutions of (3) may be represented by an integral in the form

$$r_{s,k}(N) = \int_0^1 T(\alpha)^s e(-N\alpha)\,d\alpha, \qquad T(\alpha)\sum_{x=1}^{P} e(\alpha x^k),$$

where $P = [N^{1/k}]$. Divide the unit interval (0,1] into two parts, the major arcs and the minor arcs having unions $\mathfrak{M}$ and $\mathfrak{m}$ respectively. Roughly speaking, $\mathfrak{M}$ consists of disjoint intervals centering on rational numbers h/q with small denominators q, and $\mathfrak{m}$ is made up of the remaining intervals. Hardy and Littlewood proved that the asymptotic formula

$$\int_{\mathfrak{M}} T(\alpha)^s e(-N\alpha)\,d\alpha \sim \mathfrak{S}(N)\frac{\Gamma^s(1+1/k)}{\Gamma(s/k)}N^{s/k-1}$$

holds for $s \ge 2k+1$. Hua improved this to $s \ge k+1$ in 1957 [120]; this result is best possible, and the proof is based on his estimate (2).

In the 1930s and 1940s, Hua went on to study systematically the so-called Waring-Goldbach problem. This is concerned with the solvability of (3) and its generalisations in which the variable x_i are restricted to prime numbers. For example, he proved that the number of solutions of the equation

$$N = p_1^k + \cdots + p_s^k$$

has an asymptotic formula for $s \ge 2^k + 1$ when $k \le 10$, and for $s \ge 2k^2(\log k \log\log k + 2.5)$ when $k > 10$. He also obtained analogues for prime arguments of many of the results mentioned above. The fruits of his researches are given in the well known monograph *Additive Prime Number Theory.*

TARRY'S PROBLEM

Let $N(k)$ denote the least integer t such that the system of equations

$$x_1^h + \cdots + x_t^h = y_1^h + \cdots + y_t^h, \qquad 1 \le h \le k \tag{6}$$

has a non-trivial solution, that is a solution with $x_1, \ldots, x_t, y_1, \ldots, y_t$ being positive integers, but $x_1, \ldots, x_t$ is not a permutation of $y_1, \ldots, y_t$. Let $M(k)$ be the least integer t such that (6) is solvable, and

$$x_1^{k+1} + \cdots + x_t^{k+1} \ne y_1^{k+1} + \cdots + y_t^{k+1}.$$

Evidently we have $k+1 \le N(k) \le M(k)$. In 1938 Hua [37] proved in an elementary and straightforward way that, for $k \ge 12$,

$$M(k) \le (k+1)\left(\left[\frac{\log \frac{1}{2}(k+2)}{\log(1+1/k)}\right]+1\right) \sim k^2 \log k$$

which is an improvement on an earlier result $M(k) < 7k^2(k-11)(k+3)/216$ due to E. M. Wright.

It was pointed out by Hua [104] that Vinogradov's method may be used to tackle Tarry's problem, and he used the method to obtain the following result in 1952. Let t_0 be given by the table

k	2	3	4	5	6	7	8	9	10	$k > 10$
t_0	3	8	23	55	120	207	336	540	885	$[k^2(3\log k + \log\log k + 9]$

and let $R_{k,t}(P)$ denote the number of solutions of (6) satisfying $1 \le x_i, y_i \le P$, $1 \le i \le t$. Then

$$\lim_{P\to\infty} P^{1/2k(k+1)-2t} R_{k,t}(P) = c(k,t)$$

holds for $t > t_0$ with the positive constant $c(k,t)$ depending only on k and t.

3) OTHER CONTRIBUTIONS TO NUMBER THEORY

Let $q(n)$ be the number of partitions of a positive n into unequal parts, or into odd parts. In 1942, by modifying the Hardy-Ramanujan method with Rademacher's Farey dissection of finite order, Hua [60] obtained the exact formula

$$q(n) = \frac{1}{\sqrt{2}} \sum_{\substack{k=1 \\ k \text{ odd}}}^{\infty} \sum_{\substack{(h,k)=1 \\ 0<h\le k}} \omega_{h,k} e\left(-\frac{hn}{k}\right) \frac{d}{dn} J_0\left(\frac{i\pi}{k}\left(\frac{2}{3}(n+1/24)\right)^{1/2}\right),$$

where $J_0(x)$ is Bessel's function of order 0.

Let $A(x)$ denote the number of lattice points (u,v) in the circle $u^2+v^2 \le x$. The circle problem of Gauss is the determination of the least θ such that the asymptotic formula

$$A(x) = \pi x + O(x^{\theta+\epsilon})$$

holds for every positive ϵ, with the implied constant depending on ϵ, and Gauss himself established that $\theta = 1/2$ is admissible. E. C. Titchmarsh applied the circle method to obtain $\theta = 15/46$, and in 1942 Hua [63] refined the technique to give $\theta = 13/40$. He also pointed out a gap in Vinogradov's work on the same problem.

Hua also work on the problem of the Euclidean algorithm (EA) in the real quadratic number field $\mathbb{Q}(\sqrt{d})$ where d is a square free number. He [64] proved in 1944 that there is no (EA) if $d > e^{250}$, and pointed out that 250 could be reduced to 160.

In 1959 Hua and Wang Yuan [131] wrote a note on numerical integration. They showed that if $\partial^2 f(x,y)/\partial x\, \partial y$ is continuous over $0 \leq x, y \leq 1$, then

$$\int_0^1 \int_0^1 f(x,y)\,dx\,dy - \frac{1}{F_n} \sum_{k=1}^{F_n} f\left(\frac{k}{F_n}, \frac{kF_{n-1}}{F_n}\right) \leq \frac{C \log 3F_n}{F_n}, \qquad (7)$$

where F_n is the n-th Finbonacci number. The estimate (7) is best possible apart from the improvement of the constant C. Then, in a series of papers, they generalised their method to higher dimensional cases. It is known that the ratios F_{n-1}/F_n are the best rational approximations to $(\sqrt{5}-1)/2 = 2\cos 2\pi/5$, and their method hinges on the construction of simultaneous approximations to a basis of the cyclotomic field $\mathbb{Q}(2\cos 2\pi/m)$ $(m \geq 5)$ by means of a given set of units. Their work and other contributions in this area, including the numerical information for dimensions 2 to 18, are described in their monograph *Applications of Number Theory to Numerical Analysis.*

During 1953–1957, Hua organised a seminar program on Goldbach's Problem at the Institute of Mathematics, Academia Sinica. Goldbach's problem was raised in a letter from Goldbach to Euler in which two conjectures were proposed: (A) *Every even number greater than 5 is a sum of two odd primes*, and (B) *Every odd number greater than 8 is a sum of three odd primes.* Obviously (B) is a consequence of (A). The conjecture (B) was essentially solved by Vinogradov in 1937 when he proved that all sufficiently large odd numbers are indeed sums of three odd primes. A. A. Buchstab proved that every sufficiently large even number is a sum of two numbers, each being a product of at most 4 primes, and we shall write this as (4,4) for simplicity. As fruits of the seminar program Wang first improved Buchstab's result to (3,4) in 1956, then pushed on to obtain (2,3) in 1957, while Pan Cheng Dong established (1,4) in 1963. Later Chen Jing Run established (1,2), which is the best result of this kind on the binary Golbach problem, apart from the conjecture itself, of course.

2. ALGEBRA AND GEOMETRY

4) SFIELDS

Since Hamilton's first example of non-commutative division algebra, the quaternion algebra and division algebra have received a great deal of attention. By comparison, infinite dimensional division algebras and sfields were neglected. Hua came onto the scene around 1950 and proved several theorems in this area by direct and elementary methods.

Let K be a sfield. A mapping $a \to a^\sigma$ of K onto itself is called a semi-automorphism if it satisfies

$$(a+b)^\sigma = a^\sigma + b^\sigma, \qquad (aba)^\sigma = a^\sigma b^\sigma a^\sigma \quad \text{and} \quad 1^\sigma = 1.$$

The well known examples of semi-automorphisms are automorphisms, which satisfy $(ab)^\sigma = a^\sigma b^\sigma$, and anti-automorphisms which satisfy $(ab)^\sigma = b^\sigma a^\sigma$. An outstanding problem was whether there exists a semi-automorphism which is neither an automorphism nor an anti-automorphism. Hua [8] settled this problem in 1949 by proving that every semi-automorphism is either an automorphism or an anti-automorphism. The fundamental theorem of projective geometry on a line over a sfield of characteristics $\neq 2$, namely, any one-to-one mapping carrying the projective line over a sfield of characterstics $\neq 2$ onto itself and keeping harmonic relations invariant is a semi-linear transformation induced by an automorphism or an anti-automorphism, was thus derived.

Formerly Ancochea and Kaplansky treated the problem concerning semi-automorphisms under some restrictions. Both of their methods were rooted in the structure of linear algebra and so neither of them could be extended to the general case. In 1949, Hua [71] gave a straightforward proof of the following theorem: Every proper normal subfield of a sfield is contained in its centre. This result is now called the Cartan-Brauer-Hua theorem in the literature. The earlier proof of R. Brauer and H. Cartan used the complicated devise of Galois extensions over subfields, whereas Hua's proof requires only an identity: If $ab \neq ba$, then

$$a = \left(b^{-1} - (a-1)^{-1}b^{-1}(a-1)\right)\left(a^{-1}b^{-1}a - (a-1)^{-1}b^{-1}(a-1)\right)^{-1}.$$

In 1950, Hua [96] proved that "if a sfield is not a field, then its multiplicative group is not meta-abelian".

5) GROUP THEORY AND GEOMETRY OF MATRICES

In 1946, Hua [73] published his first paper on automorphisms of classical groups, in which he determined the automorphisms of a real symplectic group. Subsequently, in 1948, he [85] determined the automorphisms of symplectic groups over any field of characteristic not equal to 2. Hua's method for the determination of the automorphisms of symplectic groups can be applied also to classical groups of other types; but since Dieudonné had already published his results on the automorphisms of classical groups in 1951, Hua [101] restricted himself to publishing only solutions, by his own method, to a series of problems left open by Dieudonné. Hua [101] determined the automorphisms of

$GL_2(K)$, $SL_4(K)$ and $PSL_4(K)$, where K is a sfield of charactersitics not equal to 2, and the automorphism of $O_4^+(K, f)$, where K is a field of charateristic not equal to 2, and f is a quadratic form of index 2. Later, Hua and Wan [105] determined the automorphisms of $SL_2(K)$ and $PSL_2(K)$, where K is a sfield of characteristic $\neq 0$, the automorphisms of $SL_4(K)$ and $PSL_4(K)$, where K is a sfield of characteristic 2, and they also established the nonisomorphism of certain linear groups. In [101] Hua wrote on the comparison of Diendonné's method and his own method as follows: "Dieudonné adopted a method which worked smoothly for n, and treated individually the cases with small n. As the author mentioned before, the difficulty increases as n diminishes; Dieudonné's method becomes very clumsy for smaller n. On the other hand the author's method starts with least possible n, which is usually the most difficult case. Therefore the reader will have little difficulty in extending the special results of this paper to the general case by meas of inductive method used in [85]. Moreover, in contrast with that of Dieudonné, the author's method uses only the calculus of matrices."

In 1951, Hua and Reiner [102, 106] determined the automorphisms of $GL_n(\mathbb{Z})$ and $PGL_n(\mathbb{Z})$ which was the start of the work on the automorphisms of classical groups over rings. They [92] also proved that $GL_n(\mathbb{Z})$ is generated by three elements, $SL_n(\mathbb{Z})$ by two elements, and $SP_{2n}(\mathbb{Z})$ by four elements for $n \geq 2$. Earlier Brahana had proved this by showing that every element of $SP_{2n}(\mathbb{Z})$ is expressible as a product of matrices taken from some finite set of matrices.

In 1940, Hua and Tuan [46] introduced the concept of the rank of a p-group. A p-group of order p^n is said to be of rank α if the maximum of the orders of its elements is $p^{n-\alpha}$. He proved, for instance, that if G is a group of order p^n and rank α ($p \geq 3$, $n \geq 2\alpha+1$), then (i) G contains one and only one subgroup of order p^n and rank α, ($2\alpha+1 \leq m \leq n$); (ii) G contains P^n cyclic subgroups of order p^n, ($\alpha < m < n-\alpha-1$); (iii) the number of elements of order $\leq p^m$, ($\alpha \leq m \leq n-\alpha$) in G is equal to $p^{m+\alpha}$. The second and third results improved on theorems of G. A. Miller and A. A. Kulakoff respectively.

The study of the geometry of matrices was initiated by Hua in 1945 and it is related to Siegel's work on fractional linear transformations. The points of the space are matrices of a certain kind, for instance, rectangular matrices, symmetric matrices or skew-symmetric matrices of the same size. There is then a group of motions in this space, and the problem is to characterise the group of motions by as few geometric invariants as possible. He discovered that the invariant "coherence" is alone sufficient to characterise the group of motions of the space. He [99] proved in 1951 the fundamental theorem of affine geometry of rectangular matrices: Let $1 < n \leq m$. Then the one-to-one mappings from the set of $n \times m$ matrices over a sfield K onto itself preserving coherence (two matrices M and N are said to be coherent if the rank of $M - N$ is 1) is necessarily of the form

$$Z_1 = PZ^{\sigma}Q + R, \tag{8}$$

where $P = P^{(n)}$ and $Q = Q^{(m)}$ are invertible matrices, R is an $n \times n$ matrix and σ is an automorphism of K; if $n = m$, then besides (8) we also have

$$Z_1 = PZ'^{\tau}Q + R,$$

where τ is an anti-automorphism of K. From this theorem he deduced the fundamental theorem of the projective geometry of rectangular matrices (the Grassmann space), and he determined the Jordan isomorphism of total matrix rings over sfields of characteristic $\neq 2$, and the Lie isomorphism of total matrix rings over sfields of characteristic $\neq 2, 3$.

The study of the geometry of matrices is very close to his research on the theory of functions of several complex variables. This urged him to study the classification problem of matrices. In 1944-46, Hua [66, 76] determined the classifications of complex symmetric and skew-symmetric matrices under the unitary group, of a pair of Hermitian matrices under congruence, and of Hermitian matrices under the orthogonal group.

In 1955, Hua organised a seminar program on algebra at the Institute of Mathematics. He and Wan Zhe Xian published a joint book *Classical Groups* which contains their results on classical groups and related problems. Besides those fields which were studied by Hua, the participants of the seminar also obtained many results in some other parts of algebra and its applications, in particular, in algebraic coding theory.

3. COMPLEX ANALYSIS

6) CLASSICAL DOMAINS

In 1935, E. Cartan proved that they are precisely six types of irreducible homogeneous bounded symmetric domains for analytic mappings. Two of these are exceptional in that they occur only in dimensions 16 and 27. The other four are the so-called "classical domains" which are defined as follows:

$$\begin{aligned}
\Re_I &= \{m \times n \text{ matrices } Z \text{ satisfying } I_m - ZZ^\star > 0\},\\
\Re_{II} &= \{\text{symmetric matrices } Z \text{ of order } n \text{ satisfying } I_n - ZZ^\star > 0\},\\
\Re_{III} &= \{\text{skew symmetric matrices } Z \text{ of order } n \text{ satisfying } I_n - ZZ^\star > 0\},\\
\Re_{IV} &= \{z = (z_1, \ldots, z_n) \in \mathbf{C}^n;\ |zz'|^2 + 1 - 2\bar{z}z' > 0, |\bar{z}z'| < 1\},
\end{aligned}$$

where the elements of Z are complex variables, I_m denotes the identity matrix of order m and $Z^\star$ is the complex conjugate of the transpose Z' of Z, while z' is the transpose of vector z.

Classical domain may be regarded as the higher-dimensional analogues of the unit disk and other domains in the complex plane. The theory of classical domains also has applications to differential equations and complex geometry.

In 1943, Siegel published his important paper on symplectic geometry, where the domain $\Re_{II}$ was studied by matrix method. In 1944, Hua [66] pointed out that the study of classical domains can be reduced to the study of the theory of geometry of matrices. He gave the matrix representations of four classical domains and their motion groups independent of Cartan and Siegel. In his paper, he gave only a brief account of those results which were related to Siegel's work. There is an editorial note to [66]: "Because (of) the poor mail between the U.S. and China, a number of minor changes in this paper have been made here, with the consent of the editors, by Prof. Hua's friend Hsio Fu Tuan and Prof. C. L. Siegel". Hua also expressed his sincere thanks in his paper to Prefessors H. Weyl, P. C. Tang and S. S. Chern for sending him the related papers by Siegel, G. Giraud and E. Cartan.

In 1953, using group representation method, Hua obtained an orthonormal system for each of the four classical domains. Roughly speaking, it is similar to the orthonormal system $e(n\theta)$ $(n = 0, \pm 1, \ldots)$ in the complex plane, by which one can easily obtain the Cauchy kernel for the unit disk. Thus Hua obtained the Cauchy kernel, Szegö kernel, Bergman kernel and Poisson kernel for the four types of classical domains with the aid of their orthonormal systems. Hua's method is distinguished by its concrete nature together with his direct style. His confidence in the being able to carry out difficult computations also comes across very clearly. Using the Cauchy kernel of a classical domain, the values of an anlytic function are determined in a classical domain if the values of the function are given in a lower dimensional manifold (characteristic manifold) of its boundary. There is an exposition of these and other results in his monograph *Harmonic analysis of Functions of Several Complex Variables in the Classical Domains*. The editor of its English edition stressed the importance of the subject matter for the representation theory of Lie groups, the theory of homogeneous spaces and the theory of automorphic functions of several complex variables. A noteworthy feature of the book is the technical machinary developed by Hua, for example, a class of algebraic identities, or the computation of integrals of functions of a matrix variable, which are of independent interest.

Using the Poisson kernel of classical domains, Hua and Q. K. Lu [128] established in 1958 the theory of harmonic functions for classical domains and solved the corresponding Dirichlet's problem for the Laplace Beltrami equation. They found some strange phenomena: (i) if a function satisfies a differential equation, it must satisfy a system of differential equations, and (ii) the Dirichlet's problem is solved if the values of the function are given in a lower dimensional manifold (characteristic manifold) of the boundary of a classical domain. Hua found here a system of differntial operators which have the similar properties of a harmonic operator. This is now called Hua's operator in the literature. Hua also studied the boundary properties of classical domains, their geometrical structures, maximum principle and proved an analogue of Schwarz's lemma for bounded domains in $\mathbf{C}^n$.

Since some classical groups may be regarded as the characteristic manifolds of classical domains, Hua [139] proved that the Fourier series on unitary group is Abel summable. This is the start of the research on Fourier analysis on classical groups. Hua's result was entended widely by Gong Sheng's work. Gong studied Abel summation, Cesàro summation, Fejér summation and various kinds of spherical summation of Fourier series on unitary groups. J. Q. Zhong extended some of the results on Fourier series on unitary group to the group of rotations.

In 1954, using elementary method, Hua [122] proved that the Riemann curvature R in the Bergman metric of a bounded domain has the following properties: (i) $2 - R$ is a sum of squares, and (ii) $R \geq -n$ under cetain restrictions. This gives an improvement on a result due to Fuchs. As a generalization of Riemann's theorem on conformal mapping, Hua proved that every noncontinuable bounded domain with constant curvature can be carried onto a unit sphere by means of an analytic mapping.

Hua was led by his work on functions of several complex variables to study some problems on partial differential equations. The result of his researches and joint work are described in his book [8] and in book [10] which was written in collaboration with Z. Q. Hu and W. Lin.

References

[1] H. Davenport, "Analytic methods for diophantine equations and diophantine inequalities", *Ann Arbor publishers*, 1962, 8.

[2] H. Halberstam, "Loo-Keng Hua: Obituary", *Acta Arith*, to appear.

[3] S. Kung and K. H. Look (S. Gong and K. Q. Lu), "Function Theory", *Loo Keng Hua; Selected papers*, Springer Verlag, 1983, 634–635.

[4] S. Salaff, "A biography of Hua Lo-Keng", *Science and Technology in East Asia, Sivin Nathan (ed)*, Watson Acad. Publ. Inc., 1977.

[5] Z. X. Wan, "Algebra and geometry", *Loo Keng Hua; Selected papers*, Springer Verlag, 1983, 281–284.

Professor Hua Loo Keng in Japan

S. IYANAGA

It is a great honour and pleasure for me to be invited to this International Symposium on Number Theory and Analysis dedicated to the memory of the late Professor Hua Loo Keng and to be given an opportunity to give a talk before you. Personally, I wonder if I am really qualified to be given this honour. This is a scientific symposium in which a talk with high scientific value should be esteemed above all. I should be very happy, if I could speak here of my findings developing Professor Hua's ideas to attain some new results on, say, the Goldbach-Waring problem or on the circle problem. My main field of research is number theory and I appreciate highly the contributions of Professor Hua and his school to these and other subjects. But I have worked in algebraic rather than in analytic number theory, and could not contribute to this latter domain. Besides, as I was born four years earlier than Professor Hua, I am no longer at an age of mathematical productivity. I hope you will excuse me in these circumstances for speaking on personal rather than on scientific matters.

In speaking of personal recollections, however, it is obvious that many of you who have carried out research by his side have much more to say than me. Professor Hua spent a considerable length of time in the Soviet Union, the United States, England, France and other European countries. He had friends in each of these countries, many of whom are attending this meeting. He also visited Japan several times, and although he did not spend a very long time in our country, he made a very deep impressions on us. Particularly, we cannot forget that he died in Tokyo just after he gave a talk at the University of Tokyo in the afternoon of 12 June, 1985.

Before speaking of these events, I should like to say something about the influence of Professor Hua's works on Japanese mathematics. If one could compile, with the aid of computers, a complete list of citations of Hua's works in papers written by Japanese mathematicians, it would make an enormous list. Unfortunately, I have not had enough time to make such a list, and I shall mention only a few of these.

In the last part of Suetuna's book on *Analytic Number Theory* (1950) which has long been known as a standard text-book in this domain in Japan, one finds Hua's result on the circle problem cited. Compared with the wide range of Hua's methods in his arithmetical works, the content of Suetuna's book was limited to special Dirichlet series, Riemann's or Dedekind's zeta-functions, Hecke's or Artin's L-functions. It uses, however, algebraic number theory and obtains general results. But it recognizes the superiority of Hua's results on the circle problem. After Suetuna, a Japanese school of analytic number theory was formed by Tutuzawa, Mitsui and other mathematicians, whose works have much in common with those of Hua's school, and Hua's results are often citied in their works.

One of Suetuna's disciples, Yoshikazu Eda, was attracted by Hua's book on *Applications of Number Theory in Numerical Analysis* (in collaboration with Professor Wang Yuan). He read it in the Chinese original soon after its publication in 1979 and translated it into Japanese. His manuscript in Japanese translation was reprinted and circulated among people interested. But an English translation of the same book was published in 1981, which hindered the Japanese translation by Eda from finding a publishing firm. Hua's books: *Additive Theory of Prime Numbers* and *Introduction to Number Theory* are mostly read in Japan in English translations. His voluminous *Selected Works* which consists of papers originally written in English, are also widely read in our country.

Hua's works on functions of matrices and on the geometry of classical domains are also well known. They are quoted in the papers by Suagawara, Morita and Satake. In the *Encyclopedic Dictionary of Mathematics* compiled by the Mathematical Society of Japan, Hua's name is quoted in the following six articles: Additive Number Theory, Lattice Point Problem, Arithmetical Functions, Discontinuous Groups and Partition of Numbers.

This does not exhaust, however, the domains of Hua's mathematical activities. In addition to the fact that he was one of the most important leaders of mathematical research in China, he was a great educator and popularizer. His last talk at the University of Tokyo was on this point and I shall speak a little about this later. Besides the domains mentioned so far, there are also topics on which he made minor but fine contributions. For example, he wrote an ingenious note on automorphisms of skew fields in 1949 while he was in Urbana, Illinois, and I had an opportunity to present it with admiration in my course on geometry at the University of Tokyo.

Although I have known him by name since the 1930s, I met him for the first time in the 1970s when he came to Japan as a member of a group on a good will visit to promote friendly relations between our two countries. The aim of such groups was not just for mathematics or science. Meetings with Japanese mathematicians were organized each time, but we could not get into deep scientific conversations. I was, however, very much impressed by his genial personality.

During these visits, I am sure he made acqaintance with a good number of Japanese personalities, including of course mathematicians. But I believe that one must count Mrs. Fumiko Shiratori among those Japanese mathematicians with whom he enjoyed the best of friendship. Mrs. Shiratori was a young student of the women's Higher Normal School in Tokyo in the 1930s (This school was renamed the Ochanomizu Women's University after the Second World War.), and her interest was in algebraic number theory. She was presented to me by Professor Kuroda who taught at the school, but we lost contact until I saw her again at the funeral of Professor Kuroda in 1972. She told me that she had been in China for more than twenty years and had continued to study mathematics by herself. However, she had the good fortune to get acquainted with Professor Hua. She had become very familiar with Chinese life and the Chinese language during her long stay, and I believe she has contributed a great deal to the mutual understanding between Chinese and Japanese scientists.

I myself visited Beijing in 1975 as a member of a scientific delegation from Japan. Although I was fortunate in being able to meet Professors Tuan Hsio-Fu, Wu Wen-Tsün and other mathematicians I was disappointed in not being able to see Professor Hua in Beijing, especially because Mrs. Shiratori had asked me to convey her best wishes to him.

One day in 1980 Mrs. Shiratori telephoned me from Beijing saying that Professor Hua was about to go to the United States, but he was in need of a certain oxygen inhaler for his heart condition, and I was asked if I could supply a good Japanese model of the apparatus which he could then collect in transit at Tokyo Airport. I am happy to report that with the help of Mr Higo, the secretary of the Association of Scientific and Technical Exchange between China and Japan, we were able to provide the service on that occasion.

In June 1985 Professor Hua was invited, as the leader of the Chinese delegation of a research group in optimization and econometrics, by the Association of Exchange between Japan and other Asian countries, and he was accompanied by Professors Chen De Quan, Ji Lei and Xu Xin Hong. On the morning of June 12, he visited the Japan Academy where I was able to welcome him together with Professors Kenjiro Kimura and Kosaku Yosida. He chatted cheerfully with us and made a point in showing me the oxygen inhaler which he kept in his car. He arrived at the University of Tokyo in the afternoon and was received by Professor Hikosaburo Komatsu, President of the Mathematical Society of Japan, Mrs. Shiratori, Professors Kosaku Yosida, Tatuzawa, Morimoto and others. Unfortunately, I myself was busy and could not attend this lecture, and I only found out from others what had happened, particularly from the report by Morimoto which appeared in the journal "Sugaku Seminar".

At four o'clock Professor Komatsu, the President, presented the speaker who was to lecture on "Some personal experiences on the popularization of mathematical methods in the People's Republic of China" to the packed audience in Room 103. Professor Hua was to speak in Chinese and was to be interpreted, but he soon asked the President if he could speak in English in order to dispense with the interpreter. With his good command of English the lecture suddenly came to life. He first showed a sheet of paper on which there were only two words: "Theory" and "Popularisation", and he then told us what he did in the 1950s in these two domains. On "Theory" he wrote the books: *Harmonic Analysis of Functions of Several Complex Variables in the Classical Domains,* and *Introduction to Number Theory.* He was happy to see Wan Zhe Xian and Lu Qi Keng developed his ideas in the domain of the first of these books, while Wang Yuan and Chen Jin Run did work on the second . On "Popularization" he introduced the Mathematics Olympiad in China and initiated the study of optimization. He spoke next on the application of number theory to numerical analysis, the subject of the book translated by Eda that I mentioned earlier. For example, if (F_n) is the sequence of Fibonacci numbers and $f(x, y)$ is a doubly periodic continuous function with (1,1) as period, then one obtains a good approximation:

$$\int_0^1 \int_0^1 f(x,y)\, dx\, dy \sim \frac{1}{F_{n+1}} \sum_{t=1}^{F_{n+1}} f\left(\frac{t}{F_{n+1}}, \frac{tF_n}{F_{n+1}}\right).$$

He also spoke on a method of estimating the area on a map, the use of non-negative matrices in econometrics and the golden ratio. He even made an experiment using tobacco on the last topic. The lecture lasted ten more minutes than the given hour, and the whole audience was charmed by his delightful presentation. Almost as soon as Professor Komatsu asked the audience if there were any questions, the speaker fell on his back. Although he was attended at once with the oxygen inhaler by Mrs. Ke Xioa-Ying, his daughter-in-law who was a doctor, and also by doctors from the university hospital Professor Hua passed away by 10 o'clock that evening at the age of 75 years.

The journal "Sugaku Seminar" published some months later several articles on Professor Hua, and Morimoto gave a report of his last talk. Tatuzawa wrote about his mathematical works and Mrs. Shiratori contributed an article on his memory. She wrote among other things that Professor Hua, for some years, had expressed his strong desire to visit Japan and to have more contact with Japanese mathematicians and the people. It is also reported that he had always said that he wished to work until his last day.

In recalling of our deep sorrow at this sad event I now see that this International Symposium dedicated to his memory is attended by his famous disciples representing today's Chinese mathematics and by his friends and admirers from all over the world. I would like to pay tribute to the illustrious memory of Professor Hua and present my congratulations to the organizers of this Symposium.

On the Least Prime in an Arithmetical Progression and Theorems Concerning the Zeros of Dirichlet's L-Functions (V)

Chen Jingrun & Liu Jianmin

Institute of Mathematics, Academia Sinica, Beijing, China

1. Introduction

Let D be a large positive integer, $(K, D)=1$, and $P(D, K)$ the least prime $p \equiv K \pmod{D}$. In 1934, S. Chowla conjectured that $P(D, K) \ll D^{1+\varepsilon}$. Three years later, P. Turan proved that under the General Riemann Conjecture, the Chowla's conjecture may hold for almost all modulo D. On the other hand, in 1949, Erdös obtained that, first, there is a constant number $C_2 = C_2(C_1)$ and an infinity of integer D, such that $P(D, K) > (1+C_1)\varphi(D)\log D$ for K's value being at least $C_2\varphi(D)$; second, there is a constant $C_4 = C_4(C_3)$, such that $P(D, K) \leqslant C_3\varphi(D)\log D$ for K's value of $C_4\varphi(D)$.

In 1944, Linnik gave a very important work. He proved that $P(D, K) \ll D^c$, where c is an absolutely constant number. Later, in 1954, Rodosskii simplified the proof of Linnik's.

The following list marks the steps in the history of the upper bound of $P(D, K)$:

Pan Chengdong [1] (1958), $C=5448$;
Chen Jingrun [2] (1964), $C=770$;
M. Jutila [3—4] (1970), $C=630, 550$;
Chen Jingrun [5] (1977), $C=168$;
M. Jutila [6] (1977), $C=80$;
S. Graham [7] (1977), $C=36$;
Chen Jingrun [9] (1979), $C=17$;
S. Graham [8] (1981), $C=20$;
Wang Wei [10] (1986), $C=16$;
Chen Jingrun & Liu Jianmin [11] (1989), $C=13.5$.

In this paper we obtain $C=11.5$.

2. Some Lemmas

Lemma 1. *For* $j=1, 2$, *let* $\rho_j = 1 - \frac{\beta_j}{\log D} + i\gamma_j$ *be a zero of* $L(s, \chi_1)$, *where* χ_1 *is a non-principal character mod* D, $0.6 \leqslant \beta_j \leqslant 0.9$, $|\gamma_1 - \gamma_2| \leqslant \frac{1}{\log D}$. *For* $k=3, 4$, *let* $\rho_k = 1 - \frac{\beta_k}{\log D} + i\gamma_k$ *be a zero of* $L(s, \chi)$, *where* χ *is a complex character mod* D, $0.6 \leqslant \beta_k \leqslant 0.9$, $|\gamma_3 - \gamma_4| \leqslant \frac{1}{\log D}$. *If* $\chi_1 \neq \chi$, $\chi_1 \neq \bar{\chi}$, *then*

$$\max_{1 \leqslant j \leqslant 4} \beta_j \geqslant 0.89.$$

Proof. Let $\sigma = 1 + \frac{a}{\log D}$, where a will be chosen later. Define

$$s(\chi_1, \chi, \chi; \rho_1, \rho_3, \rho_4) = -\mathrm{Re}\left\{\frac{\zeta'}{\zeta}(\sigma) + \frac{L'}{L}(\sigma + i\gamma_1, \chi_1) + \frac{L'}{L}(\sigma + i\gamma_3, \chi)\right.$$
$$+ \frac{L'}{L}(\sigma + i\gamma_4, \chi) + \frac{L'}{L}(\sigma + i\gamma_1 + i\gamma_3, \chi_1\chi) + \frac{L'}{L}(\sigma + i\gamma_1 + i\gamma_4, \chi_1\chi)$$
$$\left. + \frac{L'}{L}(\sigma + i\gamma_3 + i\gamma_4, \chi^2)\right\}.$$

We have $0 \leqslant s(\chi_1, \chi, \chi; \rho_1, \rho_3, \rho_4) + s(\chi_1, \chi, \bar{\chi}; \rho_1, \rho_3, \bar{\rho}_4) + s(\chi_1, \bar{\chi}, \chi; \rho_1, \bar{\rho}_3, \rho_4) + s(\chi_1, \bar{\chi}, \bar{\chi}; \rho_1, \bar{\rho}_3, \bar{\rho}_4)$.

By the method used in Lemma 1 of [11], this completes the proof.

Lemma 2. *For* $j=1, 2$, *let* $\rho_j = 1 - \frac{\beta_j}{\log D} + i\gamma_j$ *be a zero of* $L(s, \chi)$, *where* χ *is a complex character mod* D, $0.6 \leqslant \beta_j \leqslant 0.9$, $|\gamma_1 - \gamma_2| \leqslant \frac{1}{\log D}$. *For* $k=3, 4$, *let* $\rho_k = 1 - \frac{\beta_k}{\log D} + i\gamma_k$ *be a zero of* $L(s, \chi)$, *where* χ *is a complex character mod* D, $0.6 \leqslant \beta_k \leqslant 0.9$, $|\gamma_3 - \gamma_4| \leqslant \frac{1}{\log D}$. *If* $\bar{\rho}_1 \neq \rho_3$, $\bar{\rho}_1 \neq \rho_4$, $\bar{\rho}_2 \neq \rho_3$, $\bar{\rho}_2 \neq \rho_4$, *then*

$$\max_{1 \leqslant j \leqslant 4} \beta_j \geqslant 0.89.$$

Proof. See the method used in Lemma 2 of [11].

Lemma 3. *For* $j=1, 2, 3, 4$, *let* $\rho_j = 1 - \frac{\beta_j}{\log D} + i\gamma_j$ *be a zero of* $L(s, \chi)$, *where* χ *is a complex character mod* D, $0.6 \leqslant \beta_j \leqslant 0.9$, $|\gamma_j| \leqslant 1$, $|\gamma_1 - \gamma_2| \leqslant \frac{1}{\log D}$, $|\gamma_3 - \gamma_4| \leqslant \frac{1}{\log D}$. *Then* $\max_{1 \leqslant j \leqslant 4} \beta_j \geqslant 0.89$.

Proof. See the method used in Lemma 2 of [11].

Lemma 4. *Let* $\rho_1 = 1 - \frac{\beta_1}{\log D} + i\gamma_1$ *be a zero of* $L(s, \chi_1)$, *where* χ_1 *is a real non-principal character mod* D, $0.6 \leqslant \beta_1 \leqslant 0.9$, $0 < |\gamma_1| \leqslant \frac{0.5}{\log D}$. *And let* $\rho_2 = 1 - \frac{\beta_2}{\log D} + i\gamma_2$ *be a zero of* $L(s, \chi)$, *where* χ *is a real non-principal character mod* D, $0.6 \leqslant \beta_2 \leqslant 0.9$, $0 < |\gamma_2| \leqslant \frac{0.5}{\log D}$. *If* $\chi \neq \chi_1$, *then*

$$\max_{1 \leqslant j \leqslant 2} \beta_j \geqslant 0.89.$$

Proof. See the method used in Lemma 5 of [11].

Lemma 5. *For* $j = 1, 2$, *let* $\rho_j = 1 - \frac{\beta_j}{\log D} + i\gamma_j$ *be a zero of* $L(s, \chi)$, *where* χ *is a real non-principal character mod* D, $0.6 \leqslant \beta_j \leqslant 0.9$, $0 < |\gamma_j| \leqslant \frac{0.5}{\log D}$. *If* $\bar{\rho}_1 \neq \rho_2$, *then*

$$\max_{j=1,2} \beta_j \geqslant 0.89.$$

Proof. See the method used in Lemma 5. of [11].

Lemma 6. *For* $j = 1, 2$, *let* $\rho_j = 1 - \frac{\beta_j}{\log D}$ *be a zero of* $L(s, \chi_1)$, *where* χ_1 *is a real non-principal character mod* D, $0 \leqslant \beta_j \leqslant 0.9$, *and* $\rho_3 = 1 - \frac{\beta_3}{\log D} + i\gamma_3$ *be a zero of* $L(s, \chi)$, *where* χ *is a real non-principal character mod* D, $0.6 \leqslant \beta_3 \leqslant 0.9$, $0 < |\gamma_3| \leqslant \frac{0.5}{\log D}$. *If* $\chi \neq \chi_1$, *then*

$$\max_{j=1,2,3} \beta_j \geqslant 0.89.$$

Proof. See the method used in Lemma 5 of [11].

Lemma 7. *For* $j = 1, 2$, *let* $\rho_j = 1 - \frac{\beta_j}{\log D}$ *be a zero of* $L(s, \chi)$, *where* χ *is a real non-principal character mod* D, $0 < \beta_j \leqslant 0.9$, *and* $\rho_3 = 1 - \frac{\beta_3}{\log D} + i\gamma_3$ *be a zero of* $L(s, \chi)$, *where* χ *is a real non-principal character mod* D, $0.6 \leqslant \beta_3 \leqslant 0.9$, $0 < |\gamma_3| \leqslant \frac{0.5}{\log D}$. *Then*

$$\max_{j=1,2,3} \beta_j \geqslant 0.89.$$

Proof. See the method used in Lemma 5 of [11].

Lemma 8. *For* $j=1,2$, *let* $\rho_j=1-\dfrac{\beta_j}{\log D}$ *be a zero of* $L(s,\chi)$, *where* χ *is a real non-principal character mod* D, $0.6\leqslant\beta_j\leqslant0.9$. *For* $k=3,4$ *let* $\rho_k=1-\dfrac{\beta_k}{\log D}$ *be a zero of* $L(s,\chi_1)$, *where* χ_1 *is a real nonprincipal character mod* D, $0\leqslant\beta_k\leqslant0.9$. *Then*

$$\max_{1\leqslant j\leqslant4}\beta_j\geqslant0.89.$$

Proof. Let $\sigma=1+\dfrac{a}{\log D}$, where a will be chosen later. If $\chi\neq\chi_1$, then

$$0\leqslant\frac{1}{a}-\frac{4}{a+\beta_4}+0.27641\ (3).$$

If $\chi=\chi_1$, then

$$0\leqslant\frac{1}{a}-\frac{4}{a+\beta_4}+0.27641.$$

We take $a=1$, this completes the proof.

Lemma 9. *Let* $\rho_1=1-\dfrac{\beta_1}{\log D}+i\gamma_1$ *be a zero of* $L(s,\chi_1)$, *where* χ_1 *is a non-principal character mod* D, $0.1\leqslant\beta_1\leqslant0.73$, $|\gamma_1|\leqslant1$. *For* $j=2,3$, *let* $\rho_j=1-\dfrac{\beta_j}{\log D}+i\gamma_j$ *be a zero of* $L(s,\chi)$, *where* χ *is a complex character mod* D, $0.1\leqslant\beta_j\leqslant0.73$, $|\gamma_j|\leqslant1$, $|\gamma_2-\gamma_3|\leqslant\dfrac{1.194}{\log D}$. *If* $\chi_1\neq\chi$, $\chi_1\chi\neq\chi_0$, *then*

$$\max_{1\leqslant j\leqslant3}\beta_j\geqslant\min\left(0.65,\ \frac{2}{2.42813-\dfrac{1}{0.73+\beta_1}}-0.73\right).$$

Proof. By the method used in Lemma 1 of [11], we get

$$0\leqslant\frac{1}{a}-\frac{1}{a+\beta_1}-\frac{2}{a+\beta_2}+0.27641(6.5)-\frac{0.5(a+\beta_1)}{(a+\beta_1)^2+d_1^2}-\frac{a+\beta_2}{(a+\beta_2)^2+d_1^2}$$
$$-\frac{0.5a(a^2-\beta_3^2)}{(a^2+d_1^2)((a+\beta_3)^2+d_1^2)}-\frac{d_1^2}{(a^2+d_1^2)}\frac{(0.5a+\beta_3)}{((a+\beta_3)^2+d_1^2)},\qquad(2.1)$$

where a will be chosen later. $d_1=(\gamma_3-\gamma_2)\log D$, $\beta_3\leqslant a$.

Suppose that $\beta_3\leqslant0.65$, we take $a=0.73$ in (2.1). If $|d_1|\leqslant0.65$, we have

$$0\leqslant3.14841-\frac{1}{0.73+\beta_1}-\frac{2}{0.73+\beta_3}-\frac{0.5(0.73+\beta_1)}{(0.73+\beta_1)^2+0.4225}-\frac{0.73+\beta_2}{(0.73+\beta_2)^2+0.4225}.\qquad(2.2)$$

Define $A(x) = -\dfrac{0.73+x}{(0.73+x)^2+0.4225}$, we obtain

$$A(x) \leqslant A(0.65)$$

for $0.1 \leqslant x \leqslant 0.65$. By (2.2), it follows

$$0 \leqslant 2.25882 - \frac{1}{0.73+\beta_1} - \frac{2}{0.73+\beta_3}, \tag{2.3}$$

If $0.65 \leqslant |d_1| \leqslant 0.9$, we have

$$0 \leqslant 3.16653 - \frac{1}{0.73+\beta_1} - \frac{2}{0.73+\beta_3} - \frac{0.5(0.73+\beta_1)}{(0.73+\beta_1)^2+0.81} - \frac{0.73+\beta_2}{(0.73+\beta_2)^2+0.81}$$

$$- \frac{0.365((0.73)^2-0.4225)}{((0.73)^2+0.81)((0.73+0.65)^2+0.81)} - \frac{(0.65)^2}{((0.73)^2+(0.65)^2)}$$

$$\frac{0.365+\beta_3}{((0.73+\beta_3)^2+0.81)} \tag{2.4}$$

Define $A_1(x) = -\dfrac{0.365+x}{(0.73+x)^2+0.81}$, for $0.1 \leqslant x \leqslant 0.65$. We obtain $A_1(x) \leqslant \max(A_1(0.1), A_1(0.65))$. Define $B(x) = -\dfrac{0.73+x}{(0.73+x)^2+0.81}$, for $0.1 \leqslant x \leqslant 0.65$. We have $B(x) \leqslant \max(B(0.1), B(0.65))$. It follows

$$0 \leqslant 2.25573 - \frac{1}{0.73+\beta_1} - \frac{2}{0.73+\beta_3}. \tag{2.5}$$

Follow the above way, we can prove

$$0 \leqslant 2.42813 - \frac{1}{0.73+\beta_1} - \frac{2}{0.73+\beta_3} \tag{2.6}$$

if $0.9 \leqslant |d_1| \leqslant 0.97$, $0.97 \leqslant |d_1| \leqslant 1$ or $1 \leqslant |d_1| \leqslant 1.194$.

By (2.3)—(2.6), this completes this proof.

Lemma 10. *Let* $\rho_1 = 1 - \dfrac{\beta_1}{\log D} + i\gamma_1$ *be a zero of* $L(s, \overline{\chi})$, *where* χ *is a complex character mod* D, $|\gamma_1| \leqslant 1$. *For* $j = 2, 3$, *let* $\rho_j = 1 - \dfrac{\beta_j}{\log D} + i\gamma_j$ *be a zero of* $L(s, \chi)$, *where* χ *is a complex character mod* D, $|\gamma_j| \leqslant 1$, $0.1 \leqslant \beta_j \leqslant 0.73$, $|\gamma_2 - \gamma_3| \leqslant \dfrac{1.194}{\log D}$. *If* $\rho_1 \neq \overline{\rho}_2$, $\rho_1 \neq \overline{\rho}_3$, *then*

$$\max_{1 \leqslant j \leqslant 3} \beta_j \geqslant \min\left(0.65, \frac{2}{2.42813 - \dfrac{1}{0.73+\beta_1}} - 0.73\right).$$

Proof. See the method used in Lemma 2 of [11].

Lemma 11. *For* $j=1, 2, 3$, $\rho_j = 1 - \frac{\beta_j}{\log D} + i\gamma_j$ *be a zero of* $L(s, \chi)$, *where* χ *is a complex character mod* D, $|\gamma_j| \leqslant 1$, $0.1 \leqslant \beta_j \leqslant 0.73$, $|\gamma_2 - \gamma_3| \leqslant \frac{1.194}{\log D}$. *If* $\rho_1 \neq \rho_2$, $\rho_1 \neq \rho_3$, *then*

$$\max_{1\leqslant j\leqslant 3} \beta_j \geqslant \min\left(0.65, \frac{2}{2.42813 - \frac{1}{0.73+\beta_1}} - 0.73\right).$$

Proof. See the method used in Lemma 3 of [11].

Lemma 12. *Let* $\rho_1 = 1 - \frac{\beta_1}{\log D} + i\gamma_1$ *be a zero of* $L(s, \chi_1)$, *where* χ_1 *is a non-principal character mod* D, $|\gamma_1| \leqslant 1$, $0.1 \leqslant \beta_1 \leqslant 0.73$, *and let* $\rho_2 = 1 - \frac{\beta_2}{\log D} + i\gamma_2$ *be a zero of* $L(s, \chi)$, *where* χ *is a real non-principal character mod* D, $0.1 \leqslant \beta_2 \leqslant 0.73$, $0 < |\gamma_2| \leqslant \frac{0.597}{\log D}$. *If* $\chi_1 \neq \chi$, *then*

$$\max_{j=1,2} \beta_j \geqslant \min\left(0.65, \frac{1.73}{2.24 - \frac{1.248225}{0.856+\beta_1}} - 0.856\right).$$

Proof. See the method used in Lemma 5 of [11].

Lemma 13. *For* $j=1, 2$, *let* $\rho_j = 1 - \frac{\beta_j}{\log D} + i\gamma_j$ *be a zero of* $L(s, \chi)$, *where* χ *is a non-principal character mod* D, $|\gamma_j| \leqslant 1$, $0.1 \leqslant \beta_j \leqslant 0.73$, $0 < |\gamma_2| \leqslant \frac{0.597}{\log D}$. *If* $\bar{\rho}_1 \neq \rho_2$, *then*

$$\max_{j=1,2} \beta_j \geqslant \min\left(0.65, \frac{1.73}{2.24 - \frac{1.248225}{0.856+\beta_1}} - 0.856\right).$$

Proof. See the method used in Lemma 6 of [11].

Lemma 14. *For* $j=2, 3$, $\rho_j = 1 - \frac{\beta_j}{\log D}$ *be a zero of* $L(s, \chi)$, *where* χ *is a real non-principal character mod* D, $0.1 \leqslant \beta_j \leqslant 1$. *Let* $\rho_1 = 1 - \frac{\beta_1}{\log D} + i\gamma_1$ *be a zero of* $L(s, \chi_1)$, *where* χ_1 *is a non-principal character mod* D, $0.1 \leqslant \beta_1 \leqslant 1$, $|\gamma_1| \leqslant 1$. *If* $\chi_1\chi \neq \chi_0$, $\chi_1 \neq \chi$, *then*

$$\max_{1\leqslant j\leqslant 3}\beta_j\geqslant\max\left(0.646,\ \frac{2}{1.941-\dfrac{1}{0.9+\beta_1}}-0.9\right).$$

Proof. By a result (cf. [9] Lemma 10), we have

$$0\leqslant 0.82921+\frac{1}{a}-\frac{1}{a+\beta_1}-\frac{2}{a+\beta_3},$$

if $a=0.9$, then

$$\max_{1\leqslant j\leqslant 3}\beta_j\geqslant\frac{2}{1.941-\dfrac{1}{0.9+\beta_1}}\quad\text{and}\ \geqslant 0.646.$$

This completes the proof.

Lemma 15. *Let* $\rho_1=1-\dfrac{\beta_1}{\log D}+i\gamma_1$, *be a zero of* $L(s,\chi)$, *and for* $j=2,3$, *let* $\rho_j=1-\dfrac{\beta_j}{\log D}$ *be a zero of* $L(s,\chi)$, *where* χ *is a non-principal character mod* D, $0.1\leqslant\beta_1\leqslant 1$, $|\gamma_1|\leqslant 1$, $0.1\leqslant\beta_j\leqslant 1$. *Then*

$$\max_{1\leqslant j\leqslant 3}\beta_j\geqslant\max\left(0.646,\ \frac{2}{1.941-\dfrac{1}{0.9+\beta_1}}-0.9\right).$$

Proof. By the proof (cf. [9] Lemma 12), we get

$$0\leqslant 0.55281+\frac{1}{a}-\frac{1}{a+\beta_1}-\frac{2}{a+\beta_3},$$

where $\beta_3\leqslant a$, Take $a=0.9$, this completes the proof.

Lemma 16. *For* $j=1,2$, *let* $\rho_j=1-\dfrac{\beta_j}{\log D}+i\gamma_j$ *be a zero of* $L(s,\chi_j)$, *where* χ_j *is a non-principal character mod* D, $|\gamma_j|\leqslant 1$. *Assume* $\beta_2\geqslant\beta_1$, *then*

a) *If* $\chi_1\neq\bar{\chi}_2$, $\chi_1\neq\chi_2$, $\chi_1^2\neq\chi_0$, *then*

$$\beta_2\geqslant\min\left(\frac{0.1849}{\beta_1},\ \max\left(0.23469,\ \frac{1.2569}{4.51-\dfrac{1.74}{0.43+\beta_1}}-0.43\right)\right).$$

b) *If* $\chi_1\neq\bar{\chi}_2$, $\chi_1\neq\chi_2$, $\chi_1^2=\chi_0$, $\gamma_1\neq 0$, *then*

$$\beta_2\geqslant\min\left(\frac{0.2209}{\beta_1},\ \max\left(0.258,\ \frac{1.6}{3.76331-\dfrac{1.14}{0.47+\beta_1}}-0.47\right)\right).$$

c) *If* $\chi_1 \neq \overline{\chi}_2$, $\chi_1 \neq \chi_2$, $\chi_2^2 = \chi_0$, $\gamma_2 \neq 0$, *then*

$$\beta_2 \geqslant \min\left(\frac{0.2209}{\beta_1}, \max\left(0.258, \frac{1.14}{3.76331 - \frac{1.6}{0.47+\beta_1}} - 0.47\right)\right).$$

d) *If* $\gamma_1 = \gamma_2 = 0$, $\chi_1^2 = \chi_2^2 = \chi_0$, $\chi_1 \neq \chi_2$, *then*

$$\beta_2 \geqslant 0.6666 \log \frac{2}{3\beta_1}.$$

Proof. By the proof (cf. [12] Lemma 4), we notice

$$\beta_2 \geqslant \max\left(0.23469, \frac{1.2569}{4.51 - \frac{1.74}{0.43+\beta_1}} - 0.43\right).$$

for $\beta_1\beta_2 \leqslant (0.43)^2$. If $\beta_1\beta_2 \geqslant (0.43)^2$, then $\beta_2 \geqslant \frac{0.1849}{\beta_1}$. This implies the proof a). Similarly, we have b), c). By a result (cf. [8] Lemma 4), we get *d*).

Lemma 17. *Let* $\rho = 1 - \frac{\beta}{\log D} + i\frac{t}{\log D}$ *be a zero of* $L(s, \chi)$, *where* χ *is a real non-principal character mod* D, $t \neq 0$. *Then*

$$\beta \geqslant 0.158798.$$

Proof. By a result (cf. [13] Lemma 3), we have

$$\mathrm{Re}\Big(11.200472\frac{L'}{L}(\sigma, \chi_0) + 19.09712\frac{L'}{L}\left(\sigma + i\frac{t}{\log D}, \chi\right)$$

$$+ 11.6884\frac{L'}{L}\left(\sigma + 2i\frac{t}{\log D}, \chi^2\right) + 4.76\frac{L'}{L}\left(\sigma + 3i\frac{t}{\log D}, \chi^3\right)$$

$$+ \frac{L'}{L}\left(\sigma + 4i\frac{t}{\log D}, \chi^4\right)\Big) \leqslant 0. \tag{2.7}$$

Now we chose $\sigma = 1 + \frac{a}{\log D}$, $a = 0.52$. By (2.7) and results (cf. [9] Lemma 1 and Lemma 11), these imply that

$$0 \leqslant \frac{11.200472}{a} - \frac{19.09712}{a+\beta} + 0.27641(19.09712 + 4.76) + \frac{11.6884a}{a^2 + (2t)^2}$$

$$- \frac{19.09712(a+\beta)}{(a+\beta)^2 + (2t)^2} - \frac{4.76(a+\beta)}{(a+\beta)^2 + (2t)^2} + \frac{a}{a^2 + (4t)^2} - \frac{4.76(a+\beta)}{(a+\beta)^2 + (4t)^2}$$

$$\leqslant 28.133372 - \frac{19.09712}{0.52+\beta}$$

This completes the proof.

Lemma 18. *For* $j=1,\cdots,N$, *let* $\rho_j=1-\frac{\beta_j}{\log D}+i\gamma_j$ *be a zero of* $L(s,\chi_j)$, *where* χ_j *is a non-principal character mod* D, $\beta_j\geqslant\beta_{j-1}$, $\beta_0=0$, $|\gamma_j|\leqslant 1$, *Then*

$$\frac{1}{N}\sum_{j=1}^{N}\frac{\lambda_j a}{a+\beta_j}\leqslant 0.27641a+\sqrt{\frac{1}{N^2}\sum_{j=1}^{N}\lambda_j^2(1-0.27641a)+0.27641a}$$

where $\lambda_j>0$ *and* a *shall be choosen later such that* $\sum_{j=1}^{N}\lambda_j=N$, $0.36\leqslant a\leqslant 0.45$

Proof. By a result [9, Lemma 1] and Holder's inequality,

$$\left|\sum_{n=1}^{\infty}\frac{\Lambda(n)}{n^\sigma}\sum_{j=1}^{N}\frac{\lambda_j\chi_j(n)}{n^{i\gamma_j}}\right|^2\leqslant\sum_{m=1}^{\infty}\frac{\Lambda(m)\chi_0(m)}{m^\sigma}\sum_{n=1}^{\infty}\frac{\Lambda(n)}{n^\sigma}$$

$$\cdot\left|\sum_{j=1}^{N}\frac{\lambda_j\chi_j(n)}{n^{i\gamma_j}}\right|^2=\sum_{m=1}^{\infty}\frac{\Lambda(m)\chi_0(m)}{m^\sigma}\operatorname{Re}\sum_{n=1}^{\infty}\frac{\Lambda(n)}{n^\sigma}\sum_{j=1}^{N}\frac{\lambda_j\chi_j(n)}{n^{i\gamma_j}}$$

$$\cdot\sum_{j=1}^{N}\frac{\lambda_{j'}\overline{\chi}_{j'}(n)}{n^{-i\gamma_{j'}}}\leqslant\frac{\sum_{j=1}^{N}\lambda_j^2}{(\sigma-1)^2}+\varepsilon\log^2 D+\frac{(1+\varepsilon)}{\sigma-1}\sum_{j\neq j'}\operatorname{Re}\sum_{n=1}^{\infty}\frac{\Lambda(n)}{n^\sigma}$$

$$\cdot\frac{\lambda_j\lambda_{j'}\chi_j(n)\overline{\chi}_{j'}(n)}{n^{i\gamma_j-i\gamma_{j'}}}. \tag{2.8}$$

where $\sigma=1+\frac{a}{\log D}$.

First let us look at the pairs (j,j'), where $j\neq j'$, such that $\chi_j\overline{\chi}_{j'}=\chi_0$, $|\gamma_j-\gamma_{j'}|\geqslant\frac{1.194}{\log D}$. By [9, Lemma 1], it follows

$$\operatorname{Re}\sum_{n=1}^{\infty}\frac{\Lambda(n)\chi_0(n)}{n^{\sigma+i(\gamma_j-\gamma_{j'})}}\leqslant\frac{(1+\varepsilon)(\sigma-1)}{(\sigma-1)^2+(\gamma_j-\gamma_{j'})^2}\leqslant 0.27641\log D,$$

where $0.36\leqslant a\leqslant 0.45$.

Now let us look at the pairs (j,j'), where $j\neq j'$, such that $\chi_j\underline{\chi_{j'}}=\chi_p$, $|\gamma_j-\gamma_{j'}|\leqslant\frac{1.194}{\log D}$. By Lemma 9 – Lemma 15, this is available for our purpose to prove $C=11.5$ (see section 4).

If $\chi_j\overline{\chi}_{j'}\neq\chi_0$ where $j\neq j'$, by [9, Lemma 1], it follows that

$$\operatorname{Re}\sum_{n=1}^{\infty}\frac{\Lambda(n)\chi_j(n)\overline{\chi}_{j'}(n)}{n^{i\gamma_j-i\gamma_{j'}}}\leqslant 0.27641\log D.$$

Hence

$$\left|\sum_{n=1}^{\infty}\frac{\Lambda(n)}{n^\sigma}\sum_{j=1}^{N}\frac{\lambda_j\chi_j(n)}{n^{i\gamma_j}}\right|^2\leqslant\frac{\sum_{j=1}^{N}\lambda_j^2}{(\sigma-2)^2}+\frac{\sum_{j\neq j'}\lambda_j\lambda_{j'}}{\sigma-1}(0.27641)\log D,$$

i. e.

$$\left|\sum_{n=1}^{\infty}\frac{\Lambda(n)}{n^{\sigma}}\sum_{j=1}^{N}\frac{\lambda_j\chi_j(n)}{n^{i\gamma_j}}\right|\leqslant\sqrt{\frac{\sum_{j=1}^{N}\lambda_j^2}{a^2}+\frac{\sum_{j\neq j'}\lambda_j\lambda_{j'}}{a}(0.27641)(\log D)}.$$

By [9, Lemma 1], we have

$$-\operatorname{Re}\frac{L'}{L}(s,\chi)\leqslant 0.27641\log D-\sum_{\rho_k}\operatorname{Re}\frac{1}{s-\rho_k},$$

i. e.

$$\sum_{\rho_k}\operatorname{Re}\frac{1}{s-\rho_k}\leqslant 0.27641\log D+\operatorname{Re}\frac{L'}{L}(s,\chi).$$

Hence

$$\sum_{j=1}^{N}\frac{\lambda_j}{a+\beta_j}\leqslant 0.27641N+\sqrt{\frac{\sum_{j=1}^{N}\lambda_j^2}{a^2}+\frac{N^2-\sum_{j=1}^{N}\lambda_j^2}{a}(0.27641)}.$$

This implies the Lemma.

Lemma 19. *Under the assumption of Lemma* 18, *If* $\gamma_1\neq 0$, *then*

$$\beta_N\geqslant\frac{N-2B}{0.27641N+\frac{1}{a}\sqrt{\left(2B^2+\frac{(N-2B)^2}{N-2}\right)(1-0.27641a)+0.27641aN^2}-\frac{2B}{a+\beta_1}}-a$$

here $a>0$, $B>0$ shall be chosen later.

Proof. By Lemma 17, we take $\lambda_1=\lambda_2=B$, $\lambda_3=\lambda_4=\cdots=\lambda_N=\frac{N-2B}{N-2}$, then

$$\frac{1}{N^2}\sum_{j=1}^{N}\lambda_j^2=\frac{1}{N^2}\left(2B^2+(N-2)\left(\frac{N-2B}{N-2}\right)^2\right)=\frac{1}{N^2}\left(2B^2+\frac{(N-2B)^2}{N-2}\right).$$

Hence

$$\frac{a}{N}\left(\frac{2B}{a+\beta_1}+\frac{N-2B}{a+\beta_N}\right)\leqslant 0.27641a$$

$$+\sqrt{\frac{1}{N^2}\left(2B^2+\frac{(N-2B)^2}{N-2}\right)(1-0.27641a)+0.27641a}.$$

This completes the proof.
Similarly, we have

Lemma 20. *Under the assumption of Lemma* 18. *If* χ_1 *is a complex character mod* D, *then*

$$\beta_N \geqslant \frac{N-2B}{0.27641N+\frac{1}{a}\sqrt{(2B^2+\frac{(N-2B)^2}{N-2})(1-0.27641a)+0.27641aN^2}-\frac{2B}{a+\beta_1}}-a$$

here $a>0$, $B>0$ shall be chosen later.

Lemma 21. *$\prod_{\chi \bmod D} L(s,\chi)$ has at most four zeros satisfying the following conditions:*

$$1-\frac{0.2769}{\log D}<\beta<1,\ |\gamma|\leqslant 1.$$

For $j=1,2$, if $\prod_{\chi \bmod D} L(s,\chi)$ has zeros with $\rho_j=\beta_j+i\gamma_j$, where $1-\frac{0.27}{\log D}<\beta_j<1$, $|\gamma_j|\leqslant 1$ then $\prod_{\chi \bmod D} L(s,\chi)$ has at most four zeros satisfying the following conditions:

$$1-\frac{\beta_3}{\log D}\leqslant\sigma<1,\ |t|\leqslant 1, \tag{2.9}$$

where

$$\beta_3 \geqslant \max\left\{\left(4.5778-\frac{1}{0.37837+\beta_1}-\frac{1}{0.37837+\beta_2}\right)^{-1}-0.37837,\ \frac{2}{4.5778-\frac{1}{0.37837+\beta_1}}-0.37837\right\}.$$

Proof. See [11, Lemma 13] and [9, Lemma 13].

If $0.1\leqslant\lambda\leqslant 0.6$, we write $\alpha=1-\frac{\lambda}{\log D}$, considering the region

$$\alpha\leqslant\sigma<1,\ |t|\leqslant 1. \tag{2.10}$$

Suppose that $S(\lambda)=\{\chi$: there exists at least one zero of $L(s,\chi)$ in (2.10)$\}$. Let ρ_χ be zeros of $L(s,\chi)$ in the region (2.10) and $\chi\neq\chi_0$.

Put

$$H(\omega)=\frac{e^{4.75\omega\log D}(e^{\omega\log D}-1)}{\omega\log D},$$

where $\omega=\frac{-\lambda_1+i\tau}{\log D}$, $\lambda_1>0$. Set $C_1=\begin{cases}0, & \lambda\leqslant 0.545,\\ 2.061, & \lambda>0.545.\end{cases}$

Lemma 22. *If $0.1\leqslant\lambda\leqslant 0.6$, then we have*

$$\sum_{\chi\in S(\lambda)}\sum_{\rho_\chi}|H^2(\rho_\chi-1)|D^{10.5\,\mathrm{Re}(1-\rho_\chi)}\leqslant 3.642C(\lambda)e^{4.56006\lambda}+C_1.$$

$$C(\lambda)=\lambda^{-1}\left(1-\frac{(e^{3.19004\lambda}-e^{1.99002\lambda})}{1.20002\lambda}e^{-4.56006\lambda}\right).$$

Proof. See the method used in the Theorem 2 of [11].

3. Some Results Concerning the Zeros of L-Functions

For $j=1, 2$, let $\rho_j = 1 - \frac{\beta_j}{\log D} + i\frac{\gamma_j}{\log D}$ be a zero of $L(s, \chi)$, where χ is any non-principal character mod D, $|\gamma_j| \leqslant \varepsilon \log D$. Write $\Delta_{1,2} = |\gamma_1 - \gamma_2|$. Using the method used in Lemma 17 of [11], we have

Lemma 23. *If* $0 \leqslant \beta_j \leqslant 0.89$, *then*

$$\sum_{j=1}^{2} |H^2(\rho_j - 1)| D^{10.5\mathrm{Re}(1-\rho_j)} \leqslant \sum_{j=1}^{2} \frac{2.8457855 - 2\cos\gamma_j}{\gamma_j^2 + 0.7921}.$$

Define $K(x) = \frac{(C_3 - 2\cos x)}{x^2 + 0.7921}$, $C_3 = 2.8457855$, $x \geqslant 0$. Let $u = 0.7921$, we have

$$K'(x) = 2\frac{(x^2+u)\sin x - x(C_3 - 2\cos x)}{(x^2+u)^2},$$

$$K''(x) = \frac{2(x^2+u)((x^2+u)\cos x - 4x\sin x) + (C_3 - 2\cos x)(3x^2 - u)}{(x^2+u)^3}.$$

By calculation, we obtain $K''(x) \leqslant 0$ for $0 \leqslant x \leqslant 2.35$. If $0 \leqslant x \leqslant 2.35$, then $K'(x) \leqslant K'(0) = 0$ and $K(x)$ becomes a monotonous decrease function of x. Since $K''(x) \leqslant 0$, hence

$$\frac{1}{2}(K(x_1) + K(x_2)) \leqslant K\left(\frac{|x_1| + |x_2|}{2}\right)$$

for $x_1, x_2 \in [-2.35, 2.35]$.

Lemma 24.

$$\sum_{j=1}^{2} |H^2(\rho_j - 1)| D^{10.5\mathrm{Re}(1-\rho_j)} \leqslant \begin{cases} 2K\left(\frac{|\gamma_1| + |\gamma_2|}{2}\right), & -2.35 \leqslant \gamma_1, \gamma_2 \leqslant 2.35, \\ 1.783, & \textit{otherwise}. \end{cases}$$

Proof. See the proof method used in Lemma 18 of [11].

Lemma 25. *Suppose that χ is a non-principal character modulo D, and let $S(\chi)$ denote a set over all ρ with $\rho = \beta + i\gamma$, which satisfies the following conditions:*

$$L(\rho, \chi) = 0,\ 1 - \frac{0.89}{\log D} \leqslant \beta \leqslant 1 - \frac{0.6}{\log D},\ |\gamma| \leqslant 1.$$

Then, we have

$$\sum_{\rho_\chi \in S(\chi)} |H^2(\rho_\chi - 1)| D^{10.5\mathrm{Re}(1-\rho_\chi)} \leqslant 2.6851.$$

Proof. See the proof method used in Lemma 20 of [11].

Lemma 26. *Suppose that χ and χ_1 are two non-principal characters mod D, where $\chi\neq\chi_1$, $\chi\neq\bar{\chi}_1$. Let $S_1(\chi)$ and $S_1(\chi)$ denote two set over all ρ_χ, ρ_{χ_1} respectively, which satisfy the following conditions*:

$$L(\rho_\chi,\chi)=0,\ 1-\frac{1.5}{\log D}\leqslant \mathrm{Re}\rho_\chi<1-\frac{0.89}{\log D},\ |\mathrm{Im}\,\rho_\chi|\leqslant 1,$$

$$L(\rho_{\chi_1},\chi_1)=0,\ 1-\frac{1.5}{\log D}\leqslant \mathrm{Re}\,\rho_{\chi_1}<1-\frac{0.89}{\log D},\ |\mathrm{Im}\,\rho_{\chi_1}|\leqslant 1.$$

Then

$$\sum_{\rho_\chi\in S_1(\chi)}|H(\rho_\chi-1)|^2D^{9.5\mathrm{Re}(1-\rho_\chi)}+\sum_{\rho_{\chi_1}\in S_1(\chi_1)}|H(\rho_{\chi_1}-1)|^2D^{9.5\mathrm{Re}(1-\rho_{\chi_1})}\leqslant 2.24284.$$

Proof. See the proof method used in Lemma 21 of [11].

For $0.6\leqslant\lambda\leqslant 0.89$, we use the following notations: $z_1=D^{0.815}$, $z_2=D^{1.155}$, $z_3=D^{1.75}$, $z_4=D^{0.2199}$. Set $\alpha=1-\dfrac{\lambda}{\log D}$. Consider

$$\alpha\leqslant\sigma\leqslant 1,\ |t|\leqslant 1. \tag{3.1}$$

Suppose that χ is a non-principal character mod D, and set

$S_{1,D}=\{\chi$: there is, if only if, a zero $\rho_\chi=1-\dfrac{\beta_\chi}{\log D}+i\dfrac{\gamma_\chi}{\log D}$ of $L(s,\chi)$in (3.1)$\}$;

$S_{2,D}=\{\chi$: there are, if only if, two zeros of $L(s,\chi)$, $\rho_{j,\chi}=1-\dfrac{\beta_{j,\chi}}{\log D}+i\dfrac{\gamma_{j,\chi}}{\log D}$ for $j=1, 2$, in (3.1)$\}$;

$S_{3,D}=\{\chi$: there are, if only if, n zeroes of $L(s,\chi)$, $\rho_{j,\chi}=1-\dfrac{\beta_{j,\chi}}{\log D}+i\dfrac{\gamma_{j,\chi}}{\log D}$ for $j=1,\cdots,n$ with $n\geqslant 3\}$.

Assume that there are at least three element in $\bigcup_{j=1}^{3}S_{j,D}$. Otherwise, it is easily to prove our main result. Therefore, there is a non-pricipal character χ_1 mod D such that there is a zero $\rho_1=1-\dfrac{\beta_1}{\log D}+i\dfrac{\gamma_1}{\log D}$ of $L(s,\chi_1)$ in (3.1). From Lemma 1 to Lemma 8, the inequality $|\gamma_{j,\chi}-\gamma_{k,\chi}|\leqslant 1$ with $j\neq K$ holds in $S_{2,D}$ and $S_{3,D}$ except a non-principal character χ_2 mod D.

Put $E_1(\chi)=|H^2(\rho_\chi-1)|D^{10.5\mathrm{Re}(1-\rho_\chi)}$,

$$E_2(\chi)=|H^2(\rho_{1,\chi}-1)|D^{10.5\mathrm{Re}(1-\rho_{1,\chi})}+|H^2(\rho_{2,\chi}-1)|D^{10.5\mathrm{Re}(1-\rho_{2,\chi})}$$

$$E_3(\chi)=\sum_{\rho_\chi}|H^2(\rho_\chi-1)|D^{10.5\mathrm{Re}(1-\rho_\chi)}.$$

Lemma 27. *Suppose that χ is any non-principal character mod D. If $0.6\leqslant\lambda\leqslant 0.89$, then*

$$\sum_{\chi\in S_{1,D}\setminus\{\chi_2\}} E_1(\chi)+\sum_{\chi\in S_{2,D}\setminus\{\chi_2\}} E_2(\chi)+\sum_{\chi\in S_{3,D}\setminus\{\chi_2\}} E_3(\chi)\leqslant 8.6465C_1(\lambda)e^{3.5\lambda}.$$

where

$$C_1(\lambda)=\lambda^{-1}\left(1-\frac{e^{2.31\lambda}-e^{1.63\lambda}}{0.68\lambda}\,e^{-3.5\lambda}\right).$$

Proof. By the proof method used in Lemma 22 of [11], to prove Lemma 27, it suffices to estimate $\frac{1}{4}\,I_1E_2(\chi)$; $\frac{1}{9}\,E_3(\chi)I_2$ where

$$I_1=\int_{\log z_1}^{\log z_3}\left|\sum_{j=1}^{2}e^{-s_{j,\chi}t}\right|^2dt,\ I_2=\int_{\log z_1}^{\log z_3}\left|\sum_{j=1}^{3}e^{-s_{j,\chi}t}\right|^2dt.$$

By calculation, we abtain

$$\frac{1}{4}\,I_1E_2(\chi)\leqslant 1.24673,$$

$$\frac{1}{9}\,I_2E_3(\chi)\leqslant(0.4813)(2.6851)\leqslant 1.29234.$$

This completes the proof.

4. The Proof for $C=11.5$

Lemma 28. *Suppose that χ and χ_1 are non-principal character mod D, and that χ_1 is real. Suppose that $\beta_1=1-\dfrac{\delta_1}{\log D}$ is a real zero of $L(s,\chi_1)$ and that $\rho_1=1-\dfrac{\delta}{\log D}+i\gamma$ is a zero of $L(s,\chi)$ such that $\delta_1\leqslant\delta$, $|\gamma|\leqslant 1$, then*

$$\delta\geqslant\left(\frac{2}{3}-\varepsilon\right)\log\frac{2-\varepsilon}{3\delta_1}.$$

Proof. See [8].

For positive integers n, we define

$$R(n)=\frac{1}{2\pi i}\int_{2-i\infty}^{2+i\infty}H^2(\omega)e^{-\omega\log n}\,d\omega$$

and

$$J(\chi)=\frac{-1}{2\pi i}\int_{2-i\infty}^{2+i\infty}H^2(\omega)\,\frac{L'}{L}\,(\omega+1,\chi)\,d\omega.$$

Hence $R(n)=0$ if $n>D^{11.5}$ or if $n<D^{9.5}$ and $R(n)\ll\dfrac{1}{\log D}$ if $D^{9.5}\leqslant n\leqslant D^{11.5}$.

Therefore

$$J(\chi)=\sum_{D^{9.5}\leqslant n\leqslant D^{11.5}}\frac{\wedge(n)\chi(n)R(n)}{n}$$

Set

$$E_0=\begin{cases}0, & \chi\neq\chi_0,\\ 1, & \chi=\chi_0.\end{cases}$$

It follows

$$J(\chi)=E_0-\sum_{\rho_\chi}H^2(\rho_\chi-1)+0(D^{-2}),$$

where ρ_χ runs over the nontrivial zeros of $L(s,\chi)$. Hence

$$\sum_{\substack{D^{9.5}\leqslant n\leqslant D^{11.5}\\ n\equiv K(\bmod D)}}\frac{\wedge(n)R(n)}{n}=\frac{1}{\varphi(D)}\left\{1-\sum_{\chi\bmod D}\bar{\chi}(K)\sum_{\rho_\chi}H^2(\rho_\chi-1)\right\}+0\left(\frac{1}{D^2}\right).$$

Thus for the proof of $P(D,K)\ll D^{11.5}$, it is sufficient to verify

$$\sum_{\chi\bmod D}\sum_{\rho_\chi}|H^2(\rho_\chi-1)|<1-D^{-\frac{1}{2}}. \tag{4.1}$$

Set

$$Z(\chi;\xi,\eta)=\sum_{1-\frac{\xi}{\log D}\leqslant\operatorname{Re}\rho_\chi<1-\frac{\eta}{\log D}}|H^2(\rho_\chi-1)|.$$

If $0<b\leqslant 0.6$, then

$$\sum_{\chi\bmod D}\sum_{\rho_\chi}|H^2(\rho_\chi-1)|=\sum_{\chi\bmod D}Z(\chi;b,0)+\sum_{\chi\bmod D}Z(\chi;0.89,0.6)$$
$$+\sum_{\chi\bmod D}Z(\chi;\frac{1}{2}\log D,0.89)+\sum_{\chi\bmod D}Z(\chi;0.6,b). \tag{4.2}$$

Set

$$Z_1(\chi;\xi,\eta,\theta)=\sum_{1-\frac{\xi}{\log D}\leqslant\operatorname{Re}\rho_\chi<1-\frac{\eta}{\log D}}|H^2(\rho_\chi-1)|D^{\theta\operatorname{Re}(1-\rho_\chi)}.$$

We have

$$\sum_{\chi\bmod D}Z(\chi;0.6,b)=\int_b^{0.6}e^{-10.5\lambda}d_\lambda\sum_{\chi\bmod D}Z_1(\chi;\lambda,b,10.5)$$
$$=e^{-10.5(0.6)}\sum_{\chi\bmod D}Z_1(\chi;0.6,b,10.5)+10.5\int_b^{0.6}e^{-10.5\lambda}\sum_{\chi\bmod D}Z_1(\chi;\lambda,b,10.5)d\lambda \tag{4.3}$$

and

$$\sum_{\chi\bmod D}Z(\chi;0.89,0.6)=\int_{0.6}^{0.89}e^{-10.5\lambda}d_\lambda\sum_{\chi\bmod D}Z_1(\chi;\lambda,0.6,10.5)$$

$$= e^{-10.5(0.89)} \sum_{\chi \bmod D} Z_1(\chi;\ 0.89,\ 0.6,\ 10.5)$$

$$+ 10.5\int_{0.6}^{0.89} e^{-10.5\lambda} \cdot \sum_{\chi \bmod D} Z_1(\chi;\ \lambda,\ 0.6,\ 10.5)\, d\lambda. \tag{4.4}$$

On the other hand, we have

$$10.5\int_{0.6}^{0.89} e^{-10.5\lambda} \sum_{\chi \bmod D} Z_1(\chi;\ \lambda,\ 0.6,\ 10.5)\, d\lambda = 10.5\int_{0.6}^{0.89} e^{-10.5\lambda} \cdot$$

$$\sum_{\chi \bmod D} Z_1(\chi;\ \lambda,\ b,\ 10.5)\, d\lambda + \sum_{\chi \bmod D} Z_1(\chi;\ 0.6,\ b,\ 10.5)(e^{-10.5(0.89)} - e^{-10.5(0.6)}). \tag{4.5}$$

By (4.2)—(4.5), we obtain

$$\sum_{\chi \bmod D} \sum_{\rho_\chi} |H^2(\rho_\chi - 1)| = \sum_{\chi \bmod D} Z(\chi;\ b,\ 0) + 10.5\int_b^{0.6} e^{-10.5\lambda}$$

$$\sum_{\chi \bmod D} \cdot Z_1(\chi;\ \lambda,\ b,\ 10.5)\, d\lambda + e^{-10.5(0.89)} \sum_{\chi \bmod D} Z_1(\chi;\ 0.89,\ b',\ 10.5) + 10.5 \cdot$$

$$\cdot \int_{0.6}^{0.89} e^{-10.5\lambda} \sum_{\chi \bmod D} Z_1(\chi;\ \lambda,\ b,\ 10.5)\, d\lambda + \sum_{\chi \bmod D} Z(\chi;\ \frac{1}{2}\log D,\ 0.89) =$$

$$= \sum_{\chi \bmod D} Z(\chi;\ b,\ 0) - e^{-10.5b} \sum_{\chi \bmod D} Z_1(\chi;\ b,\ 0,\ 10.5) + 10.5\int_b^{0.89} e^{-10.5\lambda} \cdot$$

$$\cdot \sum_{\chi \bmod D} Z_1(\chi;\ \lambda,\ 0,\ 10.5)\, d\lambda + e^{-10.5(0.89)} \sum_{\chi \bmod D} Z_1(\chi;\ 0.89,\ 0,\ 10.5) +$$

$$+ \sum_{\chi \bmod D} Z(\chi;\ \frac{1}{2}\log D,\ 0.89).$$

Since

$$\sum_{\chi \bmod D} Z(\chi;\ \frac{1}{2}\log D,\ 0.89) = 9.5\int_{0.89}^{\infty} e^{-9.5\lambda} \sum_{\chi \bmod D} Z_1(\chi;\ \lambda,\ 0,\ 9.5)\, d\lambda$$

$$- e^{-9.5(0.89)} \sum_{\chi \bmod D} Z_1(\chi;\ 0.89,\ 0,\ 9.5).$$

We deduce that

$$\sum_{\chi \bmod D} \sum_{\rho_\chi} |H^2(\rho_\chi - 1)| = \sum_{\chi \bmod D} Z(\chi;\ b,\ 0) - e^{-10.5b} \sum_{\chi \bmod D} Z_1(\chi;\ b,\ 0,\ 10.5) +$$

$$+ 10.5\int_b^{0.89} e^{-10.5\lambda} \sum_{\chi \bmod D} Z_1(\chi;\ \lambda,\ 0,\ 10.5)\, d\lambda + 9.5\int_{0.89}^{\infty} e^{-9.5\lambda}$$

$$\cdot \sum_{\chi \bmod D} Z_1(\chi;\ \lambda,\ 0,\ 9.5)\, d\lambda \tag{4.6}$$

By the proof method used in Lemma 6 of [9] and Theorem 1 of [9], we get

$$9.5\int_{1.5}^{\infty}(2.347)\frac{6.2557}{\lambda}\left(e^{3.5\lambda}-\frac{(e^{2.31\lambda}-e^{1.63\lambda})}{0.68\lambda}\right)e^{-9.5\lambda}d\lambda\leqslant 83.20585.$$

$$\int_{1.5}^{\infty}e^{-6\lambda}d\lambda\leqslant 0.00171141.$$

By Theoem 1 of [9] and Lemma 26, we get $9.5\int_{0.89}^{1.5}(1.12142)\frac{6.2557}{\lambda}\cdot$

$$\left(e^{3.5\lambda}-\frac{(e^{2.31\lambda}-e^{1.63\lambda})}{0.68\lambda}\right)e^{-9.5\lambda}d\lambda+9.5\int_{0.89}^{1.5}(2)(2.347)e^{-9.5\lambda}d\lambda\leqslant$$

$$\leqslant 55.40127\int_{0.89}^{1.5}e^{-6\lambda}d\lambda+44.593\int_{0.89}^{1.5}e^{-9.5\lambda}d\lambda\leqslant 0.04414.$$

Hence the last term of (4.6) contributes to 0.04585141.

Put $L=\log D$ in the following discussion.

(1) Suppose that there exists a character χ mod D such that $L(s,\chi)$ has a zero $\rho_\chi=1-\frac{\beta_\chi}{L}+i\gamma_\chi$ with $1-\frac{0.44}{L}\leqslant\sigma\leqslant 1-\frac{0.103668}{L}$ and $|t|\leqslant 1$. From Lemma 28, we know that there does not exist a real non-principal character χ_1 mod D such that $\beta_1=1-\frac{\delta_1}{L}$ is a real zero of $L(s,\chi_1)=0$ and $0<\delta_1\leqslant 0.103668$.

If $0.103668\leqslant\beta_\chi\leqslant 0.12$. From Lemma 16, Lemma 19, Lemma 20, Lemma 22 and Lemma 27, we have

$$\sum_{\chi \bmod D}\sum_{\rho_\chi}|H^2(\rho_\chi-1)|\leqslant 2.02e^{-10.5(0.103668)}+\left(10.5\int_{0.50355}^{0.511902}11\cdot\right.$$

$$e^{-10.5\lambda}d\lambda+10.5\int_{0.511902}^{0.521292}12e^{-10.5\lambda}d\lambda+10.5\int_{0.521292}^{0.529472}13e^{-10.5\lambda}d\lambda+$$

$$+10.5\int_{0.529472}^{0.536717}14e^{-10.5\lambda}d\lambda+10.5\int_{0.536717}^{0.543179}15e^{-10.5\lambda}d\lambda+$$

$$+10.5\int_{0.543179}^{0.548921}16e^{-10.5\lambda}d\lambda+10.5\int_{0.548921}^{0.554059}17e^{-10.5\lambda}d\lambda+$$

$$+10.5\int_{0.554059}^{0.558697}18e^{-10.5\lambda}d\lambda+10.5\int_{0.558697}^{0.56297}19e^{-10.5\lambda}d\lambda+$$

$$+10.5\int_{0.56297}^{0.566893}20e^{-10.5\lambda}d\lambda+10.5\int_{0.566893}^{0.570473}21e^{-10.5\lambda}d\lambda+$$

$$+10.5\int_{0.570473}^{0.573752}22e^{-10.5\lambda}d\lambda+10.5\int_{0.573752}^{0.576768}23e^{-10.5\lambda}d\lambda+$$

$$+10.5\int_{0.576768}^{0.579561}24e^{-10.5\lambda}d\lambda+10.5\int_{0.579561}^{0.582175}25e^{-10.5\lambda}d\lambda+$$

$$+10.5\int_{0.582175}^{0.584606}26e^{-10.5\lambda}d\lambda+10.5\int_{0.584606}^{0.586898}27e^{-10.5\lambda}d\lambda+$$

$$+10.5\int_{0.586898}^{0.589068}28e^{-10.5\lambda}d\lambda+10.5\int_{0.589068}^{0.59109}29e^{-10.5\lambda}d\lambda+$$

$$+10.5\int_{0.59109}^{0.592991}30e^{-10.5\lambda}d\lambda+10.5\int_{0.592991}^{0.59477}31e^{-10.5\lambda}d\lambda+$$

$$+10.5\int_{0.59477}^{0.596457}32e^{-10.5\lambda}d\lambda+10.5\int_{0.596457}^{0.598055}33e^{-10.5\lambda}d\lambda+$$

$$+10.5\int_{0.598055}^{0.599562}34e^{-10.5\lambda}d\lambda+10.5\int_{0.599562}^{0.600983}35e^{-10.5\lambda}d\lambda+$$

$$+10.5\int_{0.600983}^{0.602332}36e^{-10.5\lambda}d\lambda+10.5\int_{0.602332}^{0.603625}37e^{-10.5\lambda}d\lambda+$$

$$+10.5\int_{0.603625}^{0.604851}38e^{-10.5\lambda}d\lambda+10.5\int_{0.604851}^{0.606013}39e^{-10.5\lambda}d\lambda+$$

$$+10.5\int_{0.606013}^{0.607121}40e^{-10.5\lambda}d\lambda+10.5\int_{0.607121}^{0.608204}41e^{-10.5\lambda}d\lambda+$$

$$+10.5\int_{0.608204}^{0.609242}42e^{-10.5\lambda}d\lambda+10.5\int_{0.609242}^{0.610241}43e^{-10.5\lambda}d\lambda+$$

$$+10.5\int_{0.610241}^{0.611193}44e^{-10.5\lambda}d\lambda+10.5\int_{0.611193}^{0.612102}45e^{-10.5\lambda}d\lambda+$$

$$+10.5\int_{0.612102}^{0.612977}46e^{-10.5\lambda}d\lambda+10.5\int_{0.612977}^{0.613822}47e^{-10.5\lambda}d\lambda+$$

$$+10.5\int_{0.613822}^{0.61463} 48e^{-10.5\lambda}d\lambda+10.5\int_{0.61463}^{0.6154} 49e^{-10.5\lambda}d\lambda+$$

$$+10.5\int_{0.6154}^{0.616152} 50e^{-10.5\lambda}d\lambda+10.5\int_{0.616152}^{0.616876} 51e^{-10.5\lambda}d\lambda+$$

$$+10.5\int_{0.616876}^{0.61757} 52e^{-10.5\lambda}d\lambda+10.5\int_{0.61757}^{0.618236} 53e^{-10.5\lambda}d\lambda+$$

$$+10.5\int_{0.618236}^{0.618884} 54e^{-10.5\lambda}d\lambda+10.5\int_{0.618884}^{0.619511} 55e^{-10.5\lambda}d\lambda+$$

$$+10.5\int_{0.619511}^{0.62011} 56e^{-10.5\lambda}d\lambda+10.5\int_{0.62011}^{0.62069} 57e^{-10.5\lambda}d\lambda+$$

$$+10.5\int_{0.62069}^{0.62126} 58e^{-10.5\lambda}d\lambda+10.5\int_{0.62126}^{0.621808} 59e^{-10.5\lambda}d\lambda+$$

$$+10.5\int_{0.621808}^{0.622336} 60e^{-10.5\lambda}d\lambda)(1.032694744)\leqslant 0.75913859$$

and

$$10.5\int_{0.622}^{0.89} e^{-10.5\lambda}\, 8.6465\ C_1(0.622336)e^{3.5\lambda}d\lambda\leqslant$$

$$\leqslant 89.185284\int_{0.622336}^{0.89} e^{-7\lambda}d\lambda\leqslant 0.13831.$$

Hence

$$\sum_{\chi \bmod D}\sum_{\rho_\chi}|H^2(\rho_\chi-1)|\leqslant 0.9433.$$

Similarly,

$$\sum_{\chi \bmod D}\sum_{\rho_\chi}|H^2(\rho_\chi-1)|\leqslant 0.977,$$

if $0.12\leqslant\beta_\chi\leqslant 0.14$, $0.14\leqslant\beta_\chi\leqslant 0.155$ or $0.155\leqslant\beta_\chi\leqslant 0.165$.

By Lemma 16, Lemma 21, Lemma 22 and Lemma 27, we have

$$\sum_{\chi \bmod D}\sum_{\rho_\chi}|H^2(\rho_\chi-1)|\leqslant 0.963,$$

if $0.165\leqslant\beta_\chi\leqslant 0.175$, $0.175\leqslant\beta_\chi\leqslant 0.185$, $0.185\leqslant\beta_\chi\leqslant 0.195$, and $0.195\leqslant\beta_\chi\leqslant 0.2$.

If $0.2\leqslant\beta_\chi\leqslant 0.205$, suppose that χ' is a non-principal character let

$\rho_2 = 1 - \frac{\beta_2}{L} + i\gamma_2$ be the zero of $L(s, \chi')$ such that β_2 is the closest to β_χ. From Lemma 16, Lemma 21, Lemma 22 and Lemma 27, assume $\beta_2 > 0.31$, we have

$$\sum_{\chi \bmod D} \sum_{\rho_\chi} |H^2(\rho_\chi - 1)| \leqslant 2.02e^{-10.5(0.2)} + 2.02e^{-10.5(0.31)} - 4.04\, e^{-10.5(0.32)}$$

$$+ \Big(10.5 \int_{0.32}^{0.33315} 7e^{-10.5\lambda} d\lambda + 10.5 \int_{0.33315}^{0.35317} 8e^{-10.5\lambda} d\lambda$$

$$+ 10.5 \int_{0.35317}^{0.369636} 9e^{-10.5\lambda} d\lambda + 10.5 \int_{0.369636}^{0.383382} 10e^{-10.5\lambda} d\lambda$$

$$+ 10.5 \int_{0.383382}^{0.395112} 11e^{-10.5\lambda} d\lambda + 10.5 \int_{0.395112}^{0.405246} 12e^{-10.5\lambda} d\lambda$$

$$+ 10.5 \int_{0.405246}^{0.414069} 13e^{-10.5\lambda} d\lambda + 10.5 \int_{0.414069}^{0.42187} 14e^{-10.5\lambda} d\lambda$$

$$+ 10.5 \int_{0.42187}^{0.428833} 15e^{-10.5\lambda} d\lambda + 10.5 \int_{0.428833}^{0.435045} 16e^{-10.5\lambda} d\lambda$$

$$+ 10.5 \int_{0.435045}^{0.440628} 17e^{-10.5\lambda} d\lambda + 10.5 \int_{0.440628}^{0.44573} 18e^{-10.5\lambda} d\lambda$$

$$+ 10.5 \int_{0.44573}^{0.450379} 19e^{-10.5\lambda} d\lambda + 10.5 \int_{0.450379}^{0.45462} 20e^{-10.5\lambda} d\lambda$$

$$+ 10.5 \int_{0.45462}^{0.458513} 21e^{-10.5\lambda} d\lambda + 10.5 \int_{0.458513}^{0.462092} 22e^{-10.5\lambda} d\lambda$$

$$+ 10.5 \int_{0.462092}^{0.465432} 23e^{-10.5\lambda} d\lambda + 10.5 \int_{0.465432}^{0.468541} 24e^{-10.5\lambda} d\lambda$$

$$+ 10.5 \int_{0.468541}^{0.47143} 25e^{-10.5\lambda} d\lambda + 10.5 \int_{0.47143}^{0.474121} 26e^{-10.5\lambda} d\lambda$$

$$+ 10.5 \int_{0.474121}^{0.476634} 27e^{-10.5\lambda} d\lambda + 10.5 \int_{0.476634}^{0.478987} 28e^{-10.5\lambda} d\lambda$$

$$+ 10.5 \int_{0.478987}^{0.481194} 29e^{-10.5\lambda} d\lambda + 10.5 \int_{0.481194}^{0.4833} 30e^{-10.5\lambda} d\lambda$$

$$+10.5\int_{0.4833}^{0.48529} 31e^{-10.5\lambda}d\lambda+10.5\int_{0.48529}^{0.487168} 32e^{-10.5\lambda}d\lambda$$

$$+10.5\int_{0.487168}^{0.488943} 33e^{-10.5\lambda}d\lambda+10.5\int_{0.488943}^{0.490623} 34e^{-10.5\lambda}d\lambda$$

$$+10.5\int_{0.490623}^{0.492216} 35e^{-10.5\lambda}d\lambda+10.5\int_{0.492216}^{0.493728} 36e^{-10.5\lambda}d\lambda$$

$$+10.5\int_{0.493728}^{0.495166} 37e^{-10.5\lambda}d\lambda+10.5\int_{0.495166}^{0.496534} 38e^{-10.5\lambda}d\lambda$$

$$+10.5\int_{0.496534}^{0.49783} 39e^{-10.5\lambda}d\lambda+10.5\int_{0.49783}^{0.499081} 40e^{-10.5\lambda}d\lambda$$

$$+10.5\int_{0.499081}^{0.50029} 41e^{-10.5\lambda}d\lambda+10.5\int_{0.50029}^{0.501451} 42e^{-10.5\lambda}d\lambda$$

$$+10.5\int_{0.501451}^{0.502562} 43e^{-10.5\lambda}d\lambda+10.5\int_{0.502562}^{0.503625} 44e^{-10.5\lambda}d\lambda$$

$$+10.5\int_{0.503625}^{0.50464} 45e^{-10.5\lambda}d\lambda+10.5\int_{0.50464}^{0.50624} 46e^{-10.5\lambda}d\lambda$$

$$+10.5\int_{0.505624}^{0.506564} 47e^{-10.5\lambda}d\lambda+10.5\int_{0.506564}^{0.507468} 48e^{-10.5\lambda}d\lambda$$

$$+10.5\int_{0.507468}^{0.508337} 49e^{-10.5\lambda}d\lambda+10.5\int_{0.508337}^{0.509174} 50e^{-10.5\lambda}d\lambda$$

$$+10.5\int_{0.509174}^{0.509981} 51e^{-10.5\lambda}d\lambda+10.5\int_{0.509981}^{0.510758} 52e^{-10.5\lambda}d\lambda$$

$$+10.5\int_{0.510758}^{0.511507} 53e^{-10.5\lambda}d\lambda+10.5\int_{0.511507}^{0.512231} 54e^{-10.5\lambda}d\lambda$$

$$+10.5\int_{0.512231}^{0.51293} 55e^{-10.5\lambda}d\lambda+10.5\int_{0.51293}^{0.513606} 56e^{-10.5\lambda}d\lambda$$

$$+10.5\int_{0.513606}^{0.514263} 57e^{-10.5\lambda}d\lambda+10.5\int_{0.514263}^{0.5149} 58e^{-10.5\lambda}d\lambda$$

$$+10.5\int_{0.5149}^{0.515535} 59e^{-10.5\lambda}d\lambda+10.5\int_{0.515535}^{0.516141} 60e^{-10.5\lambda}d\lambda\Bigg)$$

$$\cdot(1.032694744)\leqslant 0.622061,$$

and

$$10.5\int_{0.516141}^{0.6} e^{-10.5\lambda}3.642C(0.516141)e^{4.56006\lambda}d\lambda$$

$$\leqslant 46.8635042\int_{0.516141}^{0.6} e^{-5.93994\lambda}d\lambda\leqslant 0.143697447.$$

Hence

$$\sum_{\chi \bmod D}\sum_{\rho_\chi}|H^2(\rho_\chi-1)|\leqslant 0.98. \tag{4.7}$$

When $\beta_2\leqslant 0.31$, (4.7) also holds.

Similarly,

$$\sum_{\chi \bmod D}\sum_{\rho_\chi}|H^2(\rho_\chi-1)|\leqslant 0.9991,$$

if $0.205\leqslant\beta_\chi\leqslant 0.21$, $0.21\leqslant\beta_\chi\leqslant 0.215$, $0.215\leqslant\beta_\chi\leqslant 0.22$, $0.22\leqslant\beta_\chi\leqslant 0.225$, $0.225\leqslant\beta_\chi\leqslant 0.23$ and $0.23\leqslant\beta_\chi\leqslant 0.235$.

In these calculations, we have used the results as follows:

Denote by $S_{D,\lambda}$ the set of $L(s,\chi)$, where χ is a non-principal character modulo D, having at least zero $\rho_\chi=\beta_\chi+i\gamma_\chi$ with

$$1-\frac{\lambda}{L}\leqslant\beta_\chi\leqslant 1,\ |\gamma_\chi|\leqslant 1, \tag{4.8}$$

where $0.1\leqslant\lambda\leqslant\log\log L$. When $0.1\leqslant\lambda\leqslant\log\log L$, we donote by $Q(\lambda)$ the number of $L(s,\chi)$ in $S_{D,\lambda}$. By the method used in [9], we can obtain

$$Q(\lambda)\leqslant\frac{(h_3-h_1)}{2\lambda h_4(h_2-h_1)}\left(e^{2h_3\lambda}-\frac{(e^{2h_2\lambda}-e^{2h_1\lambda})}{2\lambda(h_2-h_1)}\right),$$

with $\frac{3}{8}+h_2+h_4<h_3$ and $\frac{3}{8}+2h_4<h_1$.

If $\lambda\geqslant 1.5$ and $1.3\leqslant\lambda\leqslant 1.5$, we take $h_4=0.09000001$, $h_2=0.7550201$, $h_3=1.22002$, $h_1=0.55501$, then

$$Q(\lambda)\leqslant\frac{18.4756513}{\lambda}\left(e^{2.44004\lambda}-\frac{(e^{1.5100402\lambda}-e^{1.11002\lambda})}{0.4000202\lambda}\right). \tag{4.9}$$

If $1.1\leqslant\lambda\leqslant 1.3$, and $1\leqslant\lambda\leqslant 1.1$, we take $h_4=0.12$, $h_2=0.9150201$, $h_3=1.41002$, $h_1=0.61501$, then

$$Q(\lambda)\leqslant\frac{11.04143383}{\lambda}\left(e^{2.82004\lambda}-\frac{(e^{1.8300402\lambda}-e^{1.23002\lambda})}{0.6000202\lambda}\right). \tag{4.10}$$

If $0.89 \leqslant \lambda \leqslant 1$, we take $h_2=0.14$, $h_2=0.9550201$, $h_3=1.47002$, $h_1=0.65501$, then

$$Q(\lambda) \leqslant 9.70217336 \left(e^{2.94004\lambda} - \frac{(e^{1.9100402\lambda} - e^{1.31002\lambda})}{0.6000202\lambda}\right). \tag{4.11}$$

By (4.9)—(4.11), we have

$$9.5\int_{1.5}^{\infty}(2.347)Q(\lambda)e^{-9.5\lambda}d\lambda \leqslant 223.3982368\int_{1.5}^{\infty}e^{-7.05996\lambda}d\lambda \leqslant 0.000796388, \tag{4.12}$$

$$9.5\int_{1.3}^{1.5}(1.12142)Q(\lambda)e^{-9.5\lambda}d\lambda \leqslant 116.1412528\int_{1.3}^{1.5}e^{-7.05996\lambda}d\lambda \leqslant 0.001285199, \tag{4.13}$$

$$9.5\int_{1}^{1.3}(1.12142)Q(\lambda)e^{-9.5\lambda}d\lambda \leqslant 84.762155771\int_{1}^{1.3}e^{-6.67996\lambda}d\lambda \leqslant 0.013787215, \tag{4.14}$$

$$9.5\int_{0.89}^{1}(1.12142)Q(\lambda)e^{-9.5\lambda}d\lambda \leqslant 71.47969735\int_{0.89}^{1}e^{-6.55996\lambda}d\lambda \leqslant 0.016319065, \tag{4.15}$$

$$9.5\int_{0.89}^{1.5}(2)(2.347)e^{-9.5\lambda}d\lambda \leqslant 0.0009^{9601} \tag{4.16}$$

From (4.12)—(4.16), the last term of (4.6) contributes to 0.033183877.

If $0.235 \leqslant \beta_\chi \leqslant 0.24$, $0.24 \leqslant \beta_\chi \leqslant 0.245$, $0.245 \leqslant \beta_\chi \leqslant 0.25$, $0.25 \leqslant \beta_\chi \leqslant 0.255$, $0.255 \leqslant \beta_\chi \leqslant 0.26$, $0.26 \leqslant \beta_\chi \leqslant 0.265$, $0.265 \leqslant \beta_\chi \leqslant 0.27$, $0.27 \leqslant \beta_\chi \leqslant 0.275$, $0.275 \leqslant \beta_\chi \leqslant 0.28$, $0.28 \leqslant \beta_\chi \leqslant 0.285$, $0.285 \leqslant \beta_\chi \leqslant 0.29$, $0.29 \leqslant \beta_\chi \leqslant 0.295$, $0.295 \leqslant \beta_\chi \leqslant 0.3$, $0.3 \leqslant \beta_\chi \leqslant 0.305$, $0.305 \leqslant \beta_\chi \leqslant 0.31$, $0.31 \leqslant \beta_\chi \leqslant 0.32$, $0.32 \leqslant \beta_\chi \leqslant 0.33$, $0.33 \leqslant \beta_\chi \leqslant 0.35$, $0.35 \leqslant \beta_\chi \leqslant 0.38$, $0.38 \leqslant \beta_\chi \leqslant 0.41$, $0.41 \leqslant \beta_\chi \leqslant 0.44$, and $\beta_\chi \geqslant 0.44$ by Lemma 19—22, Lemma 27 and very complicated calculations, we can obtain

$$\sum_{\chi \bmod D}\sum_{\rho_\chi}|H^2(\rho_\chi - 1)| \leqslant 0.99997.$$

(2) Suppose that there exists a real non-principal one such that $\rho_1 = \frac{\delta_1}{L}$ is a real zero of $L(s, \chi_1)$ with $D^{-\frac{1}{2}} L \leqslant \delta_1 \leqslant 0.103668$, then from Lemma 28 we have

$$\delta \geqslant 0.6666 \log \frac{0.6666}{0.103668} \geqslant 1.24$$

and

$$\delta \geqslant 0.6666 \log \frac{0.6666}{\delta_1}.$$

Hence
$$e^{-6.5\delta}\leqslant e^{-4.3329\log\frac{0.6666}{\delta_1}}\leqslant\left(\frac{\delta_1}{0.6666}\right)^{4.3329} \tag{4.17}$$

From Lemma 6, Theorem 1 of [9] and (4.17), we have

$$\sum_{\chi \bmod D}\sum_{\rho_\chi}|H(\rho_\chi-1)|^2\leqslant e^{-8.5\delta_1}+2.347\int_\delta^\infty e^{-9.5\lambda}dQ(\lambda)\leqslant 1-3.5\delta_1+$$

$$+22.2965\int_\delta^\infty Q(\lambda)e^{-9.5\lambda}d\lambda\leqslant 1-3.5\delta_1+300\int_\delta^\infty e^{-6.5\lambda}d\lambda\leqslant 1-3.5\delta_1+47e^{-6.5\delta}\leqslant$$

$$\leqslant 1-3.5\delta_1+47\left(\frac{\delta_1}{0.6666}\right)^{4.3329}\leqslant 1-3.5\delta_1+0.177\delta_1\leqslant 1-\delta_1.$$

Therefore, we complete the proof of (4.1).

References

[1] Pan Chengdong, On the least prime in an arithmetical progression, *Sci. Record* (N. S.), **1**(1958), 311—313.

[2] Chen Jingrun, On the least prime in an arithmetical progression, *Sci. Sin.*, **14**(1964), 1868—1871.

[3] Matti Jutila, Proc. of Sym. in Pure Math. AMS., (1971), 370.

[4] ————, A new estimate for Linnik's constant, *Ann. Acad. Sci. Fennicae* **471**(1970)8pp.

[5] Chen Jingrun, On the least prime in an arithmetical progression and two theorems concerning the zeros of Dirichlet's *L*-functions (I), *Sci. Sin.*, **20**(1977), 529—562.

[6] Matti Jutila, On the Linnik's constant, *Math. Scand.* **41**(1977), 54—62.

[7] S. Graham, Applications of Sieve Methods, Ph. D. Thesis, University of Michigan, 1977.

[8] ————, On Linnik's constant, *Acta Arith.*, **39**(1981), 163—179.

[9] Chen Jingrun, On the least prime in an arithmetical progression and two theorems concerning the zeros of Dirichlet's *L*-functions (II), *Sci. Sin.*, **22**(1979), 859—889.

[10] Wang Wei, On the least prime in an arithmetic progression, *Acta Math. Sin.*, **29**(1986), 826—836.

[11] Chen Jingrun & Liu Jianmin, On the least prime in an arithmetical progression and theorems concerning the zeros of Dirichlet's *L*-functions (III), (IV), *Sci. Sin.*, (1989) to appear.

[12] ————, The exceptional set of Goldbach numbers (III), to appear.

[13] Chen Jingrun, The exceptional set of Goldbach numbers (II), *Sci. Sin.*, **26**(1983), 714—731.

On the Even Part of BSD Conjecture for Elliptic Curves with Complex Multiplication by $\mathbb{Z}\left[\frac{1+\sqrt{-7}}{2}\right]$

Feng Keqin

University of Science and Technology of China, Hefei, Anhui, China

The even part of Birch and Swinnerton-Dyer conjecture are varified for a series of elliptic curves with complex multiplication $\mathbb{Z}\left[\frac{1+\sqrt{-7}}{2}\right]$.

1. Introduction

In the 1960 s Birch and Swinnerton-Dyer [2] conjectured certain relations between the group $E(\mathbb{Q})$ of rational points and the L-function $L_E(s)$ for elliptic curve $E=E/\mathbb{Q}$. Suppose that the elliptic curve $E/\mathbb{Q}$ is a Weil curve so that $L_E(s)$ has a meromorphic continuation to the entire complex plane. The BSD conjecture says that

(I) The rank $E(\mathbb{Q})$ equals the order of zero of $L_E(s)$ at $s=1$;

(II) If $L_E(1)\neq 0$, then

$$L_E(1)=C_E\cdot\Omega_E\cdot\ |\text{Ш}(E)|$$

where Ω_E is the real period of E, Ш(E) the Tate-Shafarevitch group of E, C_E is a certain rational number.

In 1977 Coates and Wiles [3] proved that $L_E(1)\neq 0 \Rightarrow \text{rank } E(\mathbb{Q})=0$. In 1986 Gross and Zagier [4] proved that $\text{ord}_{s=1} L_E(s)=1 \Rightarrow \text{rank } E(\mathbb{Q})\geqslant 1$. In 1987, Rubin [7] obtained the following results for elliptic curve $E/\mathbb{Q}$ with complex multiplication O_K where K is an imaginary quadratic field and O_K is the ring of integers in K.

(R1) $\text{ord}_{s=1} L_E(s)\leqslant 1 \Rightarrow \text{rank } E(\mathbb{Q})=\text{ord}_{s=1} L_E(s)$.

(R2) If $L_E(1)\neq 0$, Ш(E) is a finite abelian group (so that |Ш(E)| is a square number).

(R3) If $\text{ord}_p(L(\bar{\psi},1)/\Omega_E)=0$ where p is a prime number and $p\nmid |O_K^\times|$, then the p-primary part of Ш(E) is trivial where ψ is the Grossen character of K associated with $E/\mathbb{Q}$.

In general there are two ways to varify rank$E(\mathbb{Q})=0$. One way is 2-descent method which can also determine the order of the even part $Ш(E)_2$ of group $Ш(E)$ sometimes. Another way is to compute $L_E(1)$ directly, verify $L_E(1)\neq 0$ and use Coates-Wiles Theorem. By comparing both results we might varify the even part of the BSD conjecture (II) for some elliptic curve $E/\mathbb{Q}$. If $E/\mathbb{Q}$ has complex multiplication and the odd part of $L(\bar{\psi}, 1)/\Omega_E$ is trivial, then the whole BSD conjecture (II) might be varified by using Rubin's result (R3).

In this paper we consider the elliptic curve

$$E_D: y^2 = x^3 + 21Dx^2 + 112Dx,$$

where D is a positive square-free integer, $7 \nmid D$. Let $L_D(s)$ denote $L_{E_D}(s)$ for simplicity. In the next paragraph we shall show that $L_D(s)\neq 0$ so that rank $(E_D(\mathbb{Q}))=0$ for several kinds of D. On the other hand, we calculate the Selmer group for E_D and its dual curve $\hat{E}_D$ using 2-descent method, and show that $|Ш(E_D)|$ is odd and the even part of BSD conjecture (II) is true for E_D with several kinds of D.

2. $L_D(1)\neq 0$

For an elliptic curve $E/\mathbb{Q}$ with complex multiplication, the classical way to compute the $L_E(1)$ is using Weierstrass $\wp$-function. Thanks to the remarkable achievements in modular form theory in recent years, Tennell (1983) obtained an elementary criterion of $L_E(1)\neq 0$ for the elliptic curves $y^2=x^3-n^2x$. Lehman [6] did the same thing for the curves E_D. Before stating Lehman's result we need some notations.

For $D\equiv 1 \pmod 4$, let

$$[a, b] = \#\{(x, y, z)\in\mathbb{Z}^3 \mid x^2+y^2+28z^2=D;\ x\equiv a,\ y\equiv b \pmod{14}\},$$

$$a(D)=\begin{cases}[1,0]-[7,6] \\ [3,0]-[7,4] \\ [5,0]-[7,2] \\ [1,2]-[5,6] \\ [3,6]-[1,4] \\ [5,4]-[3,2]\end{cases} \quad \text{if } D\equiv\begin{cases}1\\2\\4\\5\\3\\6\end{cases} \pmod 7.$$

For $D\equiv 2, 3 \pmod 4$, let

$$\{a, b\} = \#\{(x, y, z)\in\mathbb{Z}^3 \mid x^2+2y^2+14z^2=D;\ x\equiv a,\ y\equiv b \pmod 7\},$$

$$b(D)=\begin{cases}\{1,0\}-\{0,2\}\\ \{3,0\}-\{0,1\}\\ \{2,0\}-\{0,3\}\\ \{2,2\}-\{1,3\}\\ \{1,1\}-\{3,2\}\\ \{3,3\}-\{2,1\}\end{cases}, \quad \text{if} \quad D\equiv\begin{cases}1\\2\\4\\5\\3\\6\end{cases} \pmod 7.$$

Let l_1 (l_2) be the number of odd prime factor p of D such that $\left[\frac{p}{7}\right]=1$, $\left[\left[\frac{p}{7}\right]=-1\right]$, $l=l_1+\left[\frac{l_2-1}{2}\right]$, and

$$S(D)=\begin{cases}(a(D)/2^l)^2, & \text{if } D\equiv 1 \pmod 4,\\ (b(D)/2^l)^2, & \text{if } D\equiv 2,3 \pmod 4.\end{cases}$$

Lehman [4] proved that

(A) $L_D(1)\neq 0 \iff S(D)\neq 0$,

(B) If $S(D)\neq 0$, then the BSD conjecture (II) for the curve E_D becomes the following form

$$|Ш(E_D)|=S(D).$$

Now we can prove $L_D(1)\neq 0$ for several kinds of curves E_D.

Theorem 1. *For the following cases of D, $S(D)$ is odd integer (so that $L_D(1)\neq 0$ and rank $(E_D(\mathbb{Q}))=0$).*

(I) *For $D=p$ case (p is an odd prime number),*

(I, 1) $p\equiv 1 \pmod 4$,

(I, 1, 1) $\left[\frac{p}{7}\right]=-1$,

(I, 1, 2) $\left[\frac{p}{7}\right]=1$ *and* $\left[\frac{-7}{p}\right]_4=-1$.

(I, 2) $p\equiv 3 \pmod 8$, $\left[\frac{p}{7}\right]=-1$.

(I, 3) $p\equiv 7 \pmod 8$, $\left[\frac{p}{7}\right]=1$ *and* $\left[\frac{1+2\sqrt{2}}{p}\right]=-1$.

(II) *For $D=2p$ case (p is an odd prime number),*

(II, 1) $p\equiv 1 \pmod 8$, $\left[\frac{p}{7}\right]=-1$.

(II, 2) $p\equiv 3 \pmod 8$, $\left[\frac{p}{7}\right]=-1$.

(II, 3) $p \equiv 5 \pmod 8$, $\left[\frac{p}{7}\right]=1$.

(II, 4) $p \equiv 7 \pmod 8$, $\left[\frac{p}{7}\right]=1$ *and* $\left[\frac{1+2\sqrt{2}}{p}\right]=-1$.

(III) *For* $n=pq$ *case*, $\left[\frac{p}{7}\right]=\left[\frac{q}{7}\right]=-1$ (p *and* q *are distinct odd prime numbers*),

(III, 1) $p \equiv q \equiv 1 \pmod 4$.
(III, 2) $p \equiv 1$, $q \equiv 3 \pmod 8$.
(III, 3) $p \equiv 3$, $q \equiv 5 \pmod 8$.

(IV) *For* $n=2pq$ *case*, $\left[\frac{p}{7}\right]=\left[\frac{q}{7}\right]=-1$ (p *and* q *are distinct odd prime numbers*),

(IV, 1) $p \equiv q \equiv 1 \pmod 8$.
(IV, 2) $p \equiv 1$, $q \equiv 3 \pmod 8$.
(IV, 3) $p \equiv q \equiv 5 \pmod 8$.
(IV, 4) $p \equiv 3$, $q \equiv 5 \pmod 8$.

Proof. For the case (I, 1, 1), $D=p\equiv 1 \pmod 4$, the equation $x^2+y^2=p$ has the integer solutions $(x, y)=(\pm x_0, \pm y_0)$. $\left[\frac{p}{7}\right]=-1$ implies $7 \nmid x_0 y_0$. Therefore the equation $x^2+y^2=p$ has a unique solution (x_0, y_0) such that $(x_0, y_0)\equiv(a, b) \pmod{14}$ and $1\leqslant a, b\leqslant 6$. The contribution of this solution for $a(D)$ is 1 or -1. On the other hand, the number of solutions of the equation $x^2+y^2+28z^2=p$ satisfying $z\neq 0$ and $(x, y)\equiv(a, b) \pmod{14}$, $1\leqslant a, b\leqslant 6$ is even. Therefore $a(D)\equiv \pm 1 \equiv 1 \pmod 2$ and $S(D)=a(D)^2$ is odd. In the same way we can prove the $S(D)=b(D)^2$ is odd for the cases (I, 2), (II, 1) and (II, 2).

For the case (I, 1, 2), $D=p\equiv 1 \pmod 4$ and $\left[\frac{p}{7}\right]=1$. The arithmetic of the principle ideal domain $\mathbb{Z}\left[\frac{1+\sqrt{-7}}{2}\right]$ implies that $x^2+28z^2=p$ has integer solution and its contribution for $a(D)$ is 2. On the other hand, the reciprocity law for power-residue says that $x^2+y^2=p$ has the solution satisfying $7 \mid xy$ if and only if $\left[\frac{-7}{p}\right]_4=1$ (see Hasse [5], p. 69). Therefore the contribution of the solution of $x^2+y^2=p$ for $a(D)$ is zero if $\left[\frac{-7}{p}\right]_4=-1$.

At last, the contribution of the solutions of $x^2+y^2+28z^2=p$ ($xyz\neq 0$) is $\equiv 0$

(mod 4). Therefore $a(D)\equiv 2 \pmod 4$ and $S(D)=\left[\dfrac{a(D)}{2}\right]^2$ is odd.

For the case (I, 3), $D=p\equiv 7 \pmod 8$ and $\left[\dfrac{p}{7}\right]=1$. Antoniadis [1, p. 18] showed that $x^2+14z^2=p$ has integer solution if and only if $\left[\dfrac{1+2\sqrt{2}}{p}\right]=-1$. Therefore $b(D)\equiv 2 \pmod 4$ and $S(D)=\left[\dfrac{b(D)}{2}\right]^2$ is odd if $\left[\dfrac{1+2\sqrt{2}}{p}\right]=-1$. By the same reason we see that $S(D)$ is odd for the case (II, 4) since $2y^2+14z^2=2p$ has integer solutions and $x^2+14z^2=2p$ has not if $D=2p$, $p\equiv 7 \pmod 8$, $\left[\dfrac{p}{7}\right]=1$ and $\left[\dfrac{1+2\sqrt{2}}{p}\right]=-1$.

For the case (II, 3), $D=2p$, $p\equiv 5 \pmod 8$ and $\left[\dfrac{p}{7}\right]=1$. The contribution of the solutions of $2y^2+14z^2=2p$ is -2. On the other hand, $x^2+14z^2=2p$ and $x^2+2y^2=2p$ has no integer solution. Therefore $b(D)\equiv 2 \pmod 4$ and $S(D)=\left[\dfrac{b(D)}{2}\right]^2$ is odd.

For the case (III, 1), $D\equiv pq\equiv 1 \pmod 4$. The assumption $\left[\dfrac{p}{7}\right]=\left[\dfrac{q}{7}\right]=-1$, $p\equiv q\equiv 1 \pmod 4$ implies that $x^2+28z^2=pq$ and $y^2+28z^2=pq$ has no solution, but $A^2+B^2=p$ and $C^2+D^2=q$ have solutions satisfying $7\mid ABCD$. Since the solutions of $x^2+y^2=pq$ come from $(x+iy)=(A+iB)(C+iD)$, it is easy to see that there is essentially one solution of $x^2+y^2=pq$ such that $7\mid xy$, so its contribution to $a(D)$ is 2 or -2. Therefore $S(D)=\left[\dfrac{a(D)}{2}\right]^2$ is odd. Similarly, $S(D)$ is also odd for the case (III, 2).

For the case (III, 3), $x^2+2y^2=pq$ has no solution since $q\equiv 5 \pmod 8$. In order to determine the solvability of $x^2+14z^2=pq$ we need more knowledge on the arithmetic of the ring $R=\mathbb{Z}[\sqrt{-14}]$. The ideal class group C of R is cyclic with order four and generated by $[\ _3]$, where $_3$ is a prime ideal in R and $_3\mid 3$, $[\ ]$ denotes the class of the ideal . Let $2=\ _2^2$, the order of $[\ _2]$ is two. The assumption $p\equiv 3 \pmod 8$ implies $\left[\dfrac{-14}{p}\right]=1$ and $p=$. The assumption $\left[\dfrac{p}{7}\right]=-1$ implies that $A^2+14B^2=p$ and $C^2+14D^2=2p$ has no integer solution. Therefore and $_2$ are not principal ideals. So the order of $[\ ]$ is four. Similarly, $q=$ and the order $[\ ]$ is four. Therefore pq or pq is principal ideal, which means that the equation $x^2+14z^2=pq$ has integer solutions and its contribution to $a(D)$ is 2. Thus $a(D)\equiv 2 \pmod 4$ and $S(D)=\left[\frac{a(D)}{2}\right]^2$ is odd.

For the case (IV), $D=2pq$, $\left[\frac{p}{7}\right]=\left[\frac{q}{7}\right]=-1$. Thus the equation $2y^2+14z^2=2pq$ has no solution. For the cases (IV, 1) and (IV, 2), $x^2+14z^2=2pq$ has no solution since $p\equiv 1 \pmod 8$ and $\left[\frac{-14}{p}\right]^2=-1$. On the other hand, $x^2+2y^2=2pq$ has essentially one solution such that $7\mid xy$ and its contribution to $b(D)$ is 2 or -2. Therefore $S(D)=\left[\frac{b(D)}{2}\right]^2$ is odd. For the cases (IV, 3) and (IV, 4), $x^2+2y^2=2pq$ has no solution since $q\equiv 5 \pmod 8$. On the other hand, as we showed in the proof of the case (III, 3), $p=$, $q=$ and the order of both [ ] and [ ] are four. Thus $_2$ or $_2$ is a principal ideal, whic means that the equation $x^2+14z^2=2pq$ has integer solutions and its contribution to $b(D)$ is 2. Therefore $S(D)=\left[\frac{b(D)}{2}\right]^2$ is odd. This complete the proof of theorem 1.

3. The 2-Descent Method

The 2-descent method can be stated briefly in the following way (for the detail, see Silverman's Book [8], Chapter X).

Suppose that $E: y^2=x^3+ax^2+bx$ is an elliptic curve over $\mathbb{Q}$, then its 2-dual curve is $E': Y^2=X^3-2aX^2+(a^2-4b)X$. We have the 2-isogeny

$$\varphi: E\to E',\ \varphi(x,y)=(y^2/x^2,\ y(b-x^2)/x^2).$$

$\operatorname{Ker}\varphi=E[\varphi]=\{0,(0,0)\}$. Let $\hat{\varphi}: E'\to E$ be the dual of φ, $\varphi\hat{\varphi}=[2]$, $\hat{\varphi}\varphi=[2]$. We have the following exact sequences

$$0\to \frac{E'(\mathbb{Q})[\hat{\varphi}]}{\varphi(E(\mathbb{Q})[2])}\to \frac{E'(\mathbb{Q})}{\varphi(E(\mathbb{Q}))}\xrightarrow{\hat{\varphi}} \frac{E(\mathbb{Q})}{2(E(\mathbb{Q}))}\to \frac{E(\mathbb{Q})}{\varphi(E'(\mathbb{Q}))}\to 0, \quad (1)$$

$$0\to E'(\mathbb{Q})/\varphi(E(\mathbb{Q}))\to S^{(\varphi)}(E/\mathbb{Q})\xrightarrow{f} Ш(E/\mathbb{Q})[\varphi]\to 0, \quad (2)$$

$$0\to E(\mathbb{Q})/\hat{\varphi}(E'(\mathbb{Q}))\to S^{(\hat{\varphi})}(E'/\mathbb{Q})\xrightarrow{\hat{f}} Ш(E'/\mathbb{Q})[\hat{\varphi}]\to 0, \quad (2')$$

$$0\to Ш(E/\mathbb{Q})[\varphi]\to Ш(E/\mathbb{Q})[2]\xrightarrow{\varphi} Ш(E'/\mathbb{Q})[\hat{\varphi}], \quad (3)$$

where Ш$(E/\mathbb{Q})$ is the Tate-Shafarevitch group of $E/\mathbb{Q}$, $S^{(\varphi)}(E/\mathbb{Q})$ is the φ-Selmer group of $E/\mathbb{Q}$, which is a finite abelian group and can be calculated by using the following isomorphism. Let

$$S=\{\infty\}\bigcup\{\text{prime factor of } 2b(a^2-4b)\}.$$

M is the subgroup of $\mathbb{Q}^*/\mathbb{Q}^{*2}$ generated by -1 and all prime factors of $2b(a^2-4b)$. For each $d\in M$, we have the curves (homogeneous spaces of $E/\mathbb{Q}$ and $E'/\mathbb{Q}$)

$$c_d: dw^2=d^2t^4-2adt^2z^2+(a^2-4b)z^4,$$

$$c'_d: dw^2=d^2t^4+adt^2z^2+bz^4.$$

Then we have the isomorphisms of groups

$$S^{(\varphi)}(E/\mathbb{Q})=\{d\in M \mid c_d(\mathbb{Q}_v)\neq\varnothing \text{ for each } v\in S\}, \tag{4}$$

$$S^{(\hat{\varphi})}(E'/\mathbb{Q})=\{d\in M \mid c'_d(\mathbb{Q}_v)\neq\varnothing \text{ for each } v\in S\}, \tag{4'}$$

where $c_d(\mathbb{Q}_v)\neq\varnothing$ means the curve c_d has a solution $(w, t, z)\neq(0, 0, 0)$ in $\mathbb{Q}_v$. With these isomorphisms, the kernel of f and $\hat{f}$ in exact sequences (2) and (2′) are

$$\operatorname{Ker} f=\{d\in M \mid c_d(\mathbb{Q})\neq\varphi\},\ \operatorname{Ker}\hat{f}=\{d\in M \mid c'_d(\mathbb{Q})\neq\varnothing\}. \tag{5}$$

For our case, the elliptic curve is

$$E_D: y^2=x^3+21Dx^2+112D^2x$$

and its dual curve is

$$E'_D: y^2=x^3-42Dx^2-7D^2x,$$

where D is a positive square-free integer and $7 \mid D$.

$$S=\{\infty, 2, 7\}\bigcup\{\text{the prime factors of } D\}.$$

M is the subgroup of $\mathbb{Q}^*/\mathbb{Q}^{*2}$ generated by -1, 2, 7 and all prime factors of D. For each $d\in M$, the homogeneous spaces are

$$c_d: dw^2=d^2t^4-42Ddt^2z^2-7D^2z^4, \tag{6}$$

$$c'_d: dw^2=d^2t^4+21Ddt^2z^2+112D^2z^4. \tag{6'}$$

From (5) we know that

$$1,\ -7\in\operatorname{Ker} f,\ 1,\ -7\in\operatorname{Ker} f. \tag{7}$$

The exact sequences (1), (2) and (2′) imply that

$$(1+\operatorname{rank} E_D(\mathbb{Q}))+1=\dim_2 E_D(\mathbb{Q})/2E_D(\mathbb{Q})+\dim_2\frac{E'_D(\mathbb{Q})[\hat{\varphi}]}{\varphi(E_D(\mathbb{Q})[2])}$$

$$(\text{since } E'_D(\mathbb{Q})[\hat{\varphi}]=\{0, (0, 0)\},\ \varphi(E_D(\mathbb{Q})[2])=0)$$

$$= \dim_2 \frac{E'_D(\mathbb{Q})}{\varphi(E_D(\mathbb{Q}))} + \dim_2 \frac{E_D(\mathbb{Q})}{\hat{\varphi}(E'_D(\mathbb{Q}))}$$

$$= \dim_2(\mathrm{Ker}\, f) + \dim_2(\mathrm{Ker}\, \hat{f})$$

$$= (\dim_2 S^{(\varphi)} - \dim_2 Ш(E_D)[\varphi]) + (\dim_2 S^{(\hat{\varphi})} - \dim_2 Ш(E'_D)[\hat{\varphi}]), \quad (8)$$

where $S^{(\varphi)} = S^{(\varphi)}(E_D/\mathbb{Q})$, $S^{(\varphi)} = S^{(\hat{\varphi})}(E'_D/\mathbb{Q})$. From (7) we know that $\dim_2(\mathrm{Ker}\, f)$, $\dim_2(\mathrm{Ker}\, \hat{f}) \geqslant 1$, and (8) implies that

$$\mathrm{rank}\ E_D(\mathbb{Q}) = 0 \Longleftrightarrow \dim_2(\mathrm{Ker}\, f) = \dim_2(\mathrm{Ker}\, \hat{f}) = 1$$

$$\Longleftrightarrow \mathrm{Ker}\, f = \{1, -7\} \text{ and } \mathrm{Ker}\, \hat{f} = \{1, 7\}.$$

Particularly, if $S^{(\varphi)} = \{1, -7\}$ and $S^{(\hat{\varphi})} = \{1, 7\}$, then rank $E_D(\mathbb{Q}) = 0$ and $Ш(E_D)[\varphi] = Ш(E')[\hat{\varphi}] = \{1\}$. Then the exact sequences (3) implies $Ш(E)[2] = \{1\}$ and $|Ш(E)|$ is odd.

From the isomorphisms (4) and (4′) we know that in order to determine the Selmer groups $S^{(\varphi)}(E_D/\mathbb{Q})$ and $S^{(\varphi)}(E'_D/\mathbb{Q})$ we need to see if the curve c_d and c'_d have the point in the local field $\mathbb{Q}_v$ for each $v \in S$ and $d \in M$. This can be done by using the Hensel lemma with the following form.

Hensel Lemma ([8], exercise 10.12) *Suppose R is a complete ring for a discret valutation v, $F(X_1, \cdots, X_N) \in R[X_1, \cdots, X_N]$, $(a_1, \cdots, a_N) \in R^N$. If*

$$v(F(a_1, \cdots, a_N)) > 2v\left[\frac{\partial F}{\partial X_i}(a_1, \cdots, a_N)\right]$$

for some i, $1 \leqslant i \leqslant N$, then $F(X_1, \cdots, X_N) = 0$ has solution in R^N.

Now we are ready to compute the Selmer group $S^{(\varphi)} = S^{(\varphi)}(E_D/\mathbb{Q})$, $S^{(\hat{\varphi})} = S^{(\hat{\varphi})}(E'_D/\mathbb{Q})$.

Lemma 1. *Suppose that $D = p$, p is a prime number and $p \neq 7$. Then*
(I) $\{1, -7\} \subset \mathrm{Ker}\, f$, *and*

$$S^{(\varphi)} = \begin{cases} \{1, -7; p, -7p\}, & \text{if } p \equiv 1 \pmod 4, \text{ and } \left[\frac{-7}{p}\right]_4 = 1, \\ \{1, -7; -p, 7p\}, & \text{if } p \equiv 7 \pmod 8, \text{ and } \left[\frac{p}{7}\right] = -1, \\ \{1, -7\}, & \text{otherwise.} \end{cases}$$

(II) $\{1, 7\} \subset \mathrm{Ker}\, \hat{f}$, *and*

$$S^{(\hat{\varphi})} = \begin{cases} \{1, 7; 2, 14; p, 7p; 2p, 14p\}, & \text{if } p \equiv 3 \pmod 4, \text{ and } \left[\frac{p}{7}\right] = 1, \\ \{1, 7; p, 7p\}, & \text{if } p \equiv 1 \pmod 4, \text{ and } \left[\frac{-7}{p}\right]_4 = 1, \\ \{1, 7; 2, 14\}, & \text{if } p \equiv 7 \pmod 8, \text{ and } \left[\frac{p}{7}\right] = -1, \\ \{1, 7\}, & \text{otherwise}. \end{cases}$$

Proof. The homogeneous space of $E = E_p$ is

$$c_d: dw^2 = d^2t^4 - 42dpz^2t^2 - 7p^2z^4$$

where d is in the subgroup of $\mathbb{Q}^*/\mathbb{Q}^{*2}$ generated by -1, 2, 7 and p. The fact $\left[\frac{-7}{p}\right] = -1$ implies $C_{-1}(\mathbb{Q}_7) = \varnothing$, so $-1 \notin S^{(\varphi)}$.

For $d=2$, the curve c_2 is $2w^2 = 4t^4 - 84z^2t^2 - 7z^4$. Let $(w, t, z) \neq (0, 0, 0)$ is a $\mathbb{Q}_2$-point on the curve c_2. We can assume that $w, t, z \in \mathbb{Z}_2$. It is easy to see that $w \neq 0$ and $v_2(2w^2)$ is odd, but $v_2(4t^4 - 84z^2t^2 - 7z^4)$ is even. This contradiction shows that $c_2(\mathbb{Q}_2) = \varnothing$ and $2 \notin S^{(\varphi)}$. With the same reason we know that 14, $2p$, $14p \notin S^{(\varphi)}$.

For $d=p$, the curve c_p is $pw^2 = t^4 - 42t^2z^2 - 7z^4$. It is easy to see that $c_p(\mathbb{Q}_7) \neq \varnothing \Longleftrightarrow \left[\frac{p}{7}\right] = 1$. So we can assume $\left[\frac{p}{7}\right] = 1$. Then

$$c_p(\mathbb{Q}_p) \neq \varnothing \Longleftrightarrow t^4 - 42t^2z^2 - 7z^4 \equiv 0 \pmod p \text{ has}$$

$$\text{non-trivial solution}$$

$$\Longleftrightarrow \left[\frac{7}{p}\right] = 1 \text{ and } \left[\frac{21+8\sqrt{7}}{p}\right] \text{ or } \left[\frac{21-8\sqrt{7}}{p}\right] = 1.$$

From $\left[\frac{p}{7}\right] = \left[\frac{7}{p}\right] = 1$ we know that $p \equiv 1 \pmod 4$. Therefore $\left[\frac{-7}{p}\right] = 1$ and $\left[\frac{21+8\sqrt{7}}{p}\right] = \left[\frac{21-8\sqrt{7}}{p}\right]$. Thus

$$c_p(\mathbb{Q}_p) \neq \varnothing \Longleftrightarrow p \equiv 1 \pmod 4, \ \left[\frac{p}{7}\right] = 1 \text{ and } \left[\frac{21+8\sqrt{7}}{p}\right] = 1.$$

$$\Longleftrightarrow p \equiv 1 \pmod 4, \ \left[\frac{p}{7}\right] = 1 \text{ and } \left[\frac{2\sqrt{7}}{p}\right] = 1$$

$$\left(\text{since } 21+8\sqrt{7}=2\sqrt{7}\left[\frac{3+\sqrt{7}}{2}\right]^2\right)$$

$$\Longleftrightarrow p\equiv 1 \pmod 4,\ \left[\frac{p}{7}\right]=1 \text{ and } \left[\frac{-7}{p}\right]_4=1.$$

On the other hand, we can show that $c_p(\mathbb{Q}_2)\neq\varnothing$ by chosing $F=pw^2-(t^4-42t-7z^4)$, $(X_1, X_2, X_3)=(w, t, z)$, $(a_1, a_2, a_3)=(4, 1, 1)$, $i=2$ in the Hensel Lemma. Therefore

$$p,\ -7p\in S^{(\varphi)}\Longleftrightarrow p\equiv 1 \pmod 4,\ \left[\frac{p}{7}\right]=1 \text{ and } \left[\frac{-7}{p}\right]_4=1.$$

For $d=-p$, the curve c_p is $-pw^2=t^4+42t^2z^2-7z^4$. We can prove the following results by the same way

$$c_{-p}(\mathbb{Q}_7)\neq\varnothing \Longleftrightarrow\left[\frac{p}{7}\right]=-1 \qquad (\text{then we assume that} \left[\frac{p}{7}\right]=-1),$$

$$c_{-p}(\mathbb{Q}_p)\neq\varnothing \Longleftrightarrow p\equiv 3 \pmod 4 \qquad (\text{then we assume that } p\equiv 3 \pmod 4),$$

$$c_{-p}(\mathbb{Q}_2)\neq\varnothing \Longleftrightarrow p\equiv 7 \pmod 8.$$

Therefore

$$-p,\ 7p\in S^{(\varphi)}\Longleftrightarrow p\equiv 7 \pmod 8 \text{ and } \left[\frac{p}{7}\right]=-1.$$

From above calculating data we obtain the structure of $S^{(\hat{\varphi})}$ described in Lemma 1. For the group $S^{(\hat{\varphi})}$, the homogeneous space of $E'=E'_p$ is

$$c'_p: dw^2=d^2t^4+21dpt^2z^2+112z^4.$$

We just list the caculating results and omit the detail for saving the space.

$$d<0\Rightarrow d\notin S^{(\hat{\varphi})} (\text{since } c'_d(\mathbb{R})=\varnothing).$$

For $d=2$, c'_2 is $w^2=2t^4+21pt^2z^2+56p^2z^4$.

$$c'_2(\mathbb{Q}_7)\neq\varnothing$$

$$c'_2(\mathbb{Q}_2)\neq\varnothing \Longleftrightarrow p\equiv 3 \pmod 4\ (\text{we assume } p\equiv 3 \pmod 4)$$

$$c'_2(\mathbb{Q}_p)\neq\varnothing \Longleftrightarrow p\equiv 7 \pmod 8;\ \text{or}$$

$$p\equiv 3 \pmod 8 \text{ and } \left[\frac{p}{7}\right]=1.$$

Therefore

$$2, 14 \in S^{(\hat\varphi)} \Leftrightarrow p \equiv 7 \pmod 8, \text{ or}$$

$$\Leftrightarrow p \equiv 3 \pmod 8 \text{ and } \left[\frac{p}{7}\right] = 1.$$

For $d=p$, c_p' is $4pw^2 = (2t^2 + 21z^2)^2 + 7z^4$,

$$c_p'(\mathbb{Q}_7) = \varnothing \Leftrightarrow \left[\frac{p}{7}\right] = 1,$$

$$c_p'(\mathbb{Q}_p) \neq \varnothing \Leftrightarrow p \equiv 3 \pmod 4, \left[\frac{p}{7}\right] = 1; \text{ or}$$

$$p \equiv 1 \pmod 4, \left[\frac{-7}{p}\right]_4 = 1,$$

$$c_p'(\mathbb{Q}_2) \neq \varnothing.$$

Therefore

$$p, 7p \in S^{(\hat\varphi)} \Leftrightarrow p \equiv 3 \pmod 4, \left[\frac{p}{7}\right] = 1; \text{ or}$$

$$p \equiv 1 \pmod 4, \left[\frac{-7}{p}\right]_4 = 1.$$

For $d=2p$, c_{2p}' is $pw^2 = 2t^4 + 21z^2t^2 + 56z^4$.

$$c_{2p}'(\mathbb{Q}_7) \neq \varnothing \Leftrightarrow \left[\frac{p}{7}\right] = 1 \text{ (we assume } \left[\frac{p}{7}\right] = 1\text{)}.$$

$$c_{2p}'(\mathbb{Q}_p) \neq \varnothing \Leftrightarrow p \equiv 3 \pmod 4.$$

$$c_{2p}'(\mathbb{Q}_p) \neq \varnothing.$$

Therefore

$$2p, 14p \in S^{(\hat\varphi)} \Leftrightarrow p \equiv 3 \pmod 4 \text{ and } \left[\frac{p}{7}\right] = 1.$$

From the above calculating data we obtain the structure of $S^{(\hat\varphi)}$ described in Lemma 1. This completes the proof of Lemma 1.

Lemma 2. *Suppose that $D=2p$, p is a prime number, $p \neq 7$. Then* (I) $\{1, -7\} \subset \operatorname{Ker} f$ and

$$S^{(\varphi)} = \begin{cases} \{1, -7; p, -7p\}, & \text{if } p \equiv 1 \pmod 8, \text{ and } \left[\frac{7}{p}\right]_4 = 1, \\ \{1, -7; 2, -14\}, & \text{if } p \equiv 5 \pmod 8, \text{ and } \left[\frac{p}{7}\right] = -1, \\ \{1, -7; -p, 7p\}, & \text{if } p \equiv 7 \pmod 8, \text{ and } \left[\frac{p}{7}\right] = -1, \\ \{1, -7\}, & \text{otherwise.} \end{cases}$$

(II) $\{1, 7\} \subset \operatorname{Ker} \hat{f}$, *and*

$$S^{(\hat{\varphi})} = \begin{cases} \{1, 7; 2, 14; p, 7p; 2p, 14p\}, & \text{if } p \equiv 3 \pmod 4, \text{ and } \left[\frac{p}{7}\right] = 1, \\ \{1, 7; p, 7p\}, & \text{if } p \equiv 1 \pmod 8 \text{ and } \left[\frac{7}{p}\right]_4 = 1, \\ \{1, 7; 2, 14\}, & \text{if } p \equiv 5 \text{ or } 7 \pmod 8, \text{ and } \left[\frac{p}{7}\right] = -1, \\ \{1, 7\}, & \text{otherwise.} \end{cases}$$

Proof. We just list the calculating results and omit the detail. The homogeneous space of E_{2p} is

$$c_d: dw^2 = d^2t^4 - 84pdt^2z^2 - 28p^2z^4,$$

where d is in the subgroup of $\mathbb{Q}^*/\mathbb{Q}^{*2}$ generated by -1, 2, 7 and p.

$$-1, -2 \notin S^{(\varphi)} \text{ (since } c_{-1}(\mathbb{Q}_7) = c_{-2}(\mathbb{Q}_7) = \varnothing),$$

$$p \in S^{(\varphi)} \Leftrightarrow p \equiv 1 \pmod 8 \text{ and } \left[\frac{7}{p}\right]_4 = 1.$$

$$-p \in S^{(\varphi)} \Leftrightarrow p \equiv 7 \pmod 8 \text{ and } \left[\frac{p}{7}\right] = -1.$$

$$2 \in S^{(\varphi)} \Leftrightarrow p \equiv 5 \pmod 8 \text{ and } \left[\frac{p}{7}\right] = -1.$$

$$\pm 2p \notin S^{(\varphi)} \text{ (since } C_{\pm 2p}(\mathbb{Q}_2) = \varnothing).$$

For the group $S^{(\hat{\varphi})}$, the homogeneous space of E'_{2p} is

$$c'_d: dw^2 = d^2 t^4 + 42dpt^2z^2 + 448p^2z^4.$$

$$d < 0 \Rightarrow d \notin S^{(\hat{\varphi})} \text{ (since } c'_d(\mathbb{R}) = \varnothing),$$

$$p \in S^{(\varphi)} \Leftrightarrow p \equiv 3 \pmod 4 \text{ and } \left[\frac{p}{7}\right] = 1; \text{ or}$$

$$p \equiv 1 \pmod 8 \text{ and } \left[\frac{7}{p}\right]_4 = 1.$$

$$2 \in S^{(\hat{\varphi})} \Leftrightarrow p \equiv 7 \pmod 8;$$

$$p \equiv 3 \pmod 8, \left[\frac{p}{7}\right] = 1; \text{ or}$$

$$p \equiv 5 \pmod 8), \quad \left[\frac{p}{7}\right] = -1.$$

$$2p \in S^{(\varphi)} \Leftrightarrow p \equiv 3 \pmod 4) \text{ and } \left[\frac{p}{7}\right] = 1.$$

Put all data together we get the results of Lemma 2.

Theorem 2. *For the cases* (I, 1, 1), (I, 1, 2), (I, 2), (II, 1), (II, 2) *and* (II, 3) *in theorem* 1, *the Tate-Shafarevitch group* $\text{III}(E_D)$ *of the elliptic curve* $E_D/\mathbb{Q}$ *has odd order and the even part of the BSD conjecture* (II) *is true.*

Proof. From the results of Lemma 1 and Lemma 2 we know that $S^{(\varphi)} = \{1, -7\} = \operatorname{Ker} f$ and $S^{(\hat{\varphi})} = \{1, 7\} = \operatorname{Ker} \hat{f}$ for all cases described in Theorem 2. Therefore $L_{E_D}(1) \neq 0$ (Theorem 1) and the order of $\text{III}(E_D)$ is odd as we explained before. The Lehmann [6] proved that if $L_{E_D}(1) \neq 0$, then the BSD conjecture (II) is equivalent to $|\text{III}(E_D)| = S(D)$. Since $S(D)$ is also odd (Theorem 1), this completes the proof of Theorem 2.

Theorem 3. *For the cases* (III) *and* (IV) *in Theorem* 1, *the Tate-Shafarevitch group* $\text{III}(E_D)$ *of the elliptic curve* $E_D/\mathbb{Q}$ *has odd order and the even part of the BSD conjecture* (*II*) *is ture.*

Proof. All we need is to show that $S^{(\varphi)} = \{1, -7\}$ and $S^{(\hat{\varphi})} = \{1, 7\}$ for the cases (III) and (IV). For the case (III), $D = pq$ and the curve c_d is

$$c_d\colon dw^2 = d^2t^4 - 42pq\, dt^2z^2 - 7p^2q^2z^4,$$

where d is an element of the subgroup of $\mathbb{Q}^*/\mathbb{Q}^{*2}$ generated by -1, 2, 7, p and q. The following statements can be checked by Hensel Lemma.

$2 \| d \Rightarrow d \notin S^{(\varphi)}$ (since $c_d(\mathbb{Q}_2) = \varnothing$),

$-1 \notin S^{(\varphi)}$ (since $c_{-1}(\mathbb{Q}_7) = \varnothing$),

$p, q \notin S^{(\varphi)}$ (since $p \in S^{(\varphi)} \Rightarrow \left[\frac{p}{7}\right] = 1$),

$$pq \notin S^{(\varphi)} \left(\text{since } pq \in S^{(\varphi)} \Leftrightarrow \begin{array}{l} \left[\frac{p}{7}\right] = \left[\frac{q}{7}\right] = 1,\ p \equiv q \equiv 1 \pmod 4);\ \text{or} \\ \left[\frac{p}{7}\right] = \left[\frac{q}{7}\right] = -1,\ p \equiv q \equiv 3 \pmod 4) \end{array}\right)$$

$-p, -q \notin S^{(\varphi)}$ (since $-p \in S^{(\varphi)} \Rightarrow p \equiv 7 \pmod 8)$),

$-pq \notin S^{(\varphi)}$ (since $-pq \in S^{(\varphi)} \Rightarrow \left[\frac{p}{7}\right] = \left[\frac{q}{7}\right] = 1$).

Therefore $S^{(\varphi)}=\{1, -7\}$ for the case (III). On the other hand, the curve c'_d is

$$c'_d: dw^2=d^2t^4+21pq\, dt^2z^2+112p^2q^2z^4,$$

$$d<0 \Rightarrow d\notin S^{(\hat{\varphi})} \text{ (since } c'_d(\mathbb{R})=\varnothing\text{)},$$

$$2\in S^{(\varphi)} \Leftrightarrow \begin{cases} (p, q)\equiv(1, 3) \pmod 8, \text{ and } \left[\frac{q}{7}\right]=1; \\ (p, q)\equiv(1, 7) \pmod 8; \\ (p, q)\equiv(5, 3) \pmod 8, \text{ and } \left[\frac{q}{7}\right]=1, \left[\frac{p}{7}\right]=-1; \\ (p, q)\equiv(5, 7) \pmod 8, \text{ and } \left[\frac{p}{7}\right]=-1. \end{cases}$$

(or: exchange the places of p and q)

$$p \text{ or } 2p\in S^{(\hat{\varphi})} \Rightarrow \left[\frac{p}{7}\right]=1.$$

$$pq \text{ or } 2pq\in S^{(\hat{\varphi})} \Rightarrow \left[\frac{p}{7}\right]=\left[\frac{q}{7}\right]=1.$$

Therefore $S^{(\hat{\varphi})}=\{1, 7\}$ for the case (III). For the case (IV), $D\equiv 2pq$ and the curve c_d is

$$c_d: dw^2=d^2t^4-84pq\, dt^2z^2-28p^2q^2z^4,$$

$$-2\notin S^{(\varphi)} \text{ (since } c_{-2}(\mathbb{Q}_7)=\varnothing\text{)},$$

$$2\in S^{(\varphi)} \Rightarrow pq\equiv 5 \pmod 8,$$

$$p \text{ or } 2p\in S^{(\varphi)} \Rightarrow \left[\frac{p}{7}\right]=1,$$

$$-p\in S^{(\varphi)} \Rightarrow p\equiv 7 \pmod 8,$$

$$-2p\in S^{(\varphi)} \Rightarrow q\equiv 3 \pmod 8 \text{ and } \left[\frac{7}{p}\right]=1,$$

$$pq \text{ or } 2pq\in S^{(\varphi)} \Rightarrow \left[\frac{7}{p}\right]=\left[\frac{7}{q}\right]=1,$$

$$-pq, \text{ and } -2pq\notin S^{(\varphi)} \text{ (since } C_{-pq}(\mathbb{Q}_7)=C_{-2pq}(\mathbb{Q}_7)=\varnothing\text{)}.$$

On the other hand, the curve c'_d is

$$c'_d: dw^2 = d^2t^4 + 42pq\ dt^2z^2 + 448p^2q^2z^4.$$

$$d<0 \Rightarrow d \notin -S^{(\hat{\varphi})} \quad (\text{since } c'_d(\mathbb{R}) = \varnothing),$$

$$2 \in S^{(\hat{\varphi})} \Rightarrow \left[\frac{2}{p}\right] = \left[\frac{2}{q}\right] = 1 \text{ and } pq \equiv 7 \ (\text{mod } 8),$$

$$p \text{ or } 2p \in S^{(\hat{\varphi})} \Rightarrow \left[\frac{p}{7}\right] = 1,$$

$$pq \text{ or } 2pq \in S^{(\varphi)} \Rightarrow \left[\frac{p}{7}\right] = \left[\frac{q}{7}\right] = 1,$$

Therefore $S^{(\varphi)} = \{1, -7\}$ and $S^{(\hat{\varphi})} = \{1, -7\}$ for the case (IV). This completes the proof of Theorem 3.

References

[1] J. A. Antoniadis, Höhere Reziprozitätsgesetze und Modulformen von Gewicht Eins, *J. reine und angew. Math.*, **2**(1985), 11—22.

[2] B. J. Birch and H. P. Swinnerton-Dyer, Notes on elliptic curves, II, *J. reine und angew. Math.*, **218** (1965), 79—108.

[3] J. Coates and A. Wiles, On the conjecture of Birch and Swinnerton-Dyer, *Invent. Math.*, **39**(1977), 223—251.

[4] B. Gross and D. Zagier, Heegner points and derivatives of L-series, *Invent. Math.*, **84**(1986), 225—320.

[5] H. Hasse, Bericht über neuere Untersuchungen und Probleme aus der Theorie der algebraischen Zahlkörper, Teil II, Würzburg, 1965.

[6] J. L. Lehmann, Rational points on elliptic curves with complex multiplication by the ring of integers in $\mathbb{Q}(\sqrt{-7})$, *Jour. of Number Theory*, **27**(1987), 253—272.

[7] K. Rubin, Tate-Shafarevich groups and L-functions of elliptic curves with complex multiplication, *Invent. Math.*, **89**(1987), 527—560.

[8] J. H. Silverman, The Arithmetic of Elliptic Curves, Springer-Verlag, New York Inc., 1986.

EIGENVALUES OF THE LAPLACIAN FOR PSL(2, $\mathbb{Z}$): SOME NEW RESULTS AND COMPUTATIONAL TECHNIQUES

Dennis A. Hejhal

In Memory of Hua Loo Keng

§ 1. Introduction

One of the most interesting problems in the Selberg trace formalism from the standpoint of *computation* is the explicit determination of discrete eigenfunctions of the automorphic Laplacian. Cf. [6, 7, 17, 20] for the necessary theoretical background.

By virtue of its distinguished role in modern number theory, it seems reasonable to place particular emphasis on the *modular group* PSL (2, $\mathbb{Z}$) and its congruence subgroups.

It is well-known that such groups [arithmetic/noncompact/finite volume] have *both* a discrete and continuous spectrum. Part of the difficulty (numerically) stems from the fact that the bulk of the discrete spectrum lies "buried" well-within the continuous spectrum.

In this paper: we'll attack *only* the "discrete side" of the spectrum.

The continuous side leads to Eisenstein series, whose computational aspects have been recently discussed elsewhere [10].

Note that there are now *two* aspects to the problem:

(A) computation of the correct eigenvalues $\lambda_n = \frac{1}{4} + R_n^2$;

(B) determination of the corresponding eigenfunctions $\varphi_n(z)$.

In the arithmetic case, part (B) is basically synonymous with computing the Fourier coefficients of φ_n. Cf. [7, pp. 256(4.7), 585, 282(3.9)].

Various papers have appeared (since the early 1970's) dealing with some combination of (A) and (B) for the case Γ=PSL (2, $\mathbb{Z}$). Cf. [4], [8], [12], [7, appendix C], [19], [23] and the *additional* references cited in [7, appendix C]. Note that [23] also treats several other groups. In all these papers: R is kept less than 25 or so.

Recent work in quantum chaos (and the Riemann zeta function) has suggested that it might prove interesting to try to attack problems (A) & (B) for the case of considerably larger R. Cf. [1, 2, 3, 15, 18, 22]. Once again: arithmetic groups seem deserving of special attention.

Because of (increasing) difficulties with the computation of the K-Bessel function $K_{iR}(X)$, the number 25 (or so!!) has remained as a kind of "barrier" to further progress.

Recently, however, Enrico Bombieri & the author have developed a *new* program for the computation of $\exp(\frac{\pi}{2} R) K_{iR}(X)$ which works quite well for R all the way out to 75000 or more. The program uses the identity

$$K_{iR}(X) = \frac{1}{2} \int_{\mathbb{R}} e^{-X\cosh t} e^{iRt} dt \qquad (1.1)$$

followed by a *deformation of contour* very similar to that used in stationary phase. [1]

This program played an essential role in our recent investigation of the distribution of zeros of Epstein zeta functions (by way of Eisenstein series). Cf. [9, 10].

Though (in that work) R was typically kept larger than 3000, there is *no* difficulty making several changes in the original CRAY code (readjusting error terms, etc.) so that $\exp(\frac{\pi}{2} R) K_{iR}(X)$ can be efficaciously computed for all $1 \leqq R \leqq 75000$, $\frac{1}{5} \leqq X < \infty$, say.

One naturally *wonders* to what extent this new code allows one to *dispense* with the "R-barrier" previously encountered [at least for $\Gamma = PSL(2, \mathbb{Z})$].

A look at [8, §6] shows that the answer to this question depends not only on the computation of $K_{iR}(X)$ but *also* (just as importantly!!) on the linear algebra associated with solving

$$\sum_{n=1}^{N} c_n I_n(z_j, R) = 0, \qquad 1 \leqq j \leqq N \qquad (1.2)$$

for various batches of test points $\{z_1, \cdots, z_N\}$. Compare §2 below.

Our *primary aim* in this paper is to report on the outcome of several new (supercomputer) experiments in this direction.

As will be seen: the results we obtain go well beyond those in [7, appendix C]. In short: one is now able to sample λ_n all the way out to $\lambda = 250000$ or so.

---

(1) The choice of contour hinges on whether $R > X$, $R \approx X$, $R < X$. The deciding factor is the size of $|R-X|^{3/2}/\sqrt{R}$. The middle case corresponds to the so-called *transitional* region. Cf. [21, pp. 235, 225, 202] for the proper perspective.

Hua Loo-Keng was quite interested in the author's original paper [8] and encouraged him to pursue further work in this area. It therefore seems especially fitting to be able to dedicate this paper to his memory.

The computer time for this work was provided by 3 sources:
(i) the Minnesota Supercomputer Institute (CRAY-2);
(ii) the National Science Foundation (XMP-48 at the San Diego Supercomputer Center);
(iii) the Swedish NFR (CRAY-1 at Linköping).

Additional support was received from NSF Grant DMS 86-07958 and the math department at Chalmers Tekniska Högskola (in Sweden). I am also grateful to Mr. Barry Rackner of the Minnesota Supercomputer Center for his assistance with the various operating systems, and in running many of the jobs [particularly those on the XMP].

§2. The Basic Procedure

The approach we use is a slight *modification* of the one outlined in [8].

We are interested in computing nonholomorphic cusp forms (i. e. Maass wave forms) for PSL(2, ℤ)\H. Such functions can be represented in the form

$$\varphi(x+iy)=\sum_{n=1}^{\infty} c_n y^{\frac{1}{2}} K_{iR}(2\pi ny)\left\{\frac{\cos(2\pi nx)}{\sin(2\pi nx)}\right\} \tag{2.1}$$

subject to the condition that:

(a) $\varphi(-\frac{1}{z})=\varphi(z)$ for $z\epsilon H$;

(b) $\sum_{n=1}^{\infty}\frac{c_n}{n^s}=\prod_p \frac{1}{1-c_p p^{-s}+p^{-2s}}$;

(c) $|c_n|\leqq d(n)n^{1/4}$.

The coefficients c_n are real; $d(n)$ is the usual divisor function; p means prime. To avoid any confusion, we explicityly remark that

$$\Delta\varphi+(\frac{1}{4}+R^2)\varphi=0. \tag{2.2}$$

In other words: $\frac{1}{4}+R^2$ is the *real* eigenvalue, not R. For later use, we also write $z^* \equiv 1/\bar{z}$.

Condition (a) reflects the fact that PSL(2, ℤ) is generated by $E(z)\equiv -\frac{1}{z}$ & $S(z)\equiv z+1$ and that φ needs to be automorphic. The condition $\varphi(z+1)=\varphi(z)$ is automatically fulfilled.

Condition (b) says that φ is an eigenfunction of the Hecke operators:

$$T_p[f] \equiv \frac{1}{\sqrt{p}}\left[f(pz)+\sum_{j=0}^{p-1} f\left(\frac{z+j}{p}\right)\right]. \tag{2.3}$$

In fact: $T_p[\varphi]=c_p\varphi$.

The generalized Ramanujan-Petersson *conjecture* asserts that $|c_n|\leq d(n)$. Condition (c) is a well-known (partial) result in this direction. Cf. [13, 14, 11] for further information.

The function φ is said to be "even" or "odd" according to whether *cos* or *sin* appears in (2.1). (Similarly for the corresponding R.)

Let F be the usual fundamental polygon for $PSL(2, \mathbb{Z})\setminus H$. For each $z\in int(F)$, *note* that condition (a) can be rewritten in the form:

$$\sum_{n=1}^{\infty} c_n I_n(z_j^!, R)=0. \tag{2.4}$$

The function $I_n(z, R)$ is an explicit combination of K-Bessel functions and sines or cosines. In the "even" case, one can take:

$$I_n(z, R)\equiv \sqrt{y^*}\, K_{iR}(2\pi ny^*)\cos(2\pi nx^*)-\sqrt{y}\, K_{iR}(2\pi ny)\cos(2\pi nx).$$

The "odd" case uses "+" and *sin*. Here $z^*\equiv x^*+iy^*$.

Since $K_{iR}(X)$ is asymptotic to $\sqrt{\frac{\pi}{2X}}\, e^{-X}$ for $X\gg R$, the terms $I_n(z, R)$ begin to decay *exponentially fast* once n exceeds $\frac{R}{2\pi y^*}$ or so. Cf. (1) *loc. cit.* and (2).

The obvious temptation is to now take N "very large" and to try to solve

$$\sum_{n=1}^{N} c_n I_n(z_j, R)=0 \tag{2.5}$$

---

(2) Two additional facts should be kept in mind here. First: that $|K_{iR}(X)|\leq K_{\frac{1}{2}}(X)=\sqrt{\frac{\pi}{2X}}\, e^{-X}$. Second that (for fixed X) the function $K_{iR}(X)$ *oscillates* ("quasitrigonometrically") in an envelope having width roughly equal to $O(e^{-\frac{\pi}{2}R})$. The *average distance* between successive zeros of $K_{iR}(X)$ is $\frac{\pi}{\omega}$ where $\omega\equiv\log\left(\frac{2R}{X}\right)$. Cf. [16, p. 315].

For later use: note that the corresponding *average* for the list of even (or odd) R_n is $\frac{12}{R}$ if $\Gamma=PSL(2, \mathbb{Z})$. See [20, §6.7] and [7, p. 511].

over a batch of N randomly chosen testpoints $\{z_1, \cdots, z_N\} \subseteqq \mathrm{int}(F)$. Cf. [8, p. 102 (top)].[3]

To allow for larger values of R [cf. [2] and (1.1)], the I_n should be *pre*multiplied by $\exp(\frac{\pi}{2} R)$ from the very start. The number N should be chosen so that

$$|I_l(z_k, R)| \leqq \begin{pmatrix} \text{something like} \\ 10^{-9} \end{pmatrix} \max_{\substack{1\leqq j\leqq N \\ 1\leqq n\leqq N}} |I_n(z_j, R)| \tag{2.6}$$

for every $l > N$ and $k \in [1, N]$. 10^{-9} can be replaced by 10^{-12} or 10^{-15} if greater accuracy is desired.

If several *precautions* are observed, it is *highly probable* that this (simple minded) procedure will actually yield correct "answers."

Four items need to be checked. Specifically:

(i) it is *essential* that the *same* R be obtained even when the points z_j are varied;
(ii) likewise for the (associated) c_n;
(iii) the (purported) c_n should satisfy the *multiplicative relations* implicit in condition (b) [e.g. $c_4 = c_2^2 - 1$, $c_6 = c_2 c_3$, $c_8 = c_2^3 - 2c_2$, $c_{12} = c_4 c_3$];
(iv) the c_n should *also* satisfy (c) and (*hopefully*) the Ramanujan-Petersson conjecture.

From the standpoint of "probability", precaution (iii) is probably the most convincing.

In practice: (iv) tends to take care of itself after (i)–(iii).

The foregoing precautions are "tempered" by the fact that (2.5) is a *truncation* and that all (real) numbers have *finite* precision on a computer. In particular:

$$\sum_{n=2}^{N} c_n I_n(z_j, R) = -I_1(z_j, R), \quad 1 \leqq j \leqq N-1 \tag{2.7}$$

typically leads to nonsensical c_n [i.e. noisy/excessively large values] once n *exceeds* a certain bound determined by R, N, and the z_j. This bound will be loosely referred to as the "c_n hump." It is quite visible in numerical experiments.[4]

This state-of-affairs plainly shows that (i)-(iv) have to be taken with a grain of salt. As precautions they should be carried out *only* so far as the basic numerical configuration allows.

In [8] and [12], R was determined by solving the equation

---

(3) The y_j are kept *bounded* [to ensure that y_j^* stays away from 0].

(4) It is reasonable to expect that the c_n "hump" will scale something like $\frac{R}{2\pi y}$, where y is *some* number betwen $\min(y_k^*)$ and $\max(y_k)$. Though this is *not* completely accurate, it is good enough for a first approximation [assuming [3] and (2.6)].

$$\det\,[I_n(z_j, R)]_{\substack{1\leq j\leq N\\ 1\leq n\leq N}} = 0. \tag{2.8}$$

This equation is *not* sufficient all by itself. Spurious R-values crop up unless precautions (i)-(iii) are observed.

An *alternative* procedure would be to replace (2.5) by (2.7) at the outset and then determine R by imposing the constraint that

$$c_4 + 1 - c_2^2 = 0 \quad (\text{say}). \tag{2.9}$$

It is *again* essential to adhere to precautions (i)-(iii).

The procedure using (2.8) is prone to *excessive* machine noise as R , so we go with (2.7) + (2.9) instead.

§3. Some Informal Remarks Concerning Implementation of the Basic Procedure

Though the basic procedure is quite transparent, its *implementation* is strewn with obstacles. In §§3-4, we briefly discuss several of the more important ones.

To begin with: observe that N has to be kept larger than something like

$$\frac{R}{2\pi y_{min}}, \text{ where } y_{min} \equiv \min(y_k^*).$$

This is good and bad. As R increases, the amount of matrix algebra implicit in solving (2.7) goes up. But, in line with (4), the c_n "hump" *also* goes up. This means that we *should* be able to "read off" significantly more coefficients (as a "reward" for our greater effort).

This rather enticing state-of-affairs was, in fact, one of the *primary* motivations for trying to implement the whole scheme.

But the "down" side is also quite apparent. At the outset, there are *two* immediate concerns.

A. Since the y_j are kept bounded, the basic matrix

$$[I_n(z_j, R)]_{\substack{1\leq j\leq N-1\\ 2\leq n\leq N}} \qquad (\text{called } J_R \text{ for short})$$

can, in principle, become more & more singular (or "ill-conditioned") as $R \to \infty$, $N \to \infty$. Cf. [5]. This causes problems in (2.7).

A machine like the CRAY2 gains *enormously* in speed by virtue of vectorization [a kind of automatic "parallel processing" of those parts of the code requiring *only* simple (*but repetitive*) arithmetic operations]. The "catch" is that the variables must all be single-precision. Passing to double-precision [i.e. 28 places instead of the usual 14-15] typically requires not 2 (or 4) but over 100 times as long. For this reason: there are obvious advantages to trying to remain with single-precision in (2.7) for as long as possible. [5]

Bear in mind (too) that $c_4+1-c_2^2=0$ is solved in effect by "hitting" (2.7) over & over again with Gauss elimination on a suitable R-grid, and then looking for changes in sign [$c_4+1-c_2^2$ being regarded as a function of R].

In proceeding along these lines: it is essential to try to arrange the z_j in some "reasonably favorable" configuration so that the *conditioning* of J_R is optimized -- and so that *single* precision does indeed suffice.

Though precaution (iii) will "let us know" when we've arrived, there are no *apriori* guarantees. One can only HOPE for the best.

We'll return to this point later.

B. The *second* difficulty is more mundane: namely the repetitive computation of *all* those Bessel functions in J_R.

To solve $c_4+1-c_2^2=0$, the basic R-grid has to be kept "finer" than a *small fraction* of the average distance between successive R_n. (Otherwise pairs of "nearby" R_n will be missed.) This means that the basic ΔR [in (2.7)] has to be kept at something like:

$$\frac{1}{50}\left(\frac{12}{R}\right) \text{ or } \frac{1}{100}\left(\frac{12}{R}\right) \text{(or less!)}. \tag{3.1}$$

Cf. [2]. This grid can (*then*) be further refined on those intervals (actually) containing a change-of-sign.

This process clearly involves a lot of computation. One naturally thinks of trying to use Lagrange interpolation (or something similar!) in an attempt to save time. [The functions $K_{iR}(X)$ are, after all, *holomorphic* with respect to R.]

This idea is actually quite fortuitous. To see this, let $X_{min}=2\pi y_{min}$. For holomorphic functions, one is inclined to say (loosely speaking) that Lagrange interpolation is *reasonable* so long as the degree is large and the grid size is less than a small fraction of the *average distance* between the sucessive "peaks and valleys" on the graph. For the Bessel functions appearing in J_R, this distance is *not less* than

$$\frac{\pi}{\log\left(\frac{2R}{X_{min}}\right)}.$$ Cf. [2]. Since

(5) Unless, of course, time and money are no object.

$$\frac{1}{10}\left[\frac{\pi}{\log\left(\frac{2R}{X_{min}}\right)}\right] >> \frac{1}{50}\left(\frac{12}{R}\right) \text{ typically,} \tag{3.2}$$

we *should* be able to reap a *very* significant savings (at least if life is reasonable). Whether life is "reasonable" or not can (of course) be *checked* by making direct "spotchecks" in various R- neighborhoods. In our case: it was determined that [for degree 11] and a coarse grid like the LHS of (3.2), we were achieving $10 \sim 12$ significant figures all the way out to $R = 1000$.

Concerns A and B suggest that there should exist *three* basic length scales

$$H1 > H2 > H3$$

in the actual program.[6]

H1 is the "coarse" grid on which the initial computation of all the Bessel functions takes place.

H2 is the "finer" scale on which —— by Lagrange interpolation —— the matrices J_R are all assembled and then "manipulated" [ala (2.7)] to locate changes-of-sign for $c_4 + 1 - c_2^2$.

H3 is the "final" scale reflecting the accuracy we *ultimately* hope to achieve for R when solving (2.9).

To describe the passage from H2 to H3, it is necessary to supply a bit more detail concerning the actual *mechanics* of the code.

Prior to doing this, we should perhaps stress that, in designing the code, our basic attitude was [and still is] that we would think of the program as being only experimental.

A number of obstacles had to be overcome in an *ad hoc* way. The "evolving" program gradually needed to take on various parameters and adjustable components. We have *not* yet made any systematic attempt to optimize *all* aspects simultaneously. [It takes nontrivial cpu time to run the necessary tests.]

On the other hand: there are *only* a finite number of eigenvalues from $R = 1$ out to $R = 1000$ (say). Once these are known to 6 decimal places (or whatever), further "streamlining of the code" could well be moot.

We have therefore sought to maintain a proper *balance*, philosophically, in our use of:

(i) educated guessing;

---

(6) at least for starters!

(ii) small-scale empirical tests;
(iii) rigorous theoretical estimates (and error analyses);
(iv) sheer luck.

§4. Further Remarks.

When designing a code that is expected to require *hours* (as opposed to minutes) of CRAY time, it is wise to make preliminary tests(7) in at least *one* TYPICAL regime. Features that are totally insignificant (or "invisible") at $R=25$ may, for instance, grow large enough to virtually "destroy" a code at $R=1000$.

In the present case, we decided to fix a modest value of R (125) and then see if we could *at least* get the code to work reasonably well in a small neighborhood of this value.

The number 125 was a compromise: large enough to be "interesting" yet small enough to allow the testing to be done fairly quickly.

Our guess was that, with a bit of "fine-tuning", the basic procedure would actually work——in *single* precision. [Verifying this *hunch* (about s. p.) was the *other* primary motivation for pursuing these experiments. The first was the *hope* of obtaining more c_n as R increased.]

The initial tests at 125 were very discouraging until (in desperation) we tried a few "last" z_j-configurations and a slightly different R-interval. We *finally* found two R-values where precaution (iii) worked beautifully.

The ensuing optimism was tempered by the *disturbing fact* that concern A was apparently very real. Depending on the z_j-batch used, the matrix J_R could either be "well-conditioned" or a "disaster." It is (to be sure) eminently reasonable to *expect* that $\det(J_R)$ will have zeros. Just like (2.8). But one would *hope* that, for generic z_j, the two zero-sets would be "reasonably disjoint."

In cases (*or regions*) where J_R is a "disaster", one will typically miss many "true" R because of all the "static."

The idea of automatically testing *several* z_j-batches to "increase the completeness probability" now becomes rather clear.

Something akin to this actually seems quite essential—*at least if one still hopes to use* only *single-precision.*

On the matter of z_j-configurations, our naive *guess* was that the z_j needed to be "spread out" as far as possible—consistent with (3). The interesting (and initially *deceptive*) feature was that this idea did *not* work—at least for rectangular configurations.

---

(7) or sample runs

In practice: the "best" J_R seemed to arise from cases having $\max(y_k)$ fairly small [e. g. 2] with a *minimal* number of vertical columns [again like 2].

There were also some difficulties with (2.9) itself which needed to be addressed in the code.

It was often observed that the zeros of $c_4+1-c_2^2$ were "fuzzy." That is: even when H2 was consistent with (3.1), there frequently appeared "strings" of 2 or more successive intervals *each* containing (*according* to the machine) a change-of-sign. The explanation for this (8) is obviously one of machine noise. Some z_j-batches were [in line with §3A] *much worse* than others in this regard.

This type of "fuzziness" was already noted by P. Cartier in connection with (2.8) — but (2.8) is generally quite a bit "noisier" than (2.9).

One naturally attempts to get a "better grip" on any R (using *only* the H2-level data) by applying a linear interpolation or something similar. Each interval containing a change-of-sign thus leads to a number called R_{temp}.

In the case of fuzzy zeros, one obtains *strings* of R_{temp}'s.

To solve the problem of eliminating the "phonies", we kept in mind the "dictum" (or basic necessity) of *always* testing several z_j batches [to maximize the probability of completeness]. There was room here to be a bit wasteful, so we simply declared as "unsafe" any case where the crossing associated with R_{temp} was *not* locally transverse [on a scale of several H2 intervals] — and then proceeded *no* further with this value (beyond printing it). This seemed to work reasonably well.

To pass from level H2 to H3, one effectively needs to "blow up" small neighborhoods of R_{temp}.

Any type of "refinement process" is bound to take time [especially if $\frac{H2}{H3}$ is large!], so it obviously pays to eliminate *beforehand* any "case" strongly suspected of being spurious. By that we mean any R_{temp} where one of the Hecke relations beyond (2.9) [e. g. $c_2c_3=c_6$] is *already* violated (9) at the level of H2.

For $R\approx 125$, this "weeding process" was already a virtual necessity in every z_j-batch.

At this point: another significant problem occurs. Prompted by certain irregularities in deciding whether the "weeding test" could actually be trusted, we performed a variety of tests on the coefficients c_n. It was found (*especially* for larger R!!) that the c_n tend to "act" like *continuous* but *nowhere* differentiable functions of

---

(8) when R is still comparatively small

(9) sufficiently badly

R as far as the *machine* goes. This is due to machine noise in solving (2.7). As n gets closer to the c_n hump, the coefficient "velocities" (wrt R) tend to grow ever larger.[10] The corresponding c_n are thus *very* sensitive.

This observation is actually relevant even at $R \approx 13.779751$ (the *first* even R). There the c_n "hump" was located at about 7. This makes testing (iii) a bit tricky. [One has to be virtually on top of the true R to see the c_6 or c_8 relations begin to work. But, since (in this case), the velocity *fluctuations* remain fairly small, one *does* eventually get there. And *all* in single precision. Cf. § 10 below.]

The "C^0-C^1 problem" was addressed by introducing a "filter" instead of employing any kind of bisection or *regula falsi* to solve (2.9).

The basic idea is to *re*compute all the Bessel functions on an appropriate H2 grid around R_{temp}, and then use Lagrange interpolation to prepare the matrices J_R along a suitable H3 grid containing R_{temp}. The "true" R is then obtained by *repeatedly* solving (2.7) and seeking to *minimize* a test functional like:

$$|c_4+1-c_2^2|+|c_6-c_2c_3|+|c_8-c_2^3+2c_2|. \tag{4.1}$$

By keeping H3 sufficiently small[11], this process seems to act as a reasonably good *filter* for whatever c_n noise [or "static"] may be present. Because regula falsi fails, one is "taking significant chances" if the filter is applied on an interval $\{|R-R_{temp}| \leqq \gamma(H2)\}$ with γ *significantly less* than 1/2 or so.

In practice, it didn't seem to make much difference which functional was used in (4.1) so long as it was reasonable. We generally used

$$\left|\frac{c_2^2-1}{c_4}-1\right|+\left|\frac{c_2c_3}{c_6}-1\right|. \tag{4.2}$$

One way to speed the process up-as regards the *repeated* solving of (2.7)—is to use the same type of code but to *assume* that the c_n can be legitimately computed at level H3 by mere *linear* interpolation of the coefficients *already* obtained at level H2. Some surprisingly good answers can be obtained in this way. Because of the C^0-C^1 problem, this method is certainly *not* rigorous—but does have the advantage of locating *some* of the true R—values (to a few less decimal places) rather quickly.[12]

The more refined version of the code can then be used if greater accuracy is desired.

One can also use this "shortcut" to help determine which cases are most "stable"—or to search for missing R-values.

This completes our remarks on the implementation of §2 except for noting one further "curiosity."

---

(10) "Velocities" like 10^5 or 10^6 are quite common (even at the level of H2).

(11) this condition is essential.

(12) at least for moderate R!! Beyond a certain point, there is simply *too much* intrinsic error in solving (2.7) for the "linear interpolation hypothesis" to be of any use at level H3. Cf. § 11.

Different machines—*and even different compilers*—can[(13)] give rise to different noise [or "garbage bits"] when the matrices J_R are prepared. For this reason: cases which were "triggered" at level H2 in *one* setting can be completely *missed* in another. This behavior reflects 2 things; namely

(i) the intrinsic variability in the c_k noise level; and

(ii) the fact that the noise is filtered only AFTER the "trigger" is "activated" at level H2 by a suitably transverse crossing of $c_4 + 1 - c_2^2$.

We saw *many* examples of this irritating *anomaly* as R went beyond 125.[(14)]

With the code structured the way it is now, the notion of trying to increase the "completeness probability" is obviously *very* fundamental. At present: one simply does not have any kind of *theoretical* guarantees [or, for that matter, any *apriori* knowledge of optimal z_j].[(15)] One can only be guided by knowledge of the average gap 12/R, testing several independent z_j-batches, and by careful exercise of precautions (i)-(iv), especially (iii).

Appendix A contains an *example* of one of our codes. Some readers may be interested in seeing what the code actually looks like. (Due to space limitations, the associated "manual" has to be omitted.)

We now turn to the experimental results···

§5. The Even Eigenvalues Less Than 50.

In this experiment, we used the CRAY−1. We considered 5 batches of z_j satisfying

$$A \leqq y_j \leqq 2 \ \& \ x_j = 0 \text{ or } \frac{1}{2} \quad \text{(i.e. } \textit{two} \text{ vertical columns)}$$

with a "control" parameter $A \in (1, 1.100]$. We then chose

$$H1 = 0.025, \ H2 = 0.001, \ H3 = 10^{-6}$$

in line with constraints (3.1) + (3.2). Since the c_n "hump" was rather small [especially for $R < 25$], we *omitted* the preliminary "weeding" of spurious R at level H2. We took $\gamma = \frac{1}{5}$ and used (4.2) as the filter.

(13) on the *same* program

(14) And were initially very perplexed by it. [Our initial *expectation* was that the results would be *identical*· or else very nearly so.]

(15) Nor does one possess any kind of computational argument principle to help with the bookkeeping [as with $\zeta(s)$].

Our results are displayed in tables 1 & 2. In table 2, the *last* digit of each c_n may be off. Compare: [7, appendix C] and [19].

Table 1

Even Eigenvalues for PSL(2, ℤ). The ordered pairs express (N, c_n "hump")

13.779751 (13, 7)	35.502349 (23, 12)	45.287438 (25, 14)
17.738563 (15, 8)	35.841677 (23, 12)	45.361613 (25, 14)
19.423481 (15, 8)	36.677553 (23, 12)	45.398470 (25, 14)
21.315796 (17, 8)	36.856349 (23, 12)	46.101456 (25, 14)
22.785909 (17, 10)	37.825072 (23, 13)	46.481402 (27, 15)
24.112353 (17, 10)	38.303276 (23, 13)	46.653318 (27, 15)
25.826244 (19, 10)	39.168085 (23, 13)	47.422896 (27, 15)
26.152085 (19, 10)	39.407532 (23, 13)	47.926558 (27, 15)
27.332708 (19, 10)	39.773623 (23, 13)	48.039331 (27, 15)
28.530747 (19, 10)	40.543351 (23, 13)	48.741666 (27, 15)
28.863394 (19, 10)	40.688666 (23, 13)	48.998308 (27, 16)
30.410679 (21, 11)	41.555578 (23, 14)	49.683520 (27, 16)
31.526582 (21, 11)	41.883003 (23, 14)	49.961697 (27, 16)
31.566275 (21, 11)	42.643489 (25, 14)	50.089705 (27, 16)
32.508118 (21, 11)	42.922228 (25, 14)	
32.891170 (21, 11)	43.267182 (25, 14)	
34.027884 (21, 11)	44.077405 (25, 14)	
34.456271 (23, 11)	44.426348 (25, 14)	

Table 2

Fourier Coefficients for the Even Eigenvalues Less than 35.

R	13.779751	17.738563	19.423481	21.315796	22.785909
c_2	1.549305	−0.765456	−0.692759	1.287534	0.267693
c_3	0.246900	−0.977777	1.562349	1.251768	−0.585496
c_5	0.737	−1.015	−0.03843	1.170	0.03834
c_7	***	***	***	***	0.99

R	24.112353	25.826244	26.152085	27.332708	28.530747
c_2	1.712436	0.258066	−1.866163	−0.209009	−1.460502
c_3	0.881068	1.333742	−0.403768	−0.114727	0.211383
c_5	−0.35537	1.276361	−0.160388	−0.700599	1.432932
c_7	1.32	0.743	−0.619	−0.252	1.1582

R	28.863394	30.410679	31.526582	31.566275	32.508118
c_2	0.770445	1.346139	−0.75896	0.530999	1.812750
c_3	−1.559404	0.186890	1.64391	−0.847997	1.171151
c_5	0.308238	1.350177	0.84987	−1.392501	−0.414880
c_7	−1.3132	0.7649	0.9736	−0.54173	0.323447

R	32.891170	34.027884	34.456271	
c_2	0.173574	1.171189	−1.235721	
c_3	−0.501554	−0.947719	−0.553993	
c_5	1.819350	−1.034348	0.900853	
c_7	−0.54885	1.14157	0.61465	

We also include 1 slightly larger example (for use as a possible reference in other experiments).

Table 3

R=47.926558	
$c_2=0.511058$	$c_7=0.028926$
$c_3=1.700681$	$c_{11}=-0.295543$
$c_5=-1.358583$	$c_{13}=-0.08748$

To convey some of the accuracy, we remark that:

$$|c_4-c_2^2+1|=2\times 10^{-8} \qquad |c_6-c_2c_3|=4\times 10^{-8}$$
$$|c_8-c_2^3+2c_2|=1.3\times 10^{-7} \qquad |c_9-c_3^2+1|=1.3\times 10^{-7}$$
$$|c_{10}-c_2c_5|=2.2\times 10^{-7} \qquad |c_{12}-c_3c_4|=3.7\times 10^{-6}$$
$$|c_{14}-c_2c_7|=5.3\times 10^{-3} \quad [\text{ for } N=27 \text{ and a } c_n \text{ hump of about } 15].$$

Ramanujan-Petersson was checked for all R-values in table 1 — at least out to the c_n "hump". There were *no* violations.

We should also mention that, once R reached 40 or so, certain R_n began to be "missed" on one batch or another. Cf. §4 paragraphs 6—7.

Sample runtimes for testing [the interval] $46 \leq R \leq 50$ are as follows:

CRAY-1	414 sec.
XMP	315 sec.
CRAY-2 (cft)	293 sec.
CRAY-2 (cft77)	278 sec.

The anomaly mentioned at the end of §4 does not occur yet. [Incidentally: *each* Bessel function takes about 0.0019/0.0002/0.0002 seconds to compute on the CRAY-1 depending on whether R>X, R≈X, R<X. The other machines are proportionately faster.]

§6. The Odd Eigenvalues Less Than 50.

Here we used the same set-up as in §5 but only 4 batches instead of 5. Since the CRAY-1 was temporarily unavailable, we switched over to the CRAY-2.

Table 4

Odd Eigenvalues for PSL (2, ℤ). The ordered pairs express (N, c_n"hump").

9.533695 (11, 6)	25.050855 (17, 10)	32.932465 (21, 12)
12.173008 (13, 7)	26.056918 (19, 10)	33.492331 (21, 12)
14.358509 (13, 7)	26.446996 (19, 11)	33.570990 (21, 12)
16.138073 (15, 8)	27.284384 (19, 10)	34.185970 (21, 12)
16.644259 (15, 8)	27.775921 (19, 10)	34.695311 (23, 12)
18.180918 (15, 8)	28.510278 (19, 10)	35.431665 (23, 12)
19.484714 (15, 8)	29.137588 (19, 11)	35.666397 (23, 12)
20.106695 (15, 8)	29.546388 (19, 10)	35.858674 (23, 12)
21.479058 (17, 9)	30.279049 (21, 10)	36.331129 (23, 12)
22.194674 (17, 9)	30.404327 (21, 10)	36.988815 (23, 12)
23.201396 (17, 9)	31.056534 (21, 11)	37.295583 (23, 12)
23.263712 (17, 10)	31.916182 (21, 11)	37.743925 (23, 12)
24.419715 (17, 10)	32.018406 (21, 12)	38.120901 (23, 12)
38.442004 (23, 12)	42.978654 (25, 13)	47.178366 (27, 14)
38.869607 (23, 12)	43.385687 (25, 13)	47.546230 (27. 14)
39.432477 (23, 12)	43.859382 (25, 13)	47.823373 (27, 14)
39.826868 (23, 12)	44.282110 (25, 13)	48.149810 (27, 15)
40.272111 (23, 12)	44.294967 (25, 13)	48.355412 (27, 15)
40.858127 (23, 12)	44.777046 (25, 13)	48.840152 (27, 15)
40.880467 (23, 12)	45.112201 (25, 14)	48.896682 (27, 15)
40.990437 (23, 13)	45.686380 (25, 14)	49.105724 (27, 15)
41.754473 (23, 13)	45.782821 (25, 14)	49.439178 (27, 16)
42.152733 (23, 13)	45.954420 (25, 14)	49.991221 (27, 15)
42.485562 (25, 13)	46.566346 (27, 14)	
42.646363 (25, 13)	46.839220 (27, 14)	

Table 5

Fourier Coefficients for the Odd Eigenvalues Less than 35·

R	9·533695	12·173008	14·358509	16·138073	16·644259
c_2	−1·06833	0·289252	−0·230915	1·161855	−1·540228
c_3	−0·4563	−1·201858	0·69560	−1·281972	0·977493
c_5	***	0·042	−1·28	−0·756	−0·105
c_7	***	***	***	***	***

R	18·180918	19·484714	20·106695	21·479058	22·194674
c_2	0·374063	−1·700188	0·858848	−0·656250	1·596844
c_3	0·101958	−0·614565	0·187279	0·226442	−1·116480
c_5	0·6372	0·8199	−1·3956	1·08229	−0·63825
c_7	***	***	***	0·423	−1·01

R	23·201396	23·263712	24·419715	25·050855	26·056918
c_2	0·169949	−1·447094	0·965541	−1·053870	1·159119
c_3	1·493056	−1·536666	−0·690260	0·552022	0·598888
c_5	−0·93998	0·10696	1·315804	−0·73366	−1·089323
c_7	−0·57	0·62	−0·545	1·546	−1·2786

R	26·446996	27·284384	27·775921	28·510278	29·137588
c_2	−0·637458	−1·20563	0·948347	−1·314095	−0·085103
c_3	−1·358607	1·66456	−0·192092	−1·410043	0·820706
c_5	1·382687	−0·45130	0·164078	−1·325850	1·031232
c_7	−0·00879	−0·80	1·131	0·0840	−1·0347

R	29·546388	30·279049	30·404327	31·056534	31·916182
c_2	1·723163	−1·781785	−1·685405	0·860109	−0·963673
c_3	−0·395634	1·005447	−1·262392	1·161835	−0·168495
c_5	−1·471689	−0·794101	0·420029	−0·165096	−0·030422
c_7	−0·1287	−0·1079	1·7351	−1·1332	−0·33567

R	32·018406	32·932465	33·492331	33·570990	34·185970
c_2	1·607004	0·601183	−1·71021	−0·576907	1·507063
c_3	−1·561208	0·755352	1·08331	−1·861226	−0·116023
c_5	0·641005	−0·238555	1·06151	−0·552746	0·363646
c_7	0·03243	1·02183	−0·8569	−0·46703	−1·79833

(continued)

R	34.695311	
c_2	−0.332079	
c_3	0.548638	
c_5	−0.065240	
c_7	1.03357	

We again include 1 slightly larger example.

Table 6

R=47.178366	
$c_2=1.314569$	$c_7=0.397455$
$c_3=0.522603$	$c_{11}=0.44840$
$c_5=-0.189176$	$c_{13}=-1.724$

Ramanujan-Petersson was checked in table 4, at least out to the c_n "hump." There were *no* violations.

§7. Even Eigenvalues Around R=125.

We did not know what to expect here. We decided to run a batch of experiments to see if we could find the 15 or so expected even eigenvalues in the interval [124.875, 126.325]. The interest centered on what sort of accuracy could be obtained (using *just* single precision), and on *how many* "trials" would have to be run. In general: we took

$$H1=0.025,\ H2=0.001,\ H3=10^{-6}$$

and used (4.2) with $\gamma=\frac{1}{3}$. Note that H1 and H2 are consistent with constraints (3.1) + (3.2).

The z_j-batches were similar to §5 except that we also used some having 3 or 4 vertical columns.

In each case: the relevant N is about 55 and the c_n "hump" is about 32.

There were 26 jobs in all.

Here is a summary of our results. [We *believe* that this listing is complete.]

Table 7

Even Eigenvalues in [124·875, 126·325] Average Gap=0·0960

R	appears in	out of (jobs)
124·898691	13	21
124·994438	7	21
125·036859	12	21
125·313840	14	21
125·347558	12	21
125·523988	20	21
125·673602	16	21
125·896473	18	21
126·018778	7	21
126·066382	9	21
126·113995	21	21
126·250406	5	5
126·313569	3	5
126·321149	1	5

21 jobs dealt with [124·875, 126·125]; 5 jobs dealt with [126·125, 126·325]. The frequency count shows that some R_n are (apparently) more "visible" than others.

The following statistics may also be of interest:

completeness probabilities for [124·875, 126·125]

jobs with 2 columns averaged 8·09 (true values) out of 11	(over 11 jobs)
jobs with 3 columns averaged 6·67 (true values) out of 11	(over 6 jobs)
jobs with 4 columns averaged 5·00 (true values) out of 11	(over 4 jobs)

[the corresponding ratios are 0·74, 0·61, 0·45]

average runtimes for [124·875, 126·125]

17 jobs on CRAY2 : 791 sec. per job
4 jobs on XMP : 695 sec. per job

[the "noisier" jobs were all run on the CRAY-2]

In most jobs: *some* spurious R-values managed to slip through all the way til the end. In the "noisiest" jobs, this number occasionally ran as high as 70% of the final listing. *True* R-values typically have their first few "Hecke differences" *vanishing* to between 4 and 7 decimal places.

For the sake of completeness, we include a coefficient listing for 3 of our better R-values. As usual: the last digit in any c_n may be off.

Table 8

R	125.313840	125.347558	125.523988
c_2	−0.332696	0.523115	−1.530691
c_3	1.056574	−1.012324	0.707551
c_5	−0.946097	0.586322	1.222769
c_7	0.324132	0.225425	0.706786
c_{11}	0.040097	0.106276	−1.132417
c_{13}	0.479454	−1.206671	−0.496056
c_{17}	1.254990	0.602756	0.056324
c_{19}	0.893140	−0.231553	1.094387
c_{23}	1.605562	0.407487	−0.725439
c_{29}	0.85671	−1.25613	−1.67296
c_{31}	−0.25088	−0.84230	1.02462

To convey some of the accuracy, we remark that:

$$|c_{10}-c_5c_2|=\begin{cases} 2.4\times 10^{-7} & \text{for} \quad 125.313840 \\ 7.4\times 10^{-7} & \text{for} \quad 125.347558 \\ 9.4\times 10^{-7} & \text{for} \quad 125.523988 \end{cases}$$

$$|c_{20}-c_5c_4|=\begin{cases} 9.8\times 10^{-7} \\ 2.0\times 10^{-6} \\ 1.6\times 10^{-6} \end{cases}$$

$$|c_{24}-c_8c_3|=\begin{cases} 1.0\times 10^{-6} \\ 4.8\times 10^{-7} \\ 1.6\times 10^{-6} \end{cases}$$

$$|c_{30}-c_6c_5|=\begin{cases} 4.6\times 10^{-6} & \text{for} \quad 125.313840 \\ 1.1\times 10^{-5} & \text{for} \quad 125.347558 \\ 1.5\times 10^{-5} & \text{for} \quad 125.523988 \end{cases}$$

Ramanujan-Petersson was checked for all 14 R-values in table 7 (at least out to the c_n "hump"). No violations were found. The "extremal" values are:

$$\left\{\begin{array}{lll} c_5=1.81457 & \text{for} & R=126.018778 \\ c_{29}=-1.95662 & \text{for} & R=126.250406 \end{array}\right\}.$$

§8. Even Eigenvalues Around R=250.

Here we used pretty much the same set-up as in §7, but did a bit of "tampering" with γ and H3. In general, we took

$$H1=0.050,\ H2=0.001,\ H3=10^{-6},\ \gamma=\frac{1}{10}.$$

The " $\frac{1}{10}$ " seemed to work fine except in a few cases. There we went to a finer and/or wider filtering grid; i.e.

$$H3=\frac{1}{4}\times 10^{-6} \text{ and/or } \gamma=\frac{1}{2}.$$

We also experimented with using LINPACK to solve (2.7)—in place of calling our own (*unoptimized*) subroutine SMAT.

There were 23 full-scale jobs plus a *similar* number of "shortcut" versions. Of the full-scale jobs: 21 had 2 columns, while 2 had 3 columns.

In each case: the relevant N was about 100, while the c_n "hump" was about 68.

Here is a summary of our results. [We *believe* this listing to be complete.] The R interval is $250 \leq R \leq 250.546$; the expected number of eigenvalues is 11.

Table 9

Even Eigenvalues in [250, 250.546] Average Gap=0.0480

R	appears in	out of (jobs)
250.014292	14	23
250.157023	18	23
250.171541	9	23
250.199849	11	23
250.220519	8	23
250.294156	4	23
250.323063	15	23
250.360005	5	23
250.512602	2	23
250.521575	3	23

The following statistics apply:

completeness probabilities for [250, 250.546]

jobs with 2 columns averaged 4.14 (true values) out of 10 (over 21 jobs)
jobs with 3 columns averaged 1.00 (true values) out of 10 (over 2 jobs)

average runtimes for [250, 250.546]

CRAY2 jobs with SMAT:	726 sec.	(8 jobs)
CRAY2 jobs with LINPACK:	734 sec.	(6 jobs)
XMP jobs with LINPACK:	789 sec.	(4 jobs)
finer and/or wider CRAY2 jobs with LINPACK:	1646 sec.	(5 jobs)

standard "short-cut" job on either machine: 200 seconds

typical runtimes for solving (2.7) with N = 100

CRAY2 with LINPACK	0.006 sec
CRAY2 with SMAT	0.086 sec

[though slower, SMAT occasionally gave *better* answers than LINPACK].

For $R \sim 250$, *true* R-values typically have their first few "Hecke differences" vanishing to between 4 and 6 decimal places.

As in §7 we include a coefficient listing for 3 of our better R-values.

Table 10

R	250.014292	250.157023	250.171541
c_2	1.963398	0.934272	−0.847139
c_3	0.966543	0.555460	0.626670
c_5	−0.572737	1.400902	−1.558957
c_7	−1.908284	−0.617719	−0.031878
c_{11}	0.258535	0.129422	−0.455558
c_{13}	0.485796	−0.875756	−0.388115
c_{17}	0.148212	−0.355089	−1.468538
c_{18}	−0.748242	−1.216197	−0.211623
c_{23}	−0.576633	−0.732391	0.234726
c_{29}	−0.542465	−0.625467	1.360191
c_{31}	−0.917826	−0.330008	−1.85193
c_{37}	−0.561364	−0.462475	1.24353
c_{41}	1.330662	−0.902229	1.43213
c_{43}	1.159073	−0.503389	0.43388
c_{47}	0.188167	0.923052	1.60671
c_{53}	−0.372197	0.499237	1.01753
c_{59}	1.832051	−1.539013	−0.23515
c_{61}	−0.27625	0.43774	−0.21213
c_{67}	0.5898	−0.864	0.0334

To convey some of the accuracy, we remark that:

$$|c_{10} - c_5 c_2| = \begin{cases} 2.0 \times 10^{-8} & \text{for } 250.014292 \\ 5.4 \times 10^{-6} & \text{for } 250.157023 \\ 6.6 \times 10^{-6} & \text{for } 250.171541 \end{cases}$$

$$|c_{30} - c_5 c_3 c_2| = \begin{cases} 2.4 \times 10^{-7} \\ 5.4 \times 10^{-6} \\ 1.5 \times 10^{-5} \end{cases}$$

$$|c_{60}-c_5c_4c_3| = \begin{cases} 2.1\times 10^{-6} \\ 6.1\times 10^{-6} \\ 5.4\times 10^{-6} \end{cases}$$

$$|c_{65}-c_{13}c_5| = \begin{cases} 1.6\times 10^{-4} & \text{for} \quad 250.014292 \\ 9.3\times 10^{-4} & \text{for} \quad 250.157023 \\ 2.5\times 10^{-4} & \text{for} \quad 250.171541 \end{cases}$$

Ramanujan-Petersson was checked for all 10 values in table 9 (at least out to the c_n "hump"). No violations were found. The extremal values were:

$$\left\{ \begin{array}{lll} c_2 = 1.96340 & \text{for} & R = 250.014292 \\ c_7 = -1.90828 & \text{for} & R = 250.014292 \end{array} \right\}.$$

§9. Even Eigenvalues Around R = 500.

The last "full-scale" search was made at 500. As in §8 we did a bit of tampering with γ and H3. In general, we took:

$$H1 = 0.050,\ H2 = 0.0005,\ (H3, \gamma) = (10^{-6}, \frac{1}{10}) \text{ or } (\frac{1}{2}\times 10^{-6}, \frac{1}{5}).$$

In several cases, however, we went with

$$H3 = 1\times 10^{-7},\ 2.5\times 10^{-7}, \text{ and/or } \gamma = \frac{1}{2}$$

to try to obtain greater accuracy.

There were 16 full-scale jobs plus a somewhat *larger* number of "short-cut" versions. All the jobs had 2 columns.

In each case: the relevant N was about 188, while the c_n "hump" was about 137.

Here is a summary of our results. [We *believe* this listing to be complete.] The chosen interval was $500 \leq R \leq 500.298$; the expected number of eigenvalues is 12.

The following statistics apply:

completeness probabilities for [500, 500.298]

jobs with 2 columns averaged 4.75 (true values) out of 12 (over 16 jobs)
[*or*: 40%]

average runtimes for [500, 500.298]

CRAY-2 jobs with SMAT:	3981 sec (3 jobs)
CRAY-2 jobs with LINPACK:	2215 sec (4 jobs)
XMP jobs with LINPACK:	2597 sec (4 jobs)

finer and/or wider jobs, either machine:	4709 sec
standard "short-cut" job on XMP, with LINPACK:	612 sec
standard "short-cut" job on CRAY-2, with LINPACK:	460 sec
standard "short-cut" job on CRAY-2, with SMAT:	730 sec

Table 11

Even Eigenvalues in [500, 500·298] Average Gap=0·0240

R	appears in	out of (jobs)
500·038825	8	16
500·048046	2	16
500·066461	14	16
500·075997	14	16
500·113941	9	16
500·138065	3	16
500·141707	2	16
500·151972	4	16
500·214756	3	16
500·232299	3	16
500·271235	1	16
500·283551	13	16

typical runtimes for solving (2·7) with N ~ 200

CRAY-2 with LINPACK	0·030 sec
CRAY-2 with SMAT	0·621 sec

For R ~ 500, true R-values typically have their first few "Hecke differences" vanishing to between 4 and 5 decimal places. If one goes out to n = 50, the number of places may *drop* to 3.

The number of *spurious* values slipping through til the very end may run as high as 50 or 60% (of the final listing) in the "noisiest" jobs.

Here are the coefficients for 2 of our better R-values.

Table 12

R	500·066461	500·283551
c_2	−0·71386	1·10979
c_3	0·82261	−0·65140
c_5	0·26012	−1·33193
c_7	1·48781	0·95336
c_{11}	0·25535	−0·29227

(continued)

c_{13}	0.80536	−0.84352
c_{17}	−0.23060	−1.24396
c_{19}	−1.47045	−1.36471
c_{23}	0.58742	−0.21683
c_{29}	−1.52386	−0.35004
c_{31}	−1.80923	1.55889
c_{37}	0.81740	1.57655
c_{41}	−1.29589	−0.11597
c_{43}	0.26127	1.35647
c_{47}	−0.87901	−1.59240
c_{53}	−0.64300	−1.54331
c_{59}	−0.15085	−0.09993
c_{61}	−0.14501	1.53694
c_{67}	−0.94672	−0.98225
c_{71}	−0.35015	−0.57396
c_{73}	−0.50796	1.18253
c_{79}	−1.73634	1.59923
c_{83}	−0.41151	−0.54689
c_{89}	−0.91326	−1.54487
c_{97}	1.60887	0.22588
c_{101}	1.78973	1.05331
c_{103}	−1.63371	1.50197
c_{107}	−0.59332	0.89831
c_{109}	−0.39324	0.47863
c_{113}	0.71463	0.90804
c_{127}	1.82748	0.87385
c_{131}	1.1836	0.5238

To convey some of the accuracy, we remark that

$$|c_{10}-c_5c_2| = \begin{cases} 8.8\times 10^{-6} & \text{for} \quad 500.066461 \\ 1.2\times 10^{-5} & \text{for} \quad 500.283551 \end{cases}$$

$$|c_{30}-c_5c_3c_2| = \begin{cases} 8.8\times 10^{-6} \\ 1.1\times 10^{-6} \end{cases}$$

$$|c_{60}-c_5c_4c_3| = \begin{cases} 7.5\times 10^{-6} \\ 4.7\times 10^{-6} \end{cases}$$

$$|c_{100}-c_{25}c_4| = \begin{cases} 4.3\times 10^{-6} \\ 4.6\times 10^{-6} \end{cases}$$

$$|c_{126}-c_{14}c_9| = \begin{cases} 2.5\times 10^{-5} \\ 6.8\times 10^{-6} \end{cases}$$

$$|c_{130}-c_{65}c_2| = \begin{cases} 6.3\times 10^{-4} & \text{for} \quad 500.066461 \\ 3.1\times 10^{-4} & \text{for} \quad 500.283551 \end{cases} .$$

Ramanujan-Petersson was checked for all 12 values in table 11 (at least out to the c_n "hump"). No violations were found. The extremal values were:

$$\begin{Bmatrix} c_2 = 1.89788 & \text{for} & R = 500.151972 \\ c_2 = -1.94412 & \text{for} & R = 500.232299 \end{Bmatrix}.$$

§ 10. Gaining Greater Accuracy.

In sections 5 and 6, we simply *took* $H3 = 10^{-6}$. As a *measure* of the intrinsic accuracy of the Bessel function (and linear algebra) routines, it is very tempting to substitute smaller & smaller values for H3 to see exactly *how much* accuracy can *ultimately* be obtained (in single precision).

"Proximity to the truth" is measured by precaution (iii) in §2.

To save time, we used a "bare-bones" form of the program which contained *only one* level (namely H3). We used the CRAY2 and looked at successively smaller neighborhoods of the (even) eigenvalues:

$$13.779751^{+} \text{ and } 41.555578^{+}$$

In the case of 13.779751, we experimented with several different z_j-batches (of 2 vertical columns) *and* several different N. The best results were obtained by taking N to be 13 [which corresponds to (only) 10^{-9} in (2.6)].

We found that:

$R = 13.7797513351890^{+}$
with
$c_2 = 1.54930447794$
$c_3 = 0.24689977245$
$c_5 = 0.737060383$
$c_7 = -0.261421$

and

$$|c_4 - c_2^2 + 1| = 1\times 10^{-10}, \qquad |c_6 - c_2c_3| = 4.8\times 10^{-8},$$
$$|c_8 - c_2^3 + 2c_2| = 1.8\times 10^{-5}, \qquad |c_9 - c_3^2 + 1| = 2.0\times 10^{-3},$$
$$|c_{10} - c_2c_5| = 5.5\times 10^{-2}.$$

The c_n "hump" has (thus) been "pushed" all the way up to N-2 or N-1 [from 7].

The other *batches* gave either 1351889⁺ or 1351890⁺, but their Hecke differences (at n = 8, 9, 10) were significantly larger…

The foregoing result compares *very* favorably with [7, appendix C] and [19].

For $R=41.555578$, we basically looked at *just one* z_j-batch but *varied* N between $23 \sim 27$. The results were pretty much the same in each case. We obtained:

$R=41.5555776736^{+}$
with
$c_2=1.0589646368$
$c_3=-0.9589822875$
$c_5=-0.6334749873$
$c_7=-0.8135527894$
$c_{11}=1.1360708$
$c_{13}=-1.11568$

and (for $N=23$)

$E_4=3\times 10^{-10}$ $\quad E_6=1\times 10^{-10}$ $\quad E_8=1\times 10^{-10}$

$E_9=1.7\times 10^{-9}$ $\quad E_{10}=2.7\times 10^{-8}$ $\quad E_{12}=1.2\times 10^{-5}$

$E_{14}=5.9\times 10^{-3}$ [in an obvious notation].

Here the c_n "hump" has been pushed from 14 out to 16 or so.

The results of these (two) "refinements" are *perfectly consistent* with an overall accuracy of $11 \sim 12$ significant figures in KBESS. Cf. §3 near (3.2).

§11. Concluding Remarks.

The results of these experiments[16] are, from one standpoint, very promising. They show that the "K-Bessel barrier" is now *all* but removed—and that the real difficulties (*mathematically*) are more in line with linear algebra. Concerning the latter: we have seen that, by employing a suitable mechanism to filter out c_k noise, it is actually possible to get *quite* far using ONLY single precision.

There are, however, *several* features that seem a bit disturbing here—especially in connection with any kind of *more systematic* (computational) analysis. Specifically:

(a) the cpu time is starting to become a bit large (as R);

(b) because of the [inevitable] regions of static, there is an increasing need to concern oneself with the notion of "completeness probability" in any particular job;

(c) the *anomaly* mentioned at the end of §4 becomes more & more noticeable as R ;

(d) we do not possess any kind of *apriori technique* for determining the number of λ_n actually present in a given interval (A, B).

(16) *viz.* §§5–10.

One is *not* really in a position to do too much about item (d) right now. [Similar difficulties occur in computing the zeros of many other (arithmetically significant functions. Compare[17] in §4.]

With regard to (a) and (b), one is *inclined* to say that we have now reached a kind of "intermediate" stage, where the *next* step should (effectively) be one "last, massive, systematic optimization."

Testing *several* z_j-batches seems inescapable — at least if one wants to use only single precision. For this reason: it is tempting to suggest that the z_j-batches should all be treated *in parallel* at levels H1 & H2 — and that any "passage" to level H3 should occur *only* in those cases judged to be most stable.

Thus far, we have been content to use local transversality of $c_4 + 1 - c_2^2$ [at the zeros of $c_4 + 1 - c_2^2$] as the basic "triggering" mechanism. This idea worked reasonably well out to $R = 500$ but there might be better ways…

This matter becomes important in connection with[12] in §4. In a parallel program of the type suggested above, one would naturally like to use the "short-cut" method as far as possible. The *problem* is that, beyond a certain point, one simply *has* to expect that the H2-level data will be plagued by sufficient machine noise to make any kind of linear interpolation down to level H3 virtually useless. Our preliminary experiments suggest that this point is already reached at $R = 1000$.

In a final series of experiments, we used the CRAY2 to do "a bit of exploration" around $R = 1000$ (with $N \approx 360$). Our *shortcut* jobs were *unable* to come up with even a *single* serious candidate for a true R_n. To make things worse, the "full-scale" jobs that we then ran (as a last resort) also failed!! The fact that H3 was 10^{-7} (instead of 10^{-6}) did *not* seem to make much difference. It appeared that the "triggering" mechanism for the filter was simply *unable* to identify *any* correct H2-intervals.

Some explanation for this failure is already "visible" in the "velocity" fluctuations for the c_k at level H2. The striking thing *at* $R = 1000$ is that these fluctuations are typically 100 times as large as the corresponding fluctuations at $R = 500$ (*even for* $k = 2$). The larger these numbers, the greater the "noise" at level H2.[17]

To treat $R \geqq 1000$, it seems *essential* to either:

(A) develop a better triggering mechanism; or

(B) somehow *reduce* the c_k noise-level at H2.

One hopeful point is this. The noise levels in the various c_k are NOT uncorrelated. There is a kind of "uniformity" with respect to k (at least when Gauss elimination is used). This uniformity partly explains why the filter itself is able to

(17) Values like 10^5 are *quite* common for $H2 = 1.25 \times 10^{-4}$ (say) and $k \leqq 6$.

function. It seems reasonable to expect that some use can be made of this fact.

On the over-all matter of further optimization, we simply mention the following points:

(1) perhaps $\exp(\frac{\pi}{2}R)K_{iR}(X)$ can be satisfactorily computed using a *wider grid* in the numerical integration; [this would certainly speed things up] (18)
(2) one should *also* experiment with using the *largest* possible H1 & H2 values;
(3) perhaps there are better methods of solving (2.7) than by standard Gauss elimination;
(4) there's an obvious need for some better "triggering" mechanisms;
(5) it would be highly advantageous to let the *machine* decide in the final listing (for each run) *which* of the R-values is "true" and which is "spurious" (possibly by referring to the results from earlier runs);
(6) with regard to parallelism (and *new types* of triggering mechanisms), *note* that any "true" R must be invariant under the process of taking arbitrary linear combinations of (2.7) with respect to distinct z_j-batches.

In conclusion: it virtually goes without saying that one *expects* that Stark's method [19] can be combined with our techniques to yield *many many* more c_k for each R_n. Stark's use of iteration in solving for c_k seems particularly suggestive. Cf. (3) above.

We hope to report on further developments in these areas in the near future.

References

1. N. L. Balazs and A. Voros, Chaos on the pseudosphere, Phys. Reports 143(3) (1986) 109 — 240.
2. M. Berry, Quantum chaology, Proc. Royal Soc. London A413(1987) 183 — 198. See also: Proc. Royal Soc. London A400(1985) 229 — 251.
3. O. Bohigas, M. J. Giannoni, and Ch. Schmit, Spectral fluctuations, random matrix theories, and chaotic motion, Springer Lecture Notes in Physics 262(1986) 118 — 138.
4. P. Cartier, Some numerical computations relating to automorphic functions, in *Computers in Number Theory* (ed. by A. O. L. Atkin and B. J. Birch), Academic Press, 1971, pp. 37 — 48.
5. G. Golub and C. Van Loan, *Matrix Computations*, Johns Hopkins Univ. Press, 1983, especially pages 25 — 27 and 71 — 72.
6. D. A. Hejhal, The Selberg trace formula and the Riemann zeta function, Duke Math. J. 43(1976) 441 — 482.
7. D. A. Hejhal, *The Selberg Trace Formula for PSL* $(2, \mathbb{R})$, volume 2, Springer Lecture Notes 1001(1983).
8. D. A. Hejhal, Some observations concerning eigenvalues of the Laplacian and Dirichlet L-series, in *Recent Progress in Analytic Number Theory* (ed. by H. Halberstam and C. Hooley), volume 2, Academic Press, 1981, pp. 95 — 110.

(18) Iterative techniques are *another* possibility. Cf. [10, p. 1369], however.

9. D. A. Hejhal and E. Bombieri, Sur les zéros des fonctions zêta d'Epstein, Comptes Rendus Acad. Sci. Paris 304(1987) 213 — 217.

10. D. A. Hejhal, Zeros of Epstein zeta functions and supercomputers, in *Proceedings of the International Congress of Mathematicians*, Berkeley, 1986, pp. 1362 — 1384.

11. D. A. Hejhal, Some remarks about cusp forms: holomorphic and non-holomorphic, Technical Report No. 1984 – 26, Chalmers Univ. of Tech. (Sweden), 1984, 33pp.

12. D. A. Hejhal and B. Berg, Some new results concerning eigenvalues of the non-Euclidean Laplacian for PSL (2, ℤ), Technical Report No. 82 — 172, University of Minnesota, 1982, 7pp.

13. H. Iwaniec, Non-holomorphic modular forms and their applications, in *Modular Forms* (ed. by R. A. Rankin), Ellis-Horwood Ltd., 1984, pp. 157 — 196.

14. N. V. Kuznecov, Petersson's conjecture for cusp forms of weight zero and Linnik's conjecture; sums of Kloosterman sums, Math. USSR Sbornik 39(1981) 299 — 342.

15. A. M. Odlyzko, On the distribution of spacings between zeros of the zeta function, Math. of Comp. 48 (1987) 273 — 308.

16. G. Pólya, Bemerkung über die Integraldarstellung der Riemannschen ξ-Funktion, Acta Math. 48(1926) 305 — 317.

17. A. Selberg, Harmonic analysis and discontinuous groups in weakly symmetric Riemannian spaces with applications to Dirichlet series, J. Indian Math. Soc. 20(1956) 47 — 87.

18. C. Se ries, Some geometrical models of chaotic dynamics, Proc. Royal Soc. London A413 (1987) 171 — 182.

19. H. Stark, Fourier coefficients of Maass waveforms, in *Modular Forms* (ed. by R. A. Rankin), Ellis-Horwood Ltd., 1984, pp. 263 — 269.

20. A. B. Venkov, *Spectral Theory of Automorphic Functions*, Proc. Steklov Inst. of Math. 153(1982). (English Translation)

21 G. N. Watson, *A Treatise on the Theory of Bessel Functions*, 2nd edition, Cambridge Univ. Press, 1944.

22. M. Wilkinson, Random matrix theory in semiclassical quantum mechanics of chaotic systems, J. Phys. A: Math. Gen. 21(1988) 1173 — 1190.

23. A. Winkler, Cusp forms and Hecke groups, J. Reine Angew. Math. 386(1988) 187 — 204.

Permanent Address of Author:
School of Mathematics
University of Minnesota
Minneapolis, Mn. 55455 USA

APPENDIX A

```
      PROGRAM DHU30A
C     experimental eigenvalue program
C     D.A.HEJHAL // October 1988
C     CRAY2 VERSION -- SINGLE PRECISION
C     this program uses an adjustable Lagrange interpolation to
C     reduce the number of KBMAT calls;
C     it also uses TRIM indices (lifted from SCALE.f);
C     and exploits ------
C     weighted Hecke indices
C     transversality controls
C     grid levels H3,H4,H5
C     full KBMAT calls at H3
C     full SMAT calls at H4
C     full SMAT calls for MAG2 -- near "transverse" R values --
C     to improve c(k) accuracy in NON-NOISY cases (cf. control #)
C     flexible program segments
C     control indices for noise/error/distortion
      IMPLICIT REAL(A-H,P-Z)
      PARAMETER(MAXDEG=51)
      PARAMETER(MG=16)
      PARAMETER(NMAX=60,NPAX=60)
      PARAMETER(NTRIM=1)
      PARAMETER(IPRINT=1)
      PARAMETER(LONG=60)
      PARAMETER(MAG=25)
      PARAMETER(LONGER=LONG*MAG)
      PARAMETER(MAG2=1000)
      PARAMETER(MTW=9+2*MAG2)
      PARAMETER(NDEG=11)
C     make certain that NDEG is odd!!
      DIMENSION XA(NMAX),YA(NMAX),XP(NMAX),YP(NMAX)
      DIMENSION R3(0:LONG)
      DIMENSION R4(0:LONGER)
      DIMENSION RZ(LONGER,NTRIM), RZT(LONGER,NTRIM)
      DIMENSION ITRIM(NTRIM), LCOUNT(NTRIM)
      DIMENSION LKOUNT(NTRIM)
      DIMENSION GIANT(NMAX,NPAX,0:NDEG)
      DIMENSION U(NMAX,NPAX)
      DIMENSION UTMP1(NMAX,NPAX), UTMP2(NMAX,NPAX)
      DIMENSION CX1(NMAX)
      DIMENSION CDH(0:LONGER,NTRIM,NMAX)
      DIMENSION CDHH(0:MTW,NMAX), WH(MTW)
      DIMENSION R97(0:MTW)
      DIMENSION CDZ(LONGER,NTRIM,NMAX)
      DIMENSION BU(NMAX)
      DIMENSION CU(-2:3,NMAX)
      DIMENSION VE(-1:1,NTRIM,NMAX)
      DIMENSION VDZ(LONGER,NTRIM,NMAX)
      DIMENSION R4T(LONGER,NTRIM), R6T(LONGER,NTRIM)
      DIMENSION VDX(LONGER,NTRIM,NMAX)
      DIMENSION VDY(LONGER,NTRIM,NMAX)
      DIMENSION A1(16),W(16)
      DIMENSION AD(MG),WD(MG)
      INTEGER*8 FAC(20)
      DIMENSION RF(20),RG(20)
      COMMON /DENNIS/ AD,WD,PI,RTM,PIH
      COMMON /DH2/ RF,RG
      COMMON /DH3/ XA,YA,XP,YP
      COMMON /DH4/ NN1R,NN1L,NN2,NN3,PI2
      COMMON /DAH/ GIANT
```

```
C
C
      UDEG=FLOAT(NDEG)
      NA97=INT((.5E0)*(UDEG-1.0E0)+(1.0E-4))
C
C
      FAC(1)=1
      DO 10  J=2,20
         FAC(J)=J*FAC(J-1)
 10   CONTINUE
      DO 12  J1=1,20
         RF(J1)=(1.0E0)/FLOAT(FAC(J1))
         RG(J1)=(1.0E0)/FLOAT(J1)
 12   CONTINUE
      PI=(4.0E0)*ATAN(1.0E0)
      PIH=(.5E0)*PI
      PI2=(2.0E0)*PI
     A1(1)= .98940 09349 91650 E0
     A1(2)= .94457 50230 73233 E0
     A1(3)= .86563 12023 87832 E0
     A1(4)= .75540 44083 55003 E0
     A1(5)= .61787 62444 02644 E0
     A1(6)= .45801 67776 57227 E0
     A1(7)= .28160 35507 79259 E0
     A1(8)= .09501 25098 376374 E0
     A1(9)=-A1(8)
     A1(10)=-A1(7)
     A1(11)=-A1(6)
     A1(12)=-A1(5)
     A1(13)=-A1(4)
     A1(14)=-A1(3)
     A1(15)=-A1(2)
     A1(16)=-A1(1)
     W(1)= .02715 24594 117541 E0
     W(2)= .06225 35239 386479 E0
     W(3)= .09515 85116 824928 E0
     W(4)= .12462 89712 55534 E0
     W(5)= .14959 59888 16577 E0
     W(6)= .16915 65193 95003 E0
     W(7)= .18260 34150 44924 E0
     W(8)= .18945 06104 55068 E0
     W(9)=W(8)
     W(10)=W(7)
     W(11)=W(6)
     W(12)=W(5)
     W(13)=W(4)
     W(14)=W(3)
     W(15)=W(2)
     W(16)=W(1)
      MM=MG/16
      RTM=(1.0E0)/FLOAT(2*MM)
      DO 23  J3=0,MM-1
      DO 24  K3=1,16
         N3=16*J3
         WD(N3+K3)=W(K3)
         AD(N3+K3)=FLOAT(1+(2*J3))-A1(K3)
 24   CONTINUE
 23   CONTINUE
C
C     INPUT DATA:
C     in addition to PARAMETERS at top
C
      R1=124.75E0
      R2=126.25E0
      NN1R=4
      NN1L=16
      NN2=4
      NN3=4
      DATA ITRIM /55/
      NR=30
      NC=2
```

```
      DO 3 J=1,NR
      DO 4 I=1,NC
        XA(I+(J-1)*NC)=(.5E0)*FLOAT(I-1)/FLOAT(NC-1)
        YA(I+(J-1)*NC)=1.10E0+(FLOAT(J)*(.9E0)/FLOAT(NR))
 4    CONTINUE
 3    CONTINUE
C
C     END-OF-DATA
C
      H3=(R2-R1)/FLOAT(LONG)
      H4=H3/FLOAT(MAG)
      H5=H4/FLOAT(MAG2)
      DO 5102 JJ=0,LONG
        R3(JJ)=R1+H3*FLOAT(JJ)
 5102 CONTINUE
      DO 5103 K=1,NMAX
        S=(XA(K)**2)+(YA(K)**2)
        XP(K)=-XA(K)/S
        YP(K)=YA(K)/S
 5103 CONTINUE
      DO 5104 JJ=0,LONGER
        R4(JJ)=R1+H4*FLOAT(JJ)
 5104 CONTINUE
C
C
C
      DO 5301 JX=0,NDEG
      R=R3(JX)
      CALL KBMAT(R,U)
      DO 5302 KO=1,NMAX
      DO 5303 K=1,NMAX
        GIANT(K,KO,JX)=U(K,KO)
 5303 CONTINUE
 5302 CONTINUE
 5301 CONTINUE
C
C     giant do-loop follows:
C
      DO 74 J=0,LONG-NDEG
C
C
      IF(J.NE.0) THEN
      RR97=R3(J+NDEG)
      CALL KBMAT(RR97,U)
CDIR$ NEXTSCALAR
      DO 5401 I=0,NDEG-1
      DO 5402 KO=1,NMAX
      DO 5403 K=1,NMAX
        S=GIANT(K,KO,I+1)
        GIANT(K,KO,I)=S
 5403 CONTINUE
 5402 CONTINUE
 5401 CONTINUE
      DO 5502 KO=1,NMAX
      DO 5503 K=1,NMAX
        S=U(K,KO)
        GIANT(K,KO,NDEG)=S
 5503 CONTINUE
 5502 CONTINUE
      ENDIF
C
C
      DO 6101 JB=0,MAG
      SX=FLOAT(NA97)+FLOAT(JB)/FLOAT(MAG)
      CALL FLAG(SX,UTMP1)
      M=JB+(J+NA97)*MAG
      DO 6102 I4=1,NTRIM
C     prepare an adjusted copy of UTMP1
C     note that the throw-away row can be modified!!!!!
      NMAT=ITRIM(I4)-1
      DO 6123 K=1,NMAT
```

```
          UTMP2(K,NMAT+1)=-UTMP1(K,1)
 6123   CONTINUE
        DO 6124 KO=1,NMAT
        DO 6125 K=1,NMAT
          UTMP2(K,KO)=UTMP1(K,KO+1)
 6125   CONTINUE
 6124   CONTINUE
        CALL SMAT(UTMP2,NMAT,CX1)
          CDH(M,I4,1)=1.0E0
        DO 6128 K=2,NMAT+1
          CDH(M,I4,K)=CX1(K-1)
 6128   CONTINUE
C
C
 6102   CONTINUE
 6101   CONTINUE
 74     CONTINUE
C
C       we now examine the Fourier coefficients
C       there is a GREAT deal of arbitrariness here;
C       fine-tuning may be necessary
C
        NII=MAG*(NA97)
        NFF=MAG*(LONG-NA97)
        NI=NII+2
        NF=NFF-2
C
C       giant do-loop follows:
C
        DO 7101 I=1,NTRIM
        NMAT=ITRIM(I)-1
        LKT=0
        LKTT=0
        DO 7102 JC=NI,NF-1
        DO 7601 N=-2,3
        DO 7602 K=2,NMAT+1
          CU(N,K)=CDH(JC+N,I,K)
 7602   CONTINUE
 7601   CONTINUE
        F40=CU(-2,4)+(1.0E0)-(CU(-2,2)**2)
        F41=CU(-1,4)+(1.0E0)-(CU(-1,2)**2)
        F42=CU(0,4)+(1.0E0)-(CU(0,2)**2)
        F43=CU(1,4)+(1.0E0)-(CU(1,2)**2)
        F44=CU(2,4)+(1.0E0)-(CU(2,2)**2)
        F45=CU(3,4)+(1.0E0)-(CU(3,2)**2)
        IF((F42*F43).LT.(0.0E0)) THEN
C       begin the detailed coefficient analysis
C       the end-statement is near 7102 (far below)
        C20=CU(0,2)
        C21=CU(1,2)
        C40=CU(0,4)
        C41=CU(1,4)
        A=C21-C20
        B=C41-C40
        E=B-(2.0E0)*C20*A
        E1=SQRT(E*E+(4.0E0)*F42*A*A)
        IF(F42.GT.(0.0E0)) THEN
          S=(E+E1)/(A*A*(2.0E0))
        ELSE
          S=(E-E1)/(A*A*(2.0E0))
        ENDIF
        SA=1.0E0-S
        RTEMP=SA*R4(JC)+S*R4(JC+1)
        DO 7603 K=2,NMAT+1
          BU(K)=SA*CU(0,K)+S*CU(1,K)
 7603   CONTINUE
        AY1=(CU(-1,2)*CU(-1,2)-(1.0E0))/CU(-1,4)
        A2=(CU(0,2)*CU(0,2)-(1.0E0))/CU(0,4)
        A3=(CU(1,2)*CU(1,2)-(1.0E0))/CU(1,4)
        A4=(CU(2,2)*CU(2,2)-(1.0E0))/CU(2,4)
        AH=(BU(2)*BU(2)-(1.0E0))/BU(4)
```

```
      B1=CU(-1,2)*CU(-1,3)/CU(-1,6)
      B2=CU(0,2)*CU(0,3)/CU(0,6)
      B3=CU(1,2)*CU(1,3)/CU(1,6)
      B4=CU(2,2)*CU(2,3)/CU(2,6)
      BH=BU(2)*BU(3)/BU(6)
C
      Z=1.0E0
      AY1=ABS(AY1-Z)
      A2=ABS(A2-Z)
      A3=ABS(A3-Z)
      A4=ABS(A4-Z)
      AH=ABS(AH-Z)
      B1=ABS(B1-Z)
      B2=ABS(B2-Z)
      B3=ABS(B3-Z)
      B4=ABS(B4-Z)
      BH=ABS(BH-Z)
C
      Q1=AMIN1(A2,AH,A3)
      Q2=AMIN1(B2,BH,B3)
      Q91=AMIN1(AY1,A2,AH,A3,A4)
      Q92=AMIN1(B1,B2,BH,B3,B4)
C
      IF(Q91.GT.(.10E0)) THEN
      GOTO 7102
      ENDIF
C
      IF(Q92.GT.(.10E0)) THEN
      GOTO 7102
      ENDIF
C
      LKT=LKT+1
      RZ(LKT,I)=RTEMP
C
C     use LOCAL transversality to trim the list
C
      FG=F43-F42
      C1=(F41-F40)/FG
      C2=(F42-F41)/FG
      C4=(F44-F43)/FG
      C5=(F45-F44)/FG
      Z89=AMIN1(C1,C2,C4,C5)
      IF(Z89.LT.(0.0E0)) THEN
C     look at B2,BH,B3 versus B1,B4
        QDE=Q2-Q92
        IF(QDE.GT.(0.0E0)) THEN
        GOTO 7102
        ENDIF
        IF(Q2.GT.(.005E0)) THEN
        GOTO 7102
        ENDIF
      ENDIF
C
      IF(Q1.GT.(.05E0)) THEN
      GOTO 7102
      ENDIF
C
      IF(Q2.GT.(.05E0)) THEN
      GOTO 7102
      ENDIF
C
      LKTT=LKTT+1
C
C
      F61=CU(-1,2)*CU(-1,3)-CU(-1,6)
      F62=CU(0,2)*CU(0,3)-CU(0,6)
      F63=CU(1,2)*CU(1,3)-CU(1,6)
      F64=CU(2,2)*CU(2,3)-CU(2,6)
      V41=(F42-F41)/H4
      V42=(F43-F42)/H4
      V43=(F44-F43)/H4
```

```
      V61=(F62-F61)/H4
      V62=(F63-F62)/H4
      V63=(F64-F63)/H4
      AT1=ABS(V42-V41)
      AT2=ABS(V42-V43)
      AT3=ABS(V42)
      AT4=ABS(V62-V61)
      AT5=ABS(V62-V63)
      AT6=ABS(V62)
      R4T(LKTT,I)=AMAX1(AT1,AT2)/AT3
      R6T(LKTT,I)=AMAX1(AT4,AT5)/AT6
C
C
C     minimize the Hecke index next;
C     use full SMAT calls for CDHH
C     NB: very noisy cases can be made WORSE here!!!
C
C
      U27=FLOAT(MAG2)+(1.0E-6)
      MRG=INT(U27/(3.0E0))+1
      H4H=(.5E0)*H4
C
C
      DO 7171 JX=0,NDEG
      R=RTEMP+FLOAT(2*JX-NDEG)*H4H
      CALL KBMAT(R,U)
      DO 7172 K0=1,NMAX
      DO 7173 K=1,NMAX
        GIANT(K,K0,JX)=U(K,K0)
 7173 CONTINUE
 7172 CONTINUE
 7171 CONTINUE
C
C
      DO 7621 JBB=0,2*MRG
        R97(JBB)=RTEMP+FLOAT(JBB-MRG)*H5
        WHN=FLOAT(NA97)+(.5E0)
        SX=WHN+FLOAT(JBB-MRG)/FLOAT(MAG2)
        CALL FLAG(SX,UTMP1)
C     prepare an adjusted copy of UTMP1
C     the throw-away row should be the same as for CDH!!!
      DO 7631 K=1,NMAT
        UTMP2(K,NMAT+1)=-UTMP1(K,1)
 7631 CONTINUE
      DO 7633 K0=1,NMAT
      DO 7635 K=1,NMAT
         UTMP2(K,K0)=UTMP1(K,K0+1)
 7635 CONTINUE
 7633 CONTINUE
      CALL SMAT(UTMP2,NMAT,CX1)
         CDHH(JBB,1)=1.0E0
      DO 7637 K=2,NMAT+1
         CDHH(JBB,K)=CX1(K-1)
 7637 CONTINUE
      C2=CDHH(JBB,2)
      C3=CDHH(JBB,3)
      C4=CDHH(JBB,4)
      C6=CDHH(JBB,6)
      C8=CDHH(JBB,8)
      C9=CDHH(JBB,9)
      WTA=(C2*C2)-1.0E0
      WTB=C2*C3
      WTC=C2*((C2*C2)-(2.0E0))
      WTD=(C3*C3)-1.0E0
      WH1=ABS(-1.0E0+(WTA/C4))
      WH2=ABS(-1.0E0+(WTB/C6))
      WH3=ABS(-1.0E0+(WTC/C8))
      WH4=ABS(-1.0E0+(WTD/C9))
C     the functional WH is adjustable!!!
      WH(1+JBB)=WH1+WH2
 7621 CONTINUE
```

```
      M98=(2*MRG)+1
      JBBM=ISMIN(M98,WH,1)
      JBBM=JBBM-1
      RZT(LKTT,I)=R97(JBBM)
      DO 7640 K=1,NMAT+1
      CDZ(LKTT,I,K)=CDHH(JBBM,K)
 7640 CONTINUE
C
C     compute velocity fluctuations
C
      DO 7641 N=-1,1
      DO 7642 K=2,NMAT+1
        VE(N,I,K)=(CU(N+1,K)-CU(N,K))/H4
 7642 CONTINUE
 7641 CONTINUE
      DO 7651 K=2,NMAT+1
        AT1=ABS(VE(0,I,K)-VE(-1,I,K))
        AT2=ABS(VE(0,I,K)-VE(1,I,K))
        AT3=ABS(VE(0,I,K))
        VDZ(LKTT,I,K)=AMAX1(AT1,AT2)
        VDX(LKTT,I,K)=AT3
        VDY(LKTT,I,K)=AMAX1(AT1,AT2)/AT3
 7651 CONTINUE
C
C
      ENDIF
C
C
 7102 CONTINUE
      LCOUNT(I)=LKT
      LKOUNT(I)=LKTT
 7101 CONTINUE
C
C     now do the final print-out
C
      PRINT*,'    '
      PRINT*,'BASIC PARAMETERS:'
      PRINT 90, R1,R2
      PRINT 91, H3
      PRINT 92, H4
      PRINT 921, H5,MRG
      PRINT 93, NDEG
      PRINT 94, R4(NI),R4(NF)
 90   FORMAT(1X,F15.10,2X,F15.10)
 91   FORMAT(1X,'H3=',F15.12)
 92   FORMAT(1X,'H4=',F15.12)
 921  FORMAT(1X,'H5=',F15.12,2X,'MRG=',I6)
 93   FORMAT(1X,'DEGREE=',I3)
 94   FORMAT(1X,'ACTIVE RANGE:',F15.10,2X,F15.10)
C
C
      PRINT*,'  '
      PRINT 9411, NR, NC
 9411 FORMAT(1X,'ROWS:',I3,3X,'COLS:',I3)
      PRINT 9412, YA(1), YA(NMAX)
 9412 FORMAT(1X,'Y RANGE:',F6.3,3X,F6.3)
C
C
      DO 8101 I=1,NTRIM
      NMAT=ITRIM(I)-1
      PRINT*,'    '
      PRINT 95, ITRIM(I)
 95   FORMAT('ITRIM=',I4)
      PRINT*,'MAXIMAL LIST (level H4):'
      PRINT 96, (RZ(L,I),L=1,LCOUNT(I))
 96   FORMAT(1X,'R=',F15.10)
C
C
      PRINT*,'TRANSVERSE LIST (level H5):'
      PRINT 961, (RZT(L,I),L=1,LKOUNT(I))
 961  FORMAT('TR=',F15.10)
```

```
C
C
         IF(IPRINT.NE.0) THEN
         DO 8201 L=1,LKOUNT(I)
         PRINT 962, RZT(L,I),R4T(L,I),R6T(L,I)
 962     FORMAT('FOR R=',F15.10,3X,E11.4,2X,E11.4)
         C2=CDZ(L,I,2)
         C3=CDZ(L,I,3)
         C4=CDZ(L,I,4)
         C5=CDZ(L,I,5)
         C6=CDZ(L,I,6)
         C7=CDZ(L,I,7)
         C8=CDZ(L,I,8)
         C9=CDZ(L,I,9)
         C10=CDZ(L,I,10)
         C12=CDZ(L,I,12)
         C14=CDZ(L,I,14)
         C15=CDZ(L,I,15)
         WTA=(C2*C2)-1.0E0
         WTB=C2*C3
         WTC=C2*((C2*C2)-(2.0E0))
         WTD=(C3*C3)-(1.0E0)
         WTAA=C2*C5
         WTBB=C3*C4
         WTCC=C2*C7
         WTDD=C3*C5
         HH4=WTA/C4
         HH6=WTB/C6
         HH8=WTC/C8
         HH9=WTD/C9
         HH10=WTAA/C10
         HH12=WTBB/C12
         HH14=WTCC/C14
         HH15=WTDD/C15
         PRINT 98, HH4,HH6,HH8,HH9
 98      FORMAT(1X,'RATIOS:',F14.8,2X,F14.8,2X,F14.8,2X,F14.8)
         PRINT 98, HH10,HH12,HH14,HH15
         PRINT 99,(K,CDZ(L,I,K),VDZ(L,I,K),VDX(L,I,K),VDY(L,I,K),
     &   K=2,NMAT+1)
 99      FORMAT(3X,'C(',I3,')=',2X,E22.15,4X,E11.4,2X,E11.4,2X,E11.4)
 8201    CONTINUE
         ENDIF
 8101    CONTINUE
         END
C
C
C
         SUBROUTINE KBMAT(R,U)
C        CRAY VERSION -- SINGLE PRECISION
         IMPLICIT REAL(A-H,P-Z)
         PARAMETER(NMAX=60,NPAX=60)
         DIMENSION XA(NMAX),YA(NMAX),XP(NMAX),YP(NMAX)
         DIMENSION U1(NMAX,NPAX)
         DIMENSION U2(NMAX,NPAX)
         DIMENSION U(NMAX,NPAX)
         COMMON /DH3/ XA,YA,XP,YP
         COMMON /DH4/ NN1R,NN1L,NN2,NN3,PI2
         S2=(5.0E0)/R
         T1=(1.0E0)/(3.0E0)
         T2=(2.0E0)/(3.0E0)
         S=AMIN1(T1,S2**T2)
         DO 6101 K=1,NMAX
         X1=XA(K)
         Y1=YA(K)
         X2=XP(K)
         Y2=YP(K)
CDIR$    NEXTSCALAR
         DO 100 N=NMAX,1,-1
           YN=(PI2)*FLOAT(N)*Y1
           C=(ABS(R-YN))/R
           IF(C.LE.S) THEN
```

```
            CALL KBES2(YN,R,RKBES)
          ELSE IF(YN.LT.R) THEN
            CALL KBES1(YN,R,RKBES)
          ELSE
            CALL KBES3(YN,R,RKBES)
          ENDIF
          U1(K,N)=SQRT(Y1)*RKBES
 100      CONTINUE
        DO 110 N=NMAX,1,-1
          YN=(PI2)*FLOAT(N)*Y2
          C=(ABS(R-YN))/R
          IF(C.LE.S) THEN
            CALL KBES2(YN,R,RKBES)
          ELSE IF(YN.LT.R) THEN
            CALL KBES1(YN,R,RKBES)
          ELSE
            CALL KBES3(YN,R,RKBES)
          ENDIF
          U2(K,N)=SQRT(Y2)*RKBES
 110    CONTINUE
 6101   CONTINUE
        DO 6102 K=1,NMAX
          X1=XA(K)
          X2=XP(K)
          DO 200 N=1,NMAX
            ZN1=(PI2)*FLOAT(N)*X1
            ZN2=(PI2)*FLOAT(N)*X2
            U11=U1(K,N)
            U22=U2(K,N)
            U(K,N)=U11*COS(ZN1)-U22*COS(ZN2)
 200      CONTINUE
 6102   CONTINUE
        RETURN
        END
C
C
      SUBROUTINE KBES1(YN,R,RKBES)
      IMPLICIT REAL(A-H,P-Z)
      PARAMETER(MG=16)
      DIMENSION AD(MG),WD(MG)
      DIMENSION RF(20),RG(20)
      COMMON /DENNIS/ AD,WD,PI,RTM,PIH
      COMMON /DH2/ RF,RG
      COMMON /DH4/ NN1R,NN1L,NN2,NN3,PI2
      DIMENSION U(MG),V(MG)
      DIMENSION VPRM(MG),T(MG),E(MG)
      DIMENSION E1(MG),E2(MG)
      DIMENSION A9(MG),B9(MG)
      DIMENSION Z(MG)
      N1=NN1R
      N2=NN1L
C     ADDITIONAL PARAMETERS:
          ZCRIT1=0.25E0
          ZCRIT2=0.25E0
          ETA=0.00E0
C
C
C     for ETA: (using cut-off) at level exp(-51)
C     .00   R<98          .33    R>492
C     .02   R>98          .50    R>1430
C     .05   R>113         .66    R>5831
C     .10   R>144         .80    R>31157
C     .20   R>238
C
      X=YN
      T8=R/X
      C25=SQRT((R-X)*(R+X))
      DD=(C25)/R
      RDD=(1.0E0)/DD
      U1=ALOG((R+C25)/X)
      C24=110.00E0/C25
```

```
        C23-SQRT(C24)
        S-T8*(U1-DD)
        C6-COS(X*S)
        S6-SIN(X*S)
        XXR-0.0E0
        XXL-0.0E0

        IF (R.LT. (40.0E0)) THEN
          DO 9381 I-2,7
          UT-U1+FLOAT(I)
          S3-SINH(UT)
          C3-COSH(UT)
          C-T8*UT-S
          S1-C/S3
          C74-SQRT((S3-C)*(S3+C))
          C1-C74/S3
          VT-(2.0E0)*ATAN(S1/(1.0E0+C1))
          TT--X*C3*C1 + R*(PIH-VT)
          IF(TT.LT.(-55.0E0)) THEN
            CC23-FLOAT(I)
            GOTO 9382
          ENDIF
 9381     CONTINUE
 9382     H1-AMIN1(C23,CC23)/FLOAT(N1)
          H2-(DD-ETA*(DD**3))/FLOAT(N2)
*
*
        ELSE
          H1-AMIN1(C23,2.0E0)/FLOAT(N1)
          H2-(DD-ETA*(DD**3))/FLOAT(N2)
        ENDIF
*
*
CDIR$   NEXTSCALAR
        DO 300  J1-1,N1
            XJ-FLOAT(-1+J1)
            AA-XJ*H1
        IF(AA.GE.ZCRIT1) THEN
            GOTO 33
        ELSE
            GOTO 34
        ENDIF
 33     DO 400  KK-1,MG
            U(KK)-U1+AA+(RTM)*H1*AD(KK)
            S3-SINH(U(KK))
            C3-COSH(U(KK))
            C-T8*U(KK)-S
            S1-C/S3
            C74-SQRT((S3-C)*(S3+C))
            C1-C74/S3
            V(KK)-(2.0E0)*ATAN(S1/(1.0E0+C1))
            T(KK)--X*C3*C1+R*(PIH-V(KK))
            E(KK)-EXP(AMAX1(T(KK),-70.0E0))
            VPRM(KK)-(T8*S3-C*C3)/(C74*S3)
            E1(KK)-E(KK)*WD(KK)
            E2(KK)-E1(KK)*VPRM(KK)
 400    CONTINUE
        GOTO 35
 34     DO 402  KK-1,MG
           Z(KK)-AA+(RTM)*H1*AD(KK)
           ZT-Z(KK)
           AQ0-(ZT**14)*(RF(16)*(DD+RG(17)*ZT)+
     &         RF(18)*(DD+RG(19)*ZT)*ZT*ZT)
           AQ1-(ZT**10)*(RF(12)*(DD+RG(13)*ZT)+
     &         RF(14)*(DD+RG(15)*ZT)*ZT*ZT)+AQ0
           AQ2-(ZT**6)*(RF(8)*(DD+RG(9)*ZT)+
     &         RF(10)*(DD+RG(11)*ZT)*ZT*ZT)+AQ1
           AQ3-(ZT**2)*(RF(4)*(DD+RG(5)*ZT)+
     &         RF(6)*(DD+RG(7)*ZT)*ZT*ZT)+AQ2
           AQ4-RF(2)*(DD+RG(3)*ZT)+AQ3
           A9(KK)-AQ4*(RDD)
```

```
          BQ0=(ZT**14)*(RF(15)*(DD+RG(16)*ZT)+
     &         RF(17)*(DD+RG(18)*ZT)*ZT*ZT)
          BQ1=(ZT**10)*(RF(11)*(DD+RG(12)*ZT)+
     &          RF(13)*(DD+RG(14)*ZT)*ZT*ZT)+BQ0
          BQ2=(ZT**6)*(RF(7)*(DD+RG(8)*ZT)+
     &          RF(9)*(DD+RG(10)*ZT)*ZT*ZT)+BQ1
          BQ3=(ZT**2)*(RF(3)*(DD+RG(4)*ZT)+
     &          RF(5)*(DD+RG(6)*ZT)*ZT*ZT)+BQ2
          BQ4=(DD+RG(2)*ZT)+BQ3
          B9(KK)=BQ4*(RDD)
          C3T=1.0E0+ZT*DD*B9(KK)
          SM=1.0E0+(RDD*ZT)
          SN1=SM+A9(KK)*ZT*ZT
          SN2=(2.0E0*SM)+A9(KK)*ZT*ZT
          CXX=SQRT(A9(KK))*SQRT(SN2)
          S1=SM/(SN1)
          C1=ZT*(CXX/SN1)
          V(KK)=(2.0E0)*ATAN(S1/(1.0E0+C1))
          VX=-B9(KK)-(B9(KK)-A9(KK))*RDD*ZT
          VPRM(KK)=VX/(CXX*SN1)
          T(KK)=R*(PIH-V(KK)-C3T*C1)
          E(KK)=EXP(AMAX1(T(KK),-70.0E0))
          E1(KK)=E(KK)*WD(KK)
          E2(KK)=E1(KK)*VPRM(KK)
 402    CONTINUE
 35     X1=SSUM(MG,E1,-1)
        X2=SSUM(MG,E2,-1)
        XXR=XXR+(RTM)*H1*(X1*C6-X2*S6)
        IF(T(MG).LT.-55.0E0) THEN
          GOTO 3
        ENDIF
 300    CONTINUE
CDIR$   NEXTSCALAR
 3      DO 600  J2=1,N2
           XJ=FLOAT(-1+J2)
           AA=-XJ*H2
        IF(AA.LE.(-ZCRIT2)) THEN
           GOTO 53
        ELSE
           GOTO 54
        ENDIF
 53     DO 700  KK=1,MG
           U(KK)=U1+AA-(RTM)*H2*AD(KK)
           S3=SINH(U(KK))
           C3=COSH(U(KK))
           C=T8*U(KK)-S
           S1=C/S3
           C74=SQRT((S3-C)*(S3+C))
           C1=-C74/S3
           V(KK)=PI-(2.0E0)*ATAN(S1/(1.0E0-C1))
           T(KK)=-X*C3*C1+R*(PIH-V(KK))
           E(KK)=EXP(AMAX1(T(KK),-70.0E0))
           VPRM(KK)=-(T8*S3-C*C3)/(C74*S3)
           E1(KK)=E(KK)*WD(KK)
           E2(KK)=E1(KK)*VPRM(KK)
 700    CONTINUE
        GOTO 55
 54     DO 702  KK=1,MG
           Z(KK)=AA-(RTM)*H2*AD(KK)
           ZT=Z(KK)
           AQ0=(ZT**14)*(RF(16)*(DD+RG(17)*ZT)+
     &          RF(18)*(DD+RG(19)*ZT)*ZT*ZT)
           AQ1=(ZT**10)*(RF(12)*(DD+RG(13)*ZT)+
     &          RF(14)*(DD+RG(15)*ZT)*ZT*ZT)+AQ0
           AQ2=(ZT**6)*(RF(8)*(DD+RG(9)*ZT)+
     &          RF(10)*(DD+RG(11)*ZT)*ZT*ZT)+AQ1
           AQ3=(ZT**2)*(RF(4)*(DD+RG(5)*ZT)+
     &          RF(6)*(DD+RG(7)*ZT)*ZT*ZT)+AQ2
           AQ4=RF(2)*(DD+RG(3)*ZT)+AQ3
           A9(KK)=AQ4*(RDD)
           BQ0=(ZT**14)*(RF(15)*(DD+RG(16)*ZT)+
```

```
     &          RF(17)*(DD+RG(18)*ZT)*ZT*ZT)
           BQ1=(ZT**10)*(RF(11)*(DD+RG(12)*ZT)+
     &          RF(13)*(DD+RG(14)*ZT)*ZT*ZT)+BQ0
           BQ2=(ZT**6)*(RF(7)*(DD+RG(8)*ZT)+
     &          RF(9)*(DD+RG(10)*ZT)*ZT*ZT)+BQ1
           BQ3=(ZT**2)*(RF(3)*(DD+RG(4)*ZT)+
     &          RF(5)*(DD+RG(6)*ZT)*ZT*ZT)+BQ2
           BQ4=(DD+RG(2)*ZT)+BQ3
           B9(KK)=BQ4*(RDD)
           C3T=1.0E0+ZT*DD*B9(KK)
           SM=1.0E0+(RDD*ZT)
           SN1=SM+A9(KK)*ZT*ZT
           SN2=(2.0E0*SM)+A9(KK)*ZT*ZT
           CXX=SQRT(A9(KK))*SQRT(SN2)
           S1=SM/(SN1)
           C1=ZT*(CXX/SN1)
           V(KK)=PI-(2.0E0)*ATAN(S1/(1.0E0-C1))
           VX=-B9(KK)-(B9(KK)-A9(KK))*RDD*ZT
           VPRM(KK)=VX/(CXX*SN1)
           T(KK)=R*(PIH-V(KK)-C3T*C1)
           E(KK)=EXP(AMAX1(T(KK),-70.0E0))
           E1(KK)=E(KK)*WD(KK)
           E2(KK)=E1(KK)*VPRM(KK)
 702     CONTINUE
 55      X1=SSUM(MG,E1,-1)
         X2=SSUM(MG,E2,-1)
         XXL=XXL+(RTM)*H2*(X1*C6-X2*S6)
         IF(T(MG).LT.-55.0E0) THEN
           GOTO 4
         ENDIF
 600     CONTINUE
 4       RKBES=XXL+XXR
C
C
         XXH=0.0E0
         IF(R.LT.(75.0E0)) THEN
         U0=U1-DD
         NH=16*(1+INT(U0))
         HH=U0/FLOAT(NH)
         DO 9711 JH=1,NH
           XJ=FLOAT(-1+JH)
           AA=XJ*HH
           DO 9722 KK=1,MG
           U(KK)=AA+(RTM)*HH*AD(KK)
           C3=COSH(U(KK))
           C8=COS(R*U(KK))
           T(KK)=X*C3-(PIH)*R
           E(KK)=EXP(AMAX1(T(KK),-70.0E0))*C8
           E1(KK)=E(KK)*WD(KK)
 9722      CONTINUE
           X1=SSUM(MG,E1,-1)
           XXH=XXH+(RTM)*HH*X1
 9711    CONTINUE
         ENDIF
         RKBES=RKBES+XXH
         RETURN
         END
C
C
         SUBROUTINE KBES2(YN,R,RKBES)
         IMPLICIT REAL(A-H,P-Z)
         PARAMETER(MG=16)
         DIMENSION AD(MG),WD(MG)
         DIMENSION RF(20),RG(20)
         COMMON /DENNIS/ AD,WD,PI,RTM,PIH
         COMMON /DH2/ RF,RG
         COMMON /DH4/ NN1R,NN1L,NN2,NN3,PI2
         DIMENSION U(MG),V(MG)
         DIMENSION VPRM(MG),T(MG),E(MG),E4(MG)
         DIMENSION A9(MG),B9(MG)
         N1=NN2
```

```
C       ADDITIONAL PARAMETER:
            UCRIT=0.125E0
        X=YN
        H=R-X
        G1=(4.0E0)*SQRT(3.0E0)
        G2=-1.0E0/3.0E0
        U12=G1*(R**G2)
        IF (R.LT.(120.0E0)) THEN
          U11=ALOG((8.0E0)+(200.0E0/R))
        ELSE IF (R.LT.(333.0E0)) THEN
          U11=1.5E0
        ELSE
          U11=U12
        ENDIF
        H1=U11/FLOAT(N1)
        XX1=0.0E0
CDIR$   NEXTSCALAR
        DO 300  J1=1,N1
             XJ=FLOAT(-1+J1)
             AA=XJ*H1
        IF(AA.GE.UCRIT) THEN
             GOTO 33
        ELSE
             GOTO 34
        ENDIF
 33     DO 400  KK=1,MG
             U(KK)=AA+(RTM)*H1*AD(KK)
             S3=SINH(U(KK))
             C3=COSH(U(KK))
             S33=S3/(U(KK))
             S1=U(KK)/S3
             C21=SQRT((S33+1.0E0)*(S33-1.0E0))
             C1=(C21)/S33
             C6=COS(H*U(KK))
             S6=SIN(H*U(KK))
             V(KK)=(2.0E0)*ATAN(S1/(1.0E0+C1))
             VPRM(KK)=(S33 - C3)/(S3*C21)
             T(KK)=-X*C3*C1 + R*(PIH-V(KK))
             E(KK)=EXP(AMAX1(T(KK),-70.0E0))
             E4(KK)=E(KK)*(C6-S6*VPRM(KK))*WD(KK)
 400    CONTINUE
        GO TO 35
 34     DO 402 KK=1,MG
              U(KK)=AA+(RTM)*H1*AD(KK)
              UT=U(KK)
              AQ0=(UT**14)*RF(16)
              AQ1=(UT**10)*(RF(12)+RF(14)*UT*UT)+AQ0
              AQ2=(UT**6)*(RF(8)+RF(10)*UT*UT) + AQ1
              AQ3=(UT**2)*(RF(4)+RF(6)*UT*UT) + AQ2
              A9(KK)=RF(2)+AQ3
              BQ0=(UT**14)*RF(17)
              BQ1=(UT**10)*(RF(13)+RF(15)*UT*UT)+BQ0
              BQ2=(UT**6)*(RF(9)+RF(11)*UT*UT) + BQ1
              BQ3=(UT**2)*(RF(5)+RF(7)*UT*UT) + BQ2
              B9(KK)=RF(3)+BQ3
              C3=1.0E0+A9(KK)*(UT*UT)
              S33=1.0E0+B9(KK)*(UT*UT)
              V11=B9(KK)*(2.0E0+B9(KK)*UT*UT)
              V12=SQRT(V11)
              VPRM(KK)=(B9(KK)-A9(KK))/(V12*S33)
              S1=(1.0E0)/(S33)
              C1=U(KK)*(V12/S33)
              C6=COS(H*U(KK))
              S6=SIN(H*U(KK))
              V(KK)=(2.0E0)*ATAN(S1/(1.0E0+C1))
              T(KK)=-X*C3*C1 + R*(PIH-V(KK))
              E(KK)=EXP(AMAX1(T(KK),-70.0E0))
              E4(KK)=(C6-S6*VPRM(KK))*E(KK)*WD(KK)
 402    CONTINUE
 35     X1=SSUM(MG,E4,-1)
        XX1=XX1+(RTM)*H1*X1
```

```
      IF (T(MG).LT.-55.0E0) THEN
         GOTO 5
      ENDIF
 300  CONTINUE
 5    RKBES=XX1
      RETURN
      END
C
C
      SUBROUTINE KBES3(YN,R,RKBES)
      IMPLICIT REAL(A-H,P-Z)
      PARAMETER(MG=16)
      DIMENSION AD(MG),WD(MG)
      COMMON /DENNIS/ AD,WD,PI,RTM,PIH
      COMMON /DH4/ NN1R,NN1L,NN2,NN3,PI2
      DIMENSION U(MG),V(MG)
      DIMENSION T(MG),E(MG),E4(MG)
      N1=NN3
      X=YN
      RPIH=R*PIH
      S=R/X
      XC=SQRT((X+R)*(X-R))
      C=(XC)/X
      AL=(2.0E0)*ATAN(S/(1.0E0+C))
      XCRAL=XC + R*AL
      IF(RPIH-XCRAL.LT.-125.0E0) THEN
           XX11=0.0E0
           GO TO 7
      END IF
      C23=110.00E0/(XC)
      U11=SQRT(C23)
      IF (R.LT.(40.0E0)) THEN
        U22=ALOG((8.0E0)+(125.0E0/X))
      ELSE
        U22=2.0E0
      ENDIF
      H1=AMIN1(U11,U22)
      H1=H1/FLOAT(N1)
      XX1=0.0E0
CDIR$ NEXTSCALAR
      DO 300 J1=1,N1
           XJ=FLOAT(-1+J1)
           AA=XJ*H1
      DO 400 KK=1,MG
           U(KK)=AA+(RTM)*H1*AD(KK)
           S3=SINH(U(KK))
           C3=COSH(U(KK))
           S33=S3/(U(KK))
           S1=S*(U(KK)/S3)
           C1=SQRT((S33+S)*(S33-S)) / S33
           V(KK)=2.0E0*ATAN(S1/(1.0E0+C1))
           T97=-X*C3*C1-R*V(KK)
           T(KK)=T97+XCRAL
           E(KK)=EXP(AMAX1(T(KK),-70.0E0))
           E4(KK)=E(KK)*WD(KK)
 400       CONTINUE
      X1=SSUM(MG,E4,-1)
      XX1=XX1+(RTM)*H1*X1
      IF(T(MG).LT.-55.0E0) THEN
           GO TO 5
      END IF
 300  CONTINUE
 5    XX11=XX1*EXP(RPIH-XCRAL)
 7    RKBES=XX11
      RETURN
      END
C
C
      SUBROUTINE FLAG(SX,SV)
      IMPLICIT REAL(A-H,P-Z)
      PARAMETER(NMAX=60,NPAX=60)
```

```
      PARAMETER(NDEG=11)
      PARAMETER(MAXDEG=51)
      DIMENSION GIANT(NMAX,NPAX,0:NDEG)
      DIMENSION SV(NMAX,NPAX)
      DIMENSION P(0:MAXDEG),D(0:MAXDEG)
      COMMON /DAH/ GIANT
      DO 2300 J=0,NDEG
        P(J)=1.0E0
        D(J)=1.0E0
      DO 2310  K8=0,NDEG
        IF(K8.NE.J) THEN
        P(J)=P(J)*(SX-FLOAT(K8))
        D(J)=D(J)*FLOAT(J-K8)
        ENDIF
 2310 CONTINUE
 2300 CONTINUE
      DO 3300 K0=1,NMAX
      DO 3310 K=1,NMAX
        SQQ=0.0E0
      DO 3320 JB=0,NDEG
        SQQ=SQQ+GIANT(K,K0,JB)*(P(JB)/D(JB))
 3320 CONTINUE
        SV(K,K0)=SQQ
 3310 CONTINUE
 3300 CONTINUE
      RETURN
      END
C
C
       SUBROUTINE SMAT(U,N,C)
       PARAMETER(NMAX=60,NPAX=60)
       DIMENSION U(NMAX,NPAX+1), C(NPAX)
       DO 1000 M=1,N
         TEMP=0.0E0
         MAXI=0
         DO 1100 J=M,N
           IF(TEMP.LT.ABS(U(J,M))) THEN
           TEMP=ABS(U(J,M))
           MAXI=J
           ENDIF
 1100    CONTINUE
*        swap rows if necessary
         IF(MAXI.NE.M) THEN
           DO 1200 J=1,N+1
           T=U(MAXI,J)
           U(MAXI,J)=U(M,J)
           U(M,J)=T
 1200      CONTINUE
         ENDIF
         TEMP=U(M,M)
         U(M,M)=1.0E0
         DO 1300 J=M+1,N+1
           U(M,J)=U(M,J)/TEMP
 1300    CONTINUE
         DO 1400 J=1,N
           IF(J.NE.M) THEN
             T=U(J,M)
             U(J,M)=0.0E0
             DO 1500 K=M+1,N+1
             U(J,K)=U(J,K)-(T*U(M,K))
 1500        CONTINUE
           ENDIF
 1400    CONTINUE
 1000  CONTINUE
       DO 2001 M=1,N
         C(M)=U(M,N+1)
 2001  CONTINUE
       RETURN
       END
```

Three Primes Theorem in a Short Interval (II)

Jia Chaohua

Department of Mathematics, Beijing University

§ 1. Introduction

In 1937, I. M. Vinogradov proved the well-known three primes theorem. Haselgrove [1], Pan Cheng-dong [2], Chen Jin-run [3] generalized it to the results in a short interval. Afterwards, Professor Pan Cheng-dong and Pan Cheng-biao pointed out that there was a defect in the proof of [2] and [3]. They provided a pure analytic proof by complex integral method. Precisely, they proved the following result:

Let N be a sufficiently large odd number, $\varepsilon \left(< \frac{1}{100} \right)$ be a small positive number,

$$U = N^{\varphi + \varepsilon}, \tag{1}$$

then the equation in prime variables

$$N = p_1 + p_2 + p_3, \tag{2}$$

$$\frac{N}{3} - U < p_j \leqslant \frac{N}{3} + U, \; j = 1, 2, 3,$$

has solutions when $\varphi = \frac{91}{96}$. The number of solutions

$$T(N) = 3 \quad (N) U^2 \log^{-3} N + O(U^2 \log^{-4} N), \tag{3}$$

where

$$(N) = \prod_{p \mid N} \left(1 - \frac{1}{(p-1)^2} \right) \prod_{p \nmid N} \left(1 + \frac{1}{(p-1)^3} \right) > \frac{1}{2}.$$

Recently, I [6] proved that φ in (1) can be taken as $\frac{13}{17}$. The basic frame in proof was taken from [2] of Pan Cheng-dong and the zero density theorem of Zhang Yi-tang [7] was used. In this paper, we use a new zero density theorem whose proof is similar to that for the zero density theorem in Huxley [8]. We obtain:

Theorem. *When* $\varphi = \frac{2}{3}$ *in* (1), *equation* (3) *still holds.*

This theorem is the same as the conclusion in Chen Jin-run [3]. Throughout the paper, we always assume $\delta=\frac{\varepsilon}{10^{40}}$, $Q_1=\log^{20}N$, $Q_2=\exp(\log\log^3 N)$, $\tau=N^{\frac{1}{3}+\frac{4}{3}\varepsilon}$, c, c_1, c_2 is positive constants which has different values in different places.

§ 2. The Translation of the Problem

$$T(N)=\int_{-\frac{1}{\tau}}^{1-\frac{1}{\tau}}\Big(\sum_{\frac{N}{3}-U<p\leqslant\frac{N}{3}+U}e(p\theta)\Big)^3 e(-N\theta)\,d\theta. \tag{4}$$

Any real number θ can be expressed as

$$\theta=\frac{a}{q}+\alpha,\ (a,q)=1,\ 0\leqslant a<q,\ 1\leqslant q\leqslant\tau,\ |\alpha|\leqslant\frac{1}{q\tau}. \tag{5}$$

All θ corresponding $q\leqslant Q_1$ make up E_1 whose supplementary set is E_2. Since $\tau>2Q$, intervals $\left[\frac{a}{q}-\frac{1}{q\tau},\ \frac{a}{q}+\frac{1}{q\tau}\right]$ $(q\leqslant Q_1)$ disjoint. In E_1, the sum aggregate of intervals $\left[\frac{a}{q}-\frac{\log^6 N}{U},\ \frac{a}{q}+\frac{\log^6 N}{U}\right]$ $(q\leqslant Q_1)$ make up $E_{1.1}$ whose supplementary set in E_1 is $E_{1.2}$. In E_2, all θ corresponding $Q_1<q\leqslant Q_2$ make up $E_{2,1}$ whose supplementary set in E_2 is $E_{2,2}$.

§ 3. The Estimation for the Integral in $E_{2,2}$

Lemma 1. (I. M. Vinogradov)

If $\theta=\frac{a}{q}+\alpha$, $|\alpha|\leqslant\frac{1}{q^2}$, $0<q\leqslant N$, $(a,q)=1$, *then*

$$\sum_{N-A<p\leqslant N}e(p\theta)$$

$$\ll A\exp(c\log\log^2 N)\left(\frac{N^{\frac{1}{2}}q^{\frac{1}{2}}}{A}+\frac{1}{q^{\frac{1}{2}}}+\frac{N^{\frac{1}{3}+\delta}}{A^{\frac{1}{2}}}\right).$$

See [2].

If $\theta\in E_{2,2}$, then $\theta=\frac{a}{q}+a$, $|a|\leqslant\frac{1}{q^2}$, $\exp(\log\log^3 N)<q\leqslant\tau$. By Lemma 1, we have

$$\sum_{\frac{N}{3}-U<p\leqslant\frac{N}{3}+U} e(p\theta)$$

$$\ll U\exp(c\log\log^2 N)\left(\frac{N^{\frac{1}{2}}\tau^{\frac{1}{2}}}{U}+\exp(-c,\log\log^3 N)+\frac{N^{\frac{1}{3}+\delta}}{U^{\frac{1}{2}}}\right)$$

$$\ll U\log^{-6}N,$$

$$\int_{\theta\in E_{2,2}}\Big(\sum_{\frac{N}{3}-U<p\leqslant\frac{N}{3}+U} e(p\theta)\Big)^3 e(-N\theta)\,d\theta$$

$$\ll \max_{\theta\in E_{2,2}}\Big|\sum_{\frac{N}{3}-U<p\leqslant\frac{N}{3}+U} e(p\theta)\Big|\int_0^1\Big|\sum_{\frac{N}{3}-U<p\leqslant\frac{N}{3}+U} e(p\theta)\Big|^2 d\theta$$

$$\ll U^2\log^{-6}N. \tag{6}$$

§ 4. Some Lemmas on the Zero of $L(s,\chi)$

Lemma 2. *If* $c_1N<x\leqslant c_2N$, $2\leqslant T\ll N$, $(q,1)=1$, *and* $\psi(x;q,1)=\sum_{\substack{n\leqslant x\\ n\equiv l\,(\mathrm{mod}\,q)}}\Lambda(n)$, *when* $q\leqslant\exp(\log\log^3 N)$, *we have*

$$\psi(x;q,1)=\frac{x}{\varphi(q)}-\frac{\tilde{\chi}(l)}{\varphi(q)}\cdot\frac{x^{\tilde{\beta}}}{\beta}-\frac{V_x}{\varphi(q)}+O\left(\frac{N}{T}\log^2 N+\frac{N^{\frac{1}{4}}\log N}{\varphi(q)}\right).$$

where

$$V_x=\sum_{x\,(\mathrm{mod}\,q)}\bar{x}(l)\sum_{|\gamma|\leqslant T}{}'\frac{x^\rho}{\rho},$$

$\rho=\beta+i\gamma$ *denotes the zero of* $L(s,\chi)$ *within* $0\leqslant\beta\leqslant1$, $\tilde{\chi}$ *is the possible exceptional character mod* q, $\tilde{\beta}$ *is the possible exceptional zero mod* q, $\sum'$ *denote summation without the exceptional zero.*

See [2].

Lemma 3. *Let* $s=\sigma+it$, *except the exceptional zero mod* q, $\prod_{\chi(\mathrm{mod}\,q)}L(s,\chi)$ *has no zero in the domain*

$$\sigma\geqslant1-c(\log q+\log^{4/5}(|t|+2))^{-1}.$$

See [5].

Lemma 4. (Siegel) *if* $\tilde{\beta}$ *is the exceptional zero mod* q, *then*

$$\mathrm{Re}\,\tilde{\beta}<1-\frac{c(\varepsilon)}{q^\varepsilon}.$$

See lemma 9 in page 255 of [9].

Lemma 5. *If $T\geqslant 4$, $q\leqslant Q=\exp(c\log\log^3 T)$, $\sigma\geqslant \frac{1}{2}$, χ is the character mod q, $N(\sigma, T, \chi)$ is the number of zeros of $L(s,\chi)$ within $\sigma\leqslant\beta<1$, $|t|\leqslant T$, then we have*

1) $N(\sigma, T+1, \chi)-N(\sigma, T, \chi)\ll \log qT$;

2) $N(\frac{1}{2}, T, \chi)\ll T\log qT$;

3) $N(\sigma, T, \chi)\ll T^{(\frac{12}{5}+\delta)(1-\sigma)}Q^{c_1}$.

1) can be found in Lemma 1 in page 76 of [9]. 2) can be deduced from 1). 3) can be found in the 3) of Lemma 3 in [6].

Lemma 6. *If $T\geqslant 2$, then*

$$\frac{1}{\varphi(q)}\sum_{\chi(\mathrm{mod}\, q)}\int_0^T |L(\frac{1}{2}+it,\chi)|^2\, dt$$

$$=\frac{\varphi(q)}{q}\{T\log T+(T-1)(2\gamma-1-\log 2\pi+\log q+\sum_{p|q}\frac{\log p}{p-1})\}$$

$$+O(T^{\frac{1}{2}}\log T),$$

where γ is the Euler constant.

See [10].

Lemma 7. $\sigma\geqslant\frac{1}{2}$, $q\leqslant Q=\exp(\log\log^3 T)$, $T_0\geqslant 2$, $T_0\gg T^{\frac{1}{2}}$, *then*

$$N(\sigma, T+T_0, \chi)-N(\sigma, T, \chi)\ll T_0^{(\frac{8}{3}+\frac{\varepsilon}{100})(1-\sigma)}Q^c,$$

Proof. Firstly we suppose $T_0\leqslant T\ll T_0^2$.

$\rho=\beta+i\gamma$ is the zero of $L(s,\chi)$ within $o\leqslant\beta<1$, $|\gamma|\leqslant T$. ρ can be divide into (I) ρ satisfies

$$\frac{1}{\log(qT)}\ll\left|\sum_{m=M}^{2M}\frac{a(m)\chi(m)}{m^\rho}\right|, \tag{7}$$

where $|a(m)|\ll d(m)$, $(qT)^\varepsilon<M\leqslant Y\ll T^c$.

(II) ρ satisfies

$$Y^{\sigma-\frac{1}{2}}(qT)^{-\delta}\ll\int_{-(\log qT)^2}^{(\log qT)^2}|L(\frac{1}{2}+it+i\gamma,\chi)|\, dt. \tag{8}$$

The process above can be found in [11, Chapter 12].

Taking $[\gamma]=t$, $\rho=\sigma+it+\delta$, we have $\mathrm{Re}(\delta)\geqslant 0$. Suppose

$$S(y)=\sum_{M<m\leqslant y}\frac{a(m)\chi(m)}{m^{\sigma+it}},$$

we have

$$\sum_{M<m\leqslant 2m}\frac{a(m)\chi(m)}{m^{\rho}}$$

$$=\frac{S(2M)}{(2M)^{\delta}}+\delta\int_M^{2M}\frac{S(y)}{y^{1+\delta}}\,dy$$

$$\ll|S(2M)|+\int_M^{2M}\frac{|S(y)|}{y}\,dy,$$

$$\frac{1}{\log(qT)}\ll\left|\sum_{m=M}^{2M}\frac{a(m)\chi(m)}{m^{\rho}}\right| \text{ equals}$$

$$\frac{1}{\log(qT)}\ll|S(2M)| \text{ or}$$

$$\frac{1}{\log(qT)}\ll\int_M^{2M}\frac{|S(y)|}{y}\,dy$$

We only consider the former. The latter can be dealt with in the same way.

$[\gamma]=t$ has $O(\log T)$ solutions. If $t=T+t_\gamma$, then $|t_\gamma|\leqslant T_0$. Thus,

$$N(\sigma,T+T_0,\chi)-N(\sigma,T,\chi) \ll\log T(R_1+R_2), \tag{9}$$

where R_1 is the number of $t_r's$, which satisfy

$$1\ll\left|\sum_{M<m\leqslant 2M}\frac{b(m)}{m^{\sigma+it_\gamma}}\right|, \tag{10}$$

where $b(m)=a(m)\chi(m)m^{-iT}\log T$, $b(m)\ll|d(m)|\log T$, $|t_r|\leqslant T_0$, $|t_r-t_s|\geqslant 1$ $(r\neq s)$; R_2 is the number of $\gamma_r's$, which satisfy

$$Y^{\sigma-\frac{1}{2}}(qT)^{-\varepsilon}\ll\int_{-(\log qT)^2}^{(\log qT)^2}|L(\frac{1}{2}+it+i\gamma_r,\chi)|\,dt. \tag{11}$$

Now we estimate R_1, R_2.

When $\frac{1}{2}\leqslant\sigma\leqslant\frac{3}{4}$, taking $Y=T_0^{4/3}$, by theorem 5 $(q=1)$ in page 40 of

[9], we know

$$\sum_{r=1}^{R}\Big|\sum_{c_1L<l\leqslant c_2L}\frac{c(l)}{l^{\sigma+it_r}}\Big|^2$$
$$\ll (T_0L^{1-2\sigma}+L^{2(1-\sigma)})\max_l|c(l)|^2. \tag{12}$$

If a is a positive integer, by (10), we get

$$1\ll\Big|\sum_{c_1M^a<m\leqslant c_2M^a}\frac{c(m)}{m^{\sigma+it_r}}\Big|^2. \quad c(m)\ll T^{\delta}.$$

By (12), we have

$$R_1\ll (T_0M^{a(1-2\sigma)}+M^{2a(1-\sigma)})T^{\delta}. \tag{13}$$

When $a\geqslant 2$, for $T_0^{\frac{1}{a+2(1-\sigma)}}\leqslant M\leqslant T_0^{\frac{1}{(a-1)+2(1-\sigma)}}$, by (13), we get

$$R_1\ll (T_0\cdot T_0^{\frac{a(1-2\sigma)}{a+2(1-\sigma)}}+T_0^{\frac{2a(1-\sigma)}{(a-1)+2(1-\sigma)}})T^{\delta}$$
$$\ll (T_0^{\frac{2(a+1)(1-\sigma)}{a+2(1-\sigma)}}+T_0^{\frac{2a(1-\sigma)}{(a-1)+2(1-\sigma)}})T_0^{2\delta}$$
$$\ll T_0^{\frac{4(1-\sigma)}{3-2\sigma}+2\delta}$$
$$\ll T_0^{\frac{8}{3}(1-\sigma)+2\delta}.$$

For $T_0^{\frac{1}{3-2\sigma}}\leqslant M\leqslant T_0^{4/3}$, taking $a=1$, we have

$$R_1\ll (T_0M^{1-2\sigma}+M^{2(1-\sigma)})T^{\delta}$$
$$\ll (T_0^{\frac{4(1-\sigma)}{3-2\sigma}}+T_0^{\frac{8}{3}(1-\sigma)})T_0^{2\delta}$$
$$\ll T_0^{\frac{8}{3}(1-\sigma)+2\delta}.$$

We always have

$$R_1\ll T_0^{\frac{8}{3}(1-\sigma)+2\delta}. \tag{14}$$

By (8), we have

$$Y^{2\sigma-1}(T_0)^{-10\delta}\ll\int_{-(\log qT)^2}^{(\log qT)^2}|L(\tfrac{1}{2}+it+i\gamma_r,\chi)|^2\,dt$$

$$R_2\ll Y^{1-2\sigma}T_0^{12\delta}\int_{T-(\log qT)^2}^{T+T_0+(\log qT)^2}|L(\tfrac{1}{2}+it,\chi)|^2\,dt.$$

By Lemma 6, we know

$$R_2 \ll Y^{1-2\sigma} T_0^{12\delta}(T_0 + T^{\frac{1}{2}})$$
$$\ll T_0^{4/3(1-2\sigma)} \cdot T_0^{12\delta} \cdot T_0$$
$$\ll T_0^{\frac{7-8\sigma}{3}+12\delta} \ll T_0^{\frac{8}{3}(1-\sigma)}. \tag{15}$$

Combining all above, we get that when $\frac{1}{2} \leqslant \sigma \leqslant \frac{3}{4}$,

$$R_1 + R_2 \ll T_0^{(\frac{8}{3}+100\delta)(1-\sigma)}. \tag{16}$$

When $\frac{3}{4} \leqslant \sigma \leqslant 1$, taking $Y = T_0^{4/3}$, by (10), we get

$$1 \ll \left| \sum_{c_1 M^a < m \leqslant c_2 M^a} \frac{c(m)}{m^{\sigma + it_r}} \right|^2, \; c(m) \ll T^{\delta}.$$

Using Halász's inequality (see § 2 in [8]), we know

$$R_1 \ll (M^{2a(1-\sigma)} + T_0 M^{2a(2-3\sigma)}) T^{\delta}. \tag{17}$$

If $a \geqslant 2$, for $T_0^{\frac{1}{(4a-2)\sigma+2-2a}} \leqslant M \leqslant T_0^{\frac{1}{(4a-6)\sigma+4-2a}}$, by (17), we get

$$R_1 \ll (T_0 \cdot T_0^{\frac{2a(2-3\sigma)}{(4a-2)\sigma+2-2a}} + T_0^{\frac{2a(1-\sigma)}{(4a-6)\sigma+4-2a}}) T_0^{2\delta}$$
$$= (T_0^{\frac{2(a+1)(1-\sigma)}{(4a-2)\sigma+2-2a}} + T_0^{\frac{2a(1-\sigma)}{(4a-6)\sigma+4-2a}}) T_0^{2\delta}$$
$$\ll T_0^{\frac{2}{\sigma}(1-\sigma)+2\delta}$$
$$\ll T_0^{\frac{8}{3}(1-\sigma)+2\delta}$$

For $T_0^{\frac{1}{2\sigma}} \leqslant M \leqslant T_0^{4/3}$, taking $a = 1$, we get

$$R_1 \ll (T_0 M^{4-6\sigma} + M^{2(1-\sigma)}) T_0^{2\delta}$$
$$\ll (T_0^{\frac{2}{\sigma}(1-\sigma)} + T_0^{\frac{8}{3}(1-\sigma)}) T_0^{2\delta}$$
$$\ll T_0^{\frac{8}{3}(1-\sigma)+2\delta}. \tag{18}$$

Using the same principle as in (15), we get again

$$R_2 \ll T_0^{\frac{8}{3}(1-\sigma)}.$$

Thus, when $\frac{3}{4} \leqslant \sigma \leqslant 1$, we have

$$R_1 + R_2 \ll T_0^{\frac{8}{3}(1-\sigma)+2\delta}. \tag{19}$$

By (6) in [6], we know that if $q \leqslant Q$, $\frac{152}{155} \leqslant \sigma < 1$, then

$$N(\sigma, T, \chi) \ll T^{1600(1-\sigma)3/2} Q^c. \tag{20}$$

When $\frac{3}{4} \leqslant \sigma \leqslant 1 - \frac{1}{10^8}$, by (19), we have

$$N(\sigma, T+T_0, \chi) - N(\sigma, T, \chi) \ll T_0^{(\frac{8}{3}+\frac{\varepsilon}{100})(1-\sigma)}$$

When $1 - \frac{1}{10^8} \leqslant \sigma < 1$, by (20), we get

$$N(\sigma, T, \chi) + N(\sigma, T+T_0, \chi)$$

$$\ll T_0^{(\frac{8}{3}+\frac{\varepsilon}{100})(1-\sigma)} Q^c. \tag{21}$$

Therefore, when $T_0 \leqslant T \ll T_0^2$, we have

$$N(\sigma, T+T_0, \chi) - N(\sigma, T, \chi)$$

$$\ll T_0^{(\frac{8}{3}+\frac{\varepsilon}{100})(1-\sigma)} Q^c. \tag{22}$$

Secondly, we suppose $T \leqslant T_0$.
By 3) of Lemma 5, we have

$$N(\sigma, T, \chi) + N(\sigma, T+T_0, \chi)$$

$$\ll T_0^{(\frac{12}{5}+\delta)(1-\sigma)} Q^c$$

$$\ll T_0^{(\frac{8}{3}+\delta)(1-\sigma)} Q^c. \tag{23}$$

From (22), (23), we know that the Lemma sets up.

§ 5. The Estimation for the Integral on $E_{2,1}$

Suppose

$$S(\theta, N) = \sum_{\frac{N}{3} - U < n \leqslant \frac{N}{3} + U} \Lambda(n) e(n\theta)$$

$$= \sum_{\frac{N}{3} - U < p \leqslant \frac{N}{3} + U} \log p \cdot e(p\theta) + O(N^{\frac{1}{2}})$$

$$=\log\frac{N}{3}\cdot\sum_{\frac{N}{3}-U<p\leqslant\frac{N}{3}+U}e(p\theta)+O(N^{\frac{1}{2}}),$$

$$\sum_{\frac{N}{3}-U<p\leqslant\frac{N}{3}+U}e(p\theta)$$

$$=(\log\frac{N}{3})^{-1}S(\theta,N)+O(N^{\frac{1}{2}}),\tag{24}$$

$$\sum_{\frac{N}{3}-U<n\leqslant\frac{N}{3}+U}\Lambda(n)e(n\theta)$$

$$=\sum_{\frac{N}{3}-U<n\leqslant\frac{N}{3}+U}\Lambda(n)e((\frac{a}{q}+\alpha)n)$$

$$=\sum_{\substack{l=1\\(l,q)=1}}^{q}e\left(\frac{al}{q}\right)\cdot\sum_{\substack{\frac{N}{3}-U<n\leqslant\frac{N}{3}+U\\n\equiv l\,(\mathrm{mod}\,q)}}\Lambda(n)e(n\alpha)+O(\log^3N).\tag{25}$$

By (9) in [6], we know

$$\sum_{\substack{\frac{N}{3}-U<n\leqslant\frac{N}{3}+U\\n\equiv l\,(\mathrm{mod}\,q)}}\Lambda(n)e(n\alpha)$$

$$=\frac{1}{\varphi(q)}\int_{\frac{N}{3}-U}^{\frac{N}{3}+U}e(\alpha x)dx-\sum_{x}x(l)\sum_{|\gamma|\leqslant T}{}'\int_{\frac{N}{3}-U}^{\frac{N}{3}+U}x^{\rho-1}e(\alpha x)dx$$

$$-\frac{\tilde{x}(l)}{\varphi(q)}\int_{\frac{N}{3}-U}^{\frac{N}{3}+U}x^{\beta-1}e(\alpha x)dx+O\left(\frac{(1+|\alpha|U)N}{T}\log^2N\right.$$

$$\left.+\frac{(1+|\alpha|U)N^{\frac{1}{4}}\log N}{\varphi(q)}\right).\tag{26}$$

When $|\alpha|\leqslant\frac{1}{N^{\frac{7}{12}+\varepsilon}}$, by (10) in [6], we know

$$\sum_{\substack{\frac{N}{3}-U<n\leqslant\frac{N}{3}+U\\n\equiv l\,(\mathrm{mod}\,q)}}\Lambda(n)e(n\alpha)=\frac{1}{\varphi(q)}\int_{\frac{N}{3}-U}^{\frac{N}{3}+U}e(\alpha x)dx$$

$$-\frac{\tilde{\chi}(l)}{\varphi(q)}\int_{\frac{N}{3}-U}^{\frac{N}{3}+U}x^{\tilde{\beta}-1}e(\alpha x)dx+O(U\exp(-c\log^{\frac{1}{5}}N)).\tag{27}$$

We might as well suppose $\frac{1}{N^{\frac{7}{12}+\varepsilon}} \leqslant \alpha \leqslant \frac{1}{q\tau}$. Let $T=\alpha N^{1+\delta}$, by (11) in [6], we know

$$\sum_{x} \bar{\chi}(l) \sum_{|\gamma| \leqslant T}{}' \int_{\frac{N}{3}-U}^{\frac{N}{3}+U} x^{\rho-1} e(\alpha x)\, dx$$

$$\ll U \log^2 N \max_{0 \leqslant T_1 \leqslant 2T} \sum_{x} \sum_{\substack{T_1 \leqslant \gamma \leqslant T_1+T_0 \\ \beta \geqslant \frac{1}{2}}}{}' N^{\beta-1}, \tag{28}$$

where $T_0=\max(\alpha U, \frac{N}{U})$.

By Lemma 3 and 7, we know

$$\sum_{x} \sum_{\substack{T_1 \leqslant \gamma \leqslant T_1+T_0 \\ \beta \geqslant \frac{1}{2}}} N^{\beta-1}$$

$$= -\int_{\frac{1}{2}}^{1} N^{\sigma-1} d(N(\sigma, T_1+T_0, q) - N(\sigma, T_1, q))$$

$$= N^{-\frac{1}{2}}(N(\frac{1}{2}, T_1+T_0, q) - N(\frac{1}{2}, T_1, q))$$

$$+ \int_{\frac{1}{2}}^{1-\frac{c}{\log^{4/5} N}} (N(\sigma, T_1+T_0, q) - N(\sigma, T_1, q)) N^{\sigma-1} \log N\, d\sigma$$

$$\ll T_0^{\frac{4}{3}+\frac{\varepsilon}{100}} N^{-\frac{1}{2}} + \int_{\frac{1}{2}}^{1-\frac{c}{\log^{4/5} N}} (T_0^{(\frac{8}{3}+\frac{\varepsilon}{100})(1-\sigma)} N^{\sigma-1}) \log N \cdot Q_2^c\, d\sigma,$$

$$T_0 = \max(\alpha U, \frac{N}{U}) \leqslant \max(\frac{U}{\tau}, \frac{N}{U}) \leqslant N^{\frac{1}{3}},$$

$$\sum_{x} \sum_{\substack{T_1 \leqslant \gamma \leqslant T_1+T_0 \\ \beta \geqslant \frac{1}{2}}} N^{\beta-1}$$

$$\ll N^{-\frac{1}{20}} + \int_{\frac{1}{2}}^{1-\frac{c}{\log^{4/5} N}} N^{-(\frac{1}{9}-\varepsilon)(1-\sigma)} Q_2^c\, d\sigma$$

$$\ll \exp(-c \log^{\frac{1}{5}} N). \tag{29}$$

From formulae above and (26), we get (27) again. By (27), we have

$$\sum_{\frac{N}{3}-U<n\leqslant\frac{N}{3}+U}\Lambda(n)e(n\theta)$$

$$=\sum_{\substack{l=1\\(l,q)=1}}^{q}e\left(\frac{a}{q}l\right)\sum_{\substack{\frac{N}{3}-U<n\leqslant\frac{N}{3}+U\\n\equiv l\,(\mathrm{mod}\ q)}}\Lambda(n)e(n\alpha)+O(\log^3 N)$$

$$=\frac{\mu(q)}{\varphi(q)}\int_{\frac{N}{3}-U}^{\frac{N}{3}+U}e(\alpha x)dx+O\left(\frac{1}{\varphi(q)}\left|\sum_{l=1}^{q}x(l)e\left(\frac{a}{q}e\right)\right|\cdot\right.$$

$$\left.\cdot\left|\int_{\frac{N}{3}-U}^{\frac{N}{3}+U}x^{\mathrm{Re}\tilde{\beta}-1}e(\alpha x)dx\right|+O\left(U\exp(-c\log^{\frac{1}{5}}N)\right)\right). \tag{30}$$

When $\theta\in E_{2,1}$, by

$$\left|\sum_{l=1}^{q}\tilde{x}(l)e\left(\frac{a}{q}l\right)\right|\ll q^{\frac{1}{2}},\ \varphi(q)\gg\frac{q}{\log q},$$

we know

$$\sum_{\frac{N}{3}-U<n\leqslant\frac{N}{3}+U}\Lambda(n)e(n\theta)$$

$$\ll\frac{U}{\varphi(q)}+\frac{U\log N}{q^{\frac{1}{2}}}+U\exp(-c\log^{\frac{1}{5}}N)$$

$$\ll U\log^{-6}N. \tag{31}$$

By (24), we get

$$\sum_{\frac{N}{3}-U<p\leqslant\frac{N}{3}+U}e(p\theta)\ll U\log^{-6}N$$

$$\int_{\theta\in E_{2,1}}\left(\sum_{\frac{N}{3}-U<p\leqslant\frac{N}{3}+U}e(p\theta)\right)^3e(-N\theta)d\theta$$

$$\ll\max_{\theta\in E_{2,1}}\left|\sum_{\frac{N}{3}-U<p\leqslant\frac{N}{3}+U}e(p\theta)\right|\int_0^1\left|\sum_{\frac{N}{3}-U<p\leqslant\frac{N}{3}+U}e(p\theta)\right|^2d\theta$$

$$\ll U^2\log^{-6}N. \tag{32}$$

§ 6. The Estimation for the Integral on $E_{1,2}$

By Lemma 4, if $q \leqslant \log^{20} N$, then

$$\operatorname{Re} \widetilde{\beta} - 1 < -\frac{c(\varepsilon)}{q^{\varepsilon}} < -\frac{1}{\log^{20\varepsilon} N},$$

$$N^{\operatorname{Re}\widetilde{\beta} - 1} = O(\exp(-\log^{1-20\varepsilon} N)) = O(\log^{-6} N).$$

$$\int_{\frac{N}{3} - U}^{\frac{N}{3} + U} e(\alpha x)\, dx \ll \frac{1}{|\alpha|}.$$

If $\theta \in E_{1,2}$, then $q \leqslant \log^{20} N$, $\frac{\log^6 N}{U} \leqslant |\alpha| \leqslant \frac{1}{q\tau}$. By (30), we get

$$\sum_{\frac{N}{3} - U < n \leqslant \frac{N}{3} + U} \Lambda(n) e(n\theta) \ll U \log^{-6} N. \tag{33}$$

By (24), we know

$$\sum_{\frac{N}{3} - U < p \leqslant \frac{N}{3} + U} e(p\theta) \ll U \log^{-6} N,$$

$$\int_{\theta \in E_{1,2}} \Bigg(\sum_{\frac{N}{3} - U < p \leqslant \frac{N}{3} + U} e(p\theta) \Bigg)^3 e(-N\theta)\, d\theta$$

$$\ll \max_{\theta \in E_{1,2}} \Bigg| \sum_{\frac{N}{3} - U < p \leqslant \frac{N}{3} + U} e(p\theta) \Bigg| \int_0^1 \Bigg| \sum_{\frac{N}{3} - U < p \leqslant \frac{N}{3} + U} e(p\theta) \Bigg|^2 d\theta$$

$$\ll U^2 \log^{-6} N. \tag{34}$$

§ 7. The Proof for the Theorem

By the discussion in §7 of [6], we know

$$T(N) = \left(\log \frac{N}{3}\right)^{-3} \sum_{q \leqslant Q_1} \frac{\mu(q)}{\varphi^3(q)} \Bigg(\sum_{\substack{a=1 \\ (a,q)=1}}^{q} e\left(-\frac{aN}{q}\right) \Bigg) 3U^2$$

$$+ O(U^2 \log^{-6} N)$$

$$= \frac{3U^2}{\log^3 N} \cdot \quad (N) + O(U^2 \log^{-4} N). \tag{35}$$

So, we prove the theorem.

REFERENCES

[1] C. B. Haselgrove, Some theorems in the analytic theory of numbers, *J. London Math. Soc.*, **26**(1951), 273—277.

[2] Pan Cheng-dong, Some new results in additive prime theory, *Acta, Math. Sin.*, **9**(1959), 315—329, Chinese.

[3] Chen Jin-run, On large odd number as sum of three almost equal primes, *Sci. Sin.*, **14**(1965), 1113—1117.

[4] Pan Cheng-dong, Pan Cheng-biao, The estimation for the trigonometric sum in prime variables in the short interval (I), *Sci. Sin.*, to appear.

[5] Pan Cheng-dong, Pan Cheng-biao, The estimation for the trigonometric sum in prime variables in the short interval (II), *Sci. Sin.*, to appear.

[6] Jia Chao-hua, Three primes theorem in a short interval (I), to appear, Chinese.

[7] Zhang Yi-tang, Two theorems on the zero density of the Riemann zeta function, *Acta Math. Sin.*, 1985, v. 1, No. 3, 274—284.

[8] M. N. Huxley, On the difference between consecutive primes, *Invent, Math.*, **15**(1972), 164—170.

[9] Pan Cheng-dong, Pan Cheng-biao, *Goldbach Conjecture*, Science Press, 1981.

[10] V. V. Rane, On the mean square value of Dirichlet L-series, *J. London Math. Soc.* **21**(1980), 203—215.

[11] H. L. Montgomery, *Topics in multiplicative number theory*, Springer, Berlin, 1971.

On the Zeros of Riemann's Zeta-Function on the Critical Line

A. A. Karatsuba

The present paper is a detailed exposition of the author's report delivered at the International Sympozium in Memory of Hua Loo Keng, Beijing, on August the 6th, 1988.

Using this opportunity, the author expresses his deep gratitude to Professor Lu Qikeng, Professor Yang Le and Professor Wang Yuan for their warm hospitality and useful conversations.

§1. On a Problem of Hua Loo Keng

Before turning to the zeros of Riemann's funcition $\zeta(s)$, I would like to dwell upon a problem, stated by Hua Loo Keng in 1938 and solved by me and my pupils G. I. Arkhipov and V. I. Tchubarikov in 1978.

This problem is closely connected with the exponential (sometimes they are also called trigonometrical or oscillatory) integrals I_r,

$$I_r=\int_{a_1}^{b_1}\cdots\int_{a_r}^{b_r}F(x_1,\cdots,x_r)e^{2\pi i f(x_1,\cdots,x_r)}dx_1\cdots dx_r,$$

where F and f are real functions of real variables and $i^2=-1$.

Consider the following system of equations

$$\begin{cases} x_1+\cdots+x_k=y_1+\cdots+y_k, \\ x_1^2+\cdots+x_k^2=y_1^2+\cdots+y_k^2, \\ \cdots\cdots\cdots\cdots\cdots\cdots\cdots\cdots\cdots \\ x_1^n+\cdots+x_k^n=y_1^n+\cdots+y_k^n. \end{cases} \quad (1)$$

$$1\leqslant x_1,\cdots,x_k,y_1,\cdots,y_k\leqslant P.$$

Here $x_1,\cdots,x_k,y_1,\cdots,y_k$ are integers. Let $J(P)=J(P;k,n)$ denote the number of solutions of that system. In 1938 Hua Loo Keng [1] proved that for $k>k_1=k_1(n)$

$$\lim_{P\to\infty} J(P)\cdot P^{-2k+\frac{n(n+1)}{2}} = \theta ,$$

where

$$\theta=\int_{-\infty}^{\infty}\cdots\int_{-\infty}^{\infty}\left|\int_0^1 e^{2\pi i(\alpha_n x^n+\cdots+\alpha_1 x)}dx\right|^{2k} d\alpha_1\cdots d\alpha_n ,$$

$$=\sum_{q_n=1}^{\infty}\cdots\sum_{q_1=1}^{\infty}\sum_{a_n=1}^{q_n}\cdots\sum_{a_1=1}^{q_1}\left|\frac{1}{q_n\cdots q_1}\sum_{x=1}^{q_n\cdots q_1} e^{2\pi i\left(\frac{a_n}{q_n}x^n+\cdots+\frac{a_1}{q_1}x\right)}\right|^{2k} .$$

$$(a_n, q_n)= 1, (a_1, q_1)=1.$$

In his paper [2] Hua Loo Keng proved that the series converges for $k>0.25\,n(n+1)$ and diverges for $k\leqslant 0.25\,n(n+1)$ while the integral θ converges for $k>0.25n(n+2)$, and formulated the problem to compute the exponent of convergence of θ i.e. that of computation of such $2k_0=2k_0(n)$, that θ converges for $k\geqslant k_0+\varepsilon$ and diverges for $k\leqslant k_0-\varepsilon$ where ε is an arbitrary small positive number.

In 1956, at the Third Congress of Mathematicians of the USSR he mentioned that problem once again (s.f. [3], p.141). In 1978 the author, G.I. Arkhipov and V.N. Tchubarikov (c.f. [4] [5]) proved that θ is convergent for

$$k>0.25\,n(n+1)+0.5$$

and divergent for $k\leqslant 0.25\,n(n+1)+0.5$ i.e. completely solved Hua's problem on the exponent of convergence of θ.

The following theorem is basic for the solution.

Theorem 1. *Let* $n\geqslant 1$, $\alpha_1, \cdots, \alpha_n$ *be real numbers* $f(x)=\alpha_n x^n+\cdots+\alpha_1 x$, $\beta_r(x)=\frac{1}{r!}f^{(r)}(x)$, $r=1,\cdots,n$, $H=H(\alpha_n,\cdots,\alpha_1)=\min\limits_{a\leqslant x\leqslant b}\sum_{r=1}^{n}\left|\beta_r(x)\right|^{\frac{1}{r}}$. *Then for the integral* J,

$$J=\int_a^b e^{2\pi i f(x)}dx ,$$

the following estimate is valid

$$|J|\leqslant\min(b-a, 6en^3H^{-1}).$$

The proof of Theorem 1 is based upon the following lemma.

Lemma 1. *Let for* $0<x<1$ *the real function* $f(x)$ *possess the derivative of n-th order,* $n>1$, *and let for some* $A>0$ *the inequality below hold*

$$|f^{(n)}(x)| \geqslant A, \qquad 0<x<1.$$

Then for the integral J_1,

$$J_1=\int_0^1 e^{2\pi i f(x)}dx,$$

the following estimate is true

$$|J_1| \leqslant \min(1, 6n A^{-\frac{1}{n}}).$$

Proof. Represent J_1 in the form of a sum of the integrals corresponding to the real and imaginary parts of the integrand

$$J_1=U+iV,$$

where

$$U=\int_0^1 \cos 2\pi f(x)dx, \qquad V=\int_0^1 \sin 2\pi f(x)dx.$$

Consider the integral U. Split the segment $0<x<1$ into two sets E_1 and E_2 where E_1 consists of those segments, whose points satisfy the condition

$$|f'(x)| \leqslant B=2^{\frac{1}{n}-1} A^{\frac{1}{n}};$$

while E_2 is the union of all the remaining segments. According to that partition U is represented as a sum of two summands

$$U=U_1+U_2,$$

where

$$U_1=\int_{E_1} \cos 2\pi f(x)dx, \qquad U_2=\int_{E_2} \cos 2\pi f(x)dx.$$

Let μ be the sum of the lengths of the segments which constitute the set E_1. Plainly

$$|U_1| \leqslant \mu.$$

We now estimate μ. We shift all the constituent segments of the set E_1 together, we get one solid segment whose length equals μ. We mark n equidistant points on the latter segment, with mutual distance equal to $\mu/(n-1)$, and then return the segments to their original positions. This gives birth to the n points $x_1, x_2, \cdots, x_n$ from E_1 such that

$$|x_k-x_j| \geqslant \frac{|k-j|\mu}{n-1}.$$

Now, consider the Lagrange interpolation polynomial $g(x)$, corresponding to the function $f'(x)$ and to the interpolation knots $x_1, x_2, \cdots, x_n$:

$$g(x)=\sum_{\nu=1}^{n} f'(x_\nu)\frac{(x-x_1)\cdots(x-x_{\nu-1})(x-x_{\nu+1})\cdots(x-x_n)}{(x_\nu-x_1)\cdots(x_\nu-x_{\nu-1})(x_\nu-x_{\nu+1})\cdots(x_\nu-x_n)}\ .$$

The difference $F(x)$,

$$F(x)=g(x)-f'(x),$$

is $(n-1)$ times differentiable and vanishes at the points $x=x_1,x_2,\cdots,x_n$. Thus, according to Rolle's theorem there exist such points $\xi_1,\xi_2,\cdots,\xi_{n-1}$ that

$$x_1<\xi_1<x_2<\xi_2<x_3<\cdots<x_{n-1}<\xi_{n-1}<x_n,$$

$$F'(\xi_1)=\cdots=F'(\xi_{n-1})=0.$$

If we apply the same speculation to $F'(x)$, $F''(x)$, etc, i. e. consequently apply $n-1$ times Rolle's theorem to $F(x)$, we arrive at a point $\xi, 0<\xi<1$ such that

$$F^{(n-1)}(\xi)=g^{(n-1)}(\xi)-f^{(n)}(\xi)=0.$$

We get:

$$\frac{f^{(n)}(\xi)}{(n-1)!}=\sum_{\nu=1}^{n}\frac{f'(x_\nu)}{(x_\nu-x_1)\cdots(x_\nu-x_{\nu-1})(x_\nu-x_{\nu+1})\cdots(x_\nu-x_n)}\ ;$$

$$\frac{A}{(n-1)!}\leqslant\frac{|f^{(n)}(\xi)|}{(n-1)!}\leqslant B\sum_{\nu=1}^{n}\frac{1}{|(x_\nu-x_1)\cdots(x_\nu-x_n)|}$$

$$\leqslant B\sum_{\nu=1}^{n}\frac{(n-1)^{n-1}}{\mu^{n-1}(\nu-1)!(n-\nu)!}=B\frac{(2n-2)^{n-1}}{(n-1)!\,\mu^{n-1}}\ ;$$

$$\mu\leqslant(2n-2)(BA^{-1})^{\frac{1}{n-1}}\ ;$$

$$|U_1|\leqslant(2n-2)(BA^{-1})^{\frac{1}{n-1}}\ .$$

We estimate $|U_2|$ from above. All the constituent segments of E_2 can be split into not more than $2n-2$ segments such that on each of the latte $f'(x)$ is monotonic and preservs its' sign. Let $x_1<x<x_2$ be some of those segments and U_3 be the corresponding part of the integral U_2. With no loss of generality we can assume that $f'(x)$ is an increasing function in that segment. If we let

$$f(x)=v,\quad f(x_1)=v_1,\quad f(x_2)=v_2,$$

we obtain:

$$U_3=\int_{v_1}^{v_2}\frac{\cos 2\pi v\,dv}{f'(x)}\ ,$$

where $f'(x)$ is considered as the function of v.

We subdivide the segment $v_1 \leqslant v \leqslant v_2$ into segments of the length $\leqslant 0.5$ by the points of the form $0.5l + 0.25$ where l is an integer. According to this partition the integral U_3 is represented by an oscillating sum whose summands are monotonically decreasing in their absolute values. Thus, we arrive at the following estimate

$$| U_3 | \leqslant \int_{v_0}^{v_0+\sigma} \frac{dv}{f'(x)} = x'' - x',$$

where

$$v_1 \leqslant v_0 \leqslant v_0 + \sigma \leqslant v_2, \qquad \sigma \leqslant 0.5,$$

$$v_0 = f(x'), \qquad v_0 + \sigma = f(x'').$$

By Lagrange's theorem on finite increments we obtain:

$$\sigma = f(x'') - f(x') = f'(\xi)(x'' - x'),$$

$$x_1 \leqslant x' \leqslant \xi \leqslant x'' \leqslant x_2;$$

$$x'' - x' = \frac{\sigma}{f'(\xi)} \leqslant \frac{1}{2B}.$$

Consequently,

$$| U_3 | \leqslant \frac{1}{2B}, \; | U_2 | \leqslant \frac{n-1}{B};$$

$$| U | \leqslant | U_1 | + | U_2 | \leqslant 2(n-1) 2^{1-\frac{1}{n}} A^{-\frac{1}{n}}.$$

Arguing in an analogous manner, we obtain the same upper estimate for $| V |$, too. Thus

$$| J_1 | \leqslant 2\sqrt{2}\, 2^{1-\frac{1}{n}} (n-1) A^{-\frac{1}{n}} < 6nA^{-\frac{1}{n}}.$$

This completes the proof of the Lemma 1.

Corollary. *Let the function* $f(x)$ *satisfy the assumptions of Lemma* 1 *on the segment* $\alpha < x < \beta$.

Then the following inequality is valid

$$\left| \int_\alpha^\beta e^{2\pi i f(x)} dx \right| \leqslant \min\left(\beta - \alpha, 6nA^{-\frac{1}{n}}\right).$$

We note here also the following I. M. Vinogradov's well-known estimate of the integral J_1 for real polynomials $f(x)$ expressed in terms of their maximal coefficient.

Lemma 2 (I. M. Vinogradov [12]). *Let* $\alpha_n, \cdots, \alpha_1$ *be real numbers*, $f(x) = \alpha_n x^n + \cdots + \alpha_1 x$, $u = \max_{1\leqslant\nu\leqslant n} |\alpha_\nu|$. *Then the following estimate is valid*

$$\left| \int_0^1 e^{2\pi i f(x)} dx \right| \leqslant \min\left(1, 32\, u^{-\frac{1}{n}}\right).$$

For the complete proof of this assertion, which is close to the above proof of lemma 1 in it's idea, the reader may be refferred to [12].

Proof of Theorem 1. We cover the segment $a<x<b$ by such nonoverlapping segments $\Delta_1, \Delta_2, \cdots, \Delta_m$, $m \leqslant 0.5(n^2+n)-1$, that for each of them there is a positive integer r, $1\leqslant r\leqslant n$ for which

$$|f^{(r)}(x)| \geqslant r!\left(\frac{1}{n} H\right)^r.$$

That partition is produced in n steps (some of the latter may be void) using the following procedure. For $k=0, 1, \cdots, n-1$ consider the functions

$$\beta_{n-k}(x) = \frac{1}{(n-k)!} f^{(n-k)}(x),$$

which are polynomials of k-th (or less) degree. The polynomials $\beta_{n-k}(x)$ possess not more than k segments of monotonicity. Thus for each $D>0$ the number of the seqments, each point of which satisfy the inequality

$$|\beta_{n-k}(x)| < D,$$

does not exceed k, while the number of the segments, each point x of which satisfies the inequality $|\beta_{n-k}(x)| \geqslant D$ does not exceed $k+1$.

Step one: $k=0$, $\beta_n(x)=\alpha_n$.

If

$$|\alpha_n| \geqslant \left(\frac{1}{n} H\right)^n,$$

then we let $\Delta_1^{(1)} = (a, b]$ and the covering process of the segment (a, b) is finished. Let it be not so, i.e. $|\alpha_n| < (\frac{1}{n} H)^n$.

Step two: $k=1$, $\beta_{n-1}(x) = n\alpha_n x + \alpha_{n-1}$. If for each $x\in(a, b)$ the inequality

$$|\beta_{n-1}(x)| \geqslant \left(\frac{1}{n} H\right)^{n-1} \tag{2}$$

is valid, then we let $\Delta_1^{(2)} = (a, b]$ and the covering process is finished. Let it be not so, i.e. there exist such $x \in (a, b)$ that

$$|\beta_{n-1}(x)| < \left(\frac{1}{n} H\right)^{n-1}.$$

The number of the segments, whose points satisfy the latter inequality does not exceed 1. We use the symbol Δ_2^1 to denote those segments, and will continue to cover them. On the other hand, the number of the segments whose satisfy the inequality (2) does not exceed two, and we use the symbols $\Delta_1^{(2)}$ and $\Delta_2^{(2)}$ to denote them (they may be void).

Let the k-th step be already acomplished. Before the step $(k+1)$ we have the set of points Δ_k^1, consisting of not more than $k-1$ segments, and each point x of Δ_k^1 satisfies the inequality

$$|\beta_{n-k+1}(x)| < \left(\frac{1}{n} H\right)^{n-k+1}.$$

We will go on with the covering of the segments of the set Δ_k^1. Moreover, the number of the segments, whose points x satisfy the inequality

$$|\beta_{n-k+1}(x)| \geqslant \left(\frac{1}{n} H\right)^{n-k+1},$$

does not exceed k, and we use the symbols $\Delta_1^{(k)}, \cdots, \Delta_k^{(k)}$ to denote those segments (some of then may be void).

Step $k+1$: the function $\beta_{n-k}(x)$ is a polynomial of the degree not exceeding k. If for each $x \in \Delta_k'$ the inequality

$$|\beta_{n-k}(x)| \geqslant \left(\frac{1}{n} H\right)^{n-k} \tag{3}$$

is valid, then the constituent intervals of Δ_k' are denoted by $\Delta_1^{(k+1)}, \cdots, \Delta_k^{(k+1)}$, and this finishes the covering process. Let it be not so, i. e. there exist $x \in \Delta_k'$ such that

$$|\beta_{n-k}(x)| < \left(\frac{1}{n} H\right)^{n-k}.$$

We denote Δ_{k+1}' the set of x's for which the above inequality is valid; Δ_{k+1}' consists of not more than k segments. We go on with the covering of the segments of the set Δ_{k+1}'. On the other hand, the number of the segments, whose points satisfy the inequality (3) does not exceed $k+1$, and to denote those segments we use the symbols $\Delta_1^{(k+1)},\cdots,\Delta_{k+1}^{(k+1)}$.

We show that after the n-th step we cover the segment (a, b) completely (if the covering is finished earlier, then we suppose all the remaining steps to be void). Let $a<\xi<b$; then due to the assumption of the theorem

$$H\leqslant\sum_{r=1}^{n} |\beta_r(\xi)|^{\frac{1}{r}},$$

i.e. there is such an r, $1\leqslant r\leqslant n$, that

$$H\leqslant n\,|\beta_r(\xi)|^{\frac{1}{r}};$$

consequently

$$|f^{(r)}(\xi)|\geqslant r!\left(\frac{1}{n}H\right)^r.$$

If we choose among such r's the maximal one and denote it k we see that ξ belong to some of the segments defined by the inequality

$$|\beta_k(x)|\geqslant\left(\frac{1}{n}H\right)^k.$$

This completes the proof of the fact that the segments $\Delta_j^{(k)}$ cover (a,b) completely, and from now on we use the symbol Δ_j to denote those segments, $j\leqslant m$. The number m of the segments of that covering does not exceed

$$2+3+\cdots+n=\frac{1}{2}(n^2+n)-1.$$

Next we estimate J. We have

$$|J|\leqslant\sum_{j=1}^{m}\left|\int_{\Delta_j} e^{2\pi i f(x)}dx\right|.$$

For each of the segments Δ_j there is such an r with $1\leqslant r\leqslant n$ that

$$|f^{(r)}(x)|\geqslant r!\left(\frac{1}{n}H\right)^r.$$

If we apply the Corollary of the Lemma 1 (note that for $r=1$ Lemma 1 and its' corollary remain valid in our case, since $f(x)$ is a polynomial of n-th degree, and consequently $f'(x)$ possesses not more than $2n-2$ segments of monotonicity) we arrive at the following estimate:

$$\left|\int_{\Delta_j} e^{2\pi i f(x)}dx\right|\leqslant 6r(r!\,n^{-r}H^r)^{-\frac{1}{r}}\leqslant 6e(n^{-1}H)^{-1};$$

$$|J| \leqslant 6menH^{-1} \leqslant 6en^3H^{-1}.$$

This completes the proof of the theorem.

We note that the theorem above provides an estimate of the integral which is exact in the order of the quantity H for each fixed polynomial

Theorem 1 can be generalized to the case of J_1 with arbitrary functions $F(x)$ and $f(x)$ (see [4] — [7]).

Below we prove the following theorem 2, which solves the problem of Hua Loo Keng.

Theorem 2. *The integral* θ *converges for* $k > 0.25n(n+1)+0.5$ *and diverges for* $k \leqslant 0.25n(n+1)+0.5$.

Proof. We start with the first assertion of the theorem. First of all we estimate the volume of the set $\Omega = \Omega(\alpha_n, \cdots, \alpha_1)$ of those points $(\alpha_n, \cdots, \alpha_1)$, where the quantity H, defined in Theorem 1, does not exceed P – a given positive integer. For $r = 1, 2, \cdots, P$, let

$$u_r = \frac{r}{P},$$

and consider the sets $\Omega_r = \Omega_r(\alpha_n, \cdots, \alpha_1)$ of those points $(\alpha_n, \cdots, \alpha_1)$ where the following estimates are valid:

$$|\beta_s(u_r)| \leqslant 2^n P^s, \qquad s = 1, 2, \cdots, n.$$

Estimate from above $\mu(\Omega_r)$ the volume of Ω_r. We have

$$\mu(\Omega_r) = \int_{\Omega_r} \cdots \int d\alpha_n \cdots d\alpha_1.$$

Introduce the following substitute of the variables in that integral

$$\alpha_s = \beta_s - \binom{s+1}{s}\beta_{s+1}u_r + \cdots + (-1)^{n-s}\binom{n}{s}\beta_n u_r^{n-s},$$

where β_s are new independent variables.

The Jacobian of that transformation equals 1. Thus we have

$$\mu(\Omega_r) = \int_{|\beta_n| \leqslant 2^n P^n} \cdots \int_{|\beta_1| \leqslant 2^n P} d\beta_n \cdots d\beta_1 = 2^{n^2+n} P^{\frac{n^2+n}{2}}.$$

It is easy to see that if at some point $(\alpha_n, \cdots, \alpha_1)$ the inequality

$$H = H(\alpha_n, \cdots, \alpha_1) \leqslant P, \tag{4}$$

holds true, then $(\alpha_n, \cdots, \alpha_1)$ belong to a domain Ω_r for some r, $1 \leqslant r \leqslant P$. In fact, if (4) holds, then there is such a ξ, $0 \leqslant \xi \leqslant 1$, that for each $s = 1, \cdots, n$ we have:

$$|\beta_s(\xi)|^{\frac{1}{s}} \leqslant P; \quad |\beta_s(\xi)| \leqslant P^s.$$

Take $r = [\xi P]$; we prove that

$$(\alpha_n, \cdots, \alpha_1) \in \Omega_r.$$

If we let $y = u_r - \xi$ we see that

$$|\beta_s(u_r)| = |\beta_s(\xi + y)| = |\beta_s(\xi) + \frac{1}{1!}\beta_s^{1}(\xi)y$$

$$+ \cdots + \frac{1}{(n-s)!}\beta_s^{(n-s)}(\xi)y^{n-s}| \leqslant \sum_{k=0}^{n-s}\binom{s+k}{k}P^s \leqslant 2^n P^s.$$

Thus we obtained that each point $(\alpha_n, \cdots, \alpha_1)$ which belongs to the domain Ω of the points satisfying inequality (4), belongs for some r, $1 \leqslant r \leqslant P$ to the domain Ω and moreover

$$\mu(\Omega_r) = 2^{n^2+n} P^{\frac{n^2+n}{2}}.$$

Hence

$$\mu(\Omega) \leqslant \sum_{r=1}^{P} \mu(\Omega_r) = 2^{n^2+n} P^{\frac{n^2+n}{2}+1}.$$

Denote $\pi(P)$ the set of those points $(\alpha_n, \cdots, \alpha_1)$, which satisfy the inequalities

$$P < H = H(\alpha_n, \cdots, \alpha_1) \leqslant 2P.$$

Then for the integral θ we have following estimate:

$$\theta \leqslant \sum_{m=0}^{\infty} \int \cdots \int_{\pi(2^m)} \left| \int_0^1 e^{2\pi i(\alpha_n x^n + \cdots + \alpha_1 x)} dx \right|^{2k} d\alpha_n \cdots d\alpha_1$$

$$+ \int \cdots \int_{\Omega(1)} \left| \int_0^1 e^{2\pi i(\alpha_n x^n + \cdots + \alpha_1 x)} dx \right|^{2k} d\alpha_n \cdots d\alpha_1.$$

To estimate the integral J,

$$J = \int_0^1 e^{2\pi i(\alpha_n x^n + \cdots + \alpha_1 x)} dx,$$

for

$$(\alpha_n, \cdots, \alpha_1) \in \pi(2^m),$$

we use theorem 1, while the integral over the set $\Omega(1)$ we estimate trivially. If, in addition, we use the estimate of the volume of $\pi(P)$, we see that

$$\theta \leqslant (12en^3)^{2k} 2^{2n^2+4} \sum_{m=0}^{\infty} 2^{m(\frac{1}{2}(n^2+n)+1-2k)} + 2^{n^2+n}.$$

The series at the right-hand side is convergent if

$$2k > \frac{n^2+n}{2} + 1,$$

which completes the proof of the first assertion of the theorem.

Now we prove the second asseryion of the theorem. For integral $P \geqslant (600\,n)^{15n}$, $r=1,2,\cdots,P$, consider the polynomials

$$f(x\,;r) = \alpha_n x^n + \alpha_{n-1} x^{n-1} + \cdots + \alpha_1 x$$

whose senior coefficients α_n belong to the segment

$$P^n < \alpha_n \leqslant (2P)^n$$

and the coefficients $\alpha_{n-1}, \cdots, \alpha_1$ are defined by the relation

$$\alpha_n x^n + \alpha_{n-1} x^{n-1} + \cdots + \alpha_1 x + \alpha_0$$
$$= \alpha_n (x-x_r)^n + \beta_{n-1}(x-x_r)^{n-1} + \cdots + \beta_1 (x-x_r),$$

where

$x_r = \dfrac{1}{4} + \dfrac{r}{2P}$; $\beta_{n-1}, \cdots, \beta_2, \beta_1$ is arbitrary numbers satisfying

$$|\beta_{n-1}| \leqslant (c_1 P)^{n-1}, \cdots, |\beta_2| \leqslant (c_1 P)^2,$$
$$|\beta_1| \leqslant c_1 P, \quad c_1 = (600\,n)^{-9n}.$$

Next we prove that if $r_1 \neq r_2$, then

$$f(x\,;r_1) \neq f(x\,;r_2).$$

In fact, the coefficient α_{n-1} of the polynomial $f(x\,;r)$ equals

$$\alpha_{n-1} = -n\left(\frac{1}{4} + \frac{r}{2P}\right)\alpha_n + \beta_{n-1},$$

and if α'_{n-1} and α''_{n-1} stand, respectively, for the coefficients α_{n-1} of the polynomial $f(x\,;r_1)$ and $f(x\,;r_2)$, then we see that:

$$|\alpha'_{n-1} - \alpha''_{n-1}| = \left|\frac{n}{2P}(r_2 - r_1)\alpha_n + \beta'_{n-1} - \beta''_{n-1}\right| \geqslant \frac{n}{2P} P^n - 2(c_1 P)^{n-1} > 0,$$

and it follows that $f(x\,;r_1) \neq f(x\,;r_2)$.

If we let $P_m=(600\,n)^{15m}$, we arrive at the following estimate of θ from below:

$$\theta>\sum_{m=n}^{\infty}\sum_{r=1}^{P_m}\int_{P_m^n}^{(2P_m)^n}\int_{-(c_1P_m)^{n-1}}^{(c_1P_m)^{n-1}}\cdots\int_{-c_1P_m}^{c_1P_m}|J|^{2k}\,d\alpha_n d\beta_{n-1}\cdots d\beta_1, \tag{5}$$

where

$$J=J(\alpha_n,\beta_{n-1},\cdots,\beta_1)=\int_0^1 e^{2\pi i f(x)}dx,$$

$$f(x)=\alpha_n(x-x_r)^n+\beta_{n-1}(x-x_r)^{n-1}+\cdots+\beta_1(x-x_r).$$

Estimate $|J|$ from below. Let

$$\Delta=\frac{c}{P}, \quad c=(600\,n)^{3n},\ P=P_m.$$

We have

$$J=\int_{-x_r}^{1-x_r}e^{2\pi i(\alpha_nx^n+\beta_{n-1}x^{n-1}+\cdots+\beta_1x)}dx=J_1+J_2+J_3,$$

where

$$J_1=\int_{-\Delta}^{\Delta}e^{2\pi i(\alpha_nx^n+\beta_{n-1}x^{n-1}+\cdots+\beta_1x)}dx,$$

$$J_2=\int_{-x_r}^{-\Delta}e^{2\pi i(\alpha_nx^n+\beta_{n-1}x^{n-1}+\cdots+\beta_1x)}dx,$$

$$J_3=\int_{\Delta}^{1-x_r}e^{2\pi i(\alpha_nx^n+\beta_{n-1}x^{n-1}+\cdots+\beta_1x)}dx.$$

We estimate the integrals J_2 and J_3 from above, using Theorem 1. For each x from the segments $-x_r\leqslant x\leqslant-\Delta$, $\Delta\leqslant x\leqslant 1-x_r$, the following inequality holds true:

$$\left|\frac{d^{n-1}}{dx^{n-1}}(\alpha_nx^n+\beta_{n-1}x^{n-1}+\cdots+\beta_1x)\right|=|n!\,\alpha_nx+(n-1)!\,\beta_{n-1}|$$

$$\geqslant n!\,\alpha_n|x|-(n-1)!\,|\beta_{n-1}|\geqslant n!\,P^n\Delta-(n-1)!\,(c_1P)^{n-1}>0.5\,cn!\,P^{n-1}.$$

Consequently, the quantity H, defined in Theorem 1, is not less than

$$(0.5\,cn\,P^{n-1})^{\frac{1}{n-1}}\geqslant c^{\frac{1}{n-1}}P,$$

and hence

$$|J_2|<6en^3c^{-\frac{1}{n-1}}P^{-1};\ |J_3|<6en^3c^{-\frac{1}{n-1}}P^{-1}.$$

We estimate the integral J_1 from below. We have:

$$J_1=\int_{-\Delta}^{\Delta} e^{2\pi i\alpha_n x^n}\,dx+\int_{-\Delta}^{\Delta}\Phi(x)e^{2\pi i\alpha_n x^n}dx,$$

where

$$\Phi(x)=e^{2\pi i(\beta_{n-1}x^{n-1}+\cdots+\beta_1 x)}-1.$$

For $|x|\leqslant\Delta$ we estimate $|\Phi(x)|$ trivially:

$$|\Phi(x)|=2\,|\sin\pi(\beta_{n-1}x^{n-1}+\cdots+\beta_1 x)|\leqslant 2\pi\,(|\beta_{n-1}|\Delta^{n-1}+\cdots+|\beta_1|\Delta)$$

$$\leqslant 2\pi((c_1\Delta P)^{n-1}+\cdots+c_1\Delta P)<2\pi(cc_1+(cc_1)^2+\cdots)=\frac{2\pi cc_1}{1-cc_1}.$$

Hence

$$\left|\int_{-\Delta}^{\Delta}\Phi(x)e^{2\pi i\alpha_n x^n}\,dx\right|<\frac{4\pi c^2c_1}{1-cc_1}P^{-1}.$$

Furthermore

$$\int_{-\Delta}^{\Delta}e^{2\pi i\alpha_n x^n}\,dx=\int_{-\infty}^{\infty}e^{2\pi i\alpha_n x^n}dx+R,$$

where

$$|R|\leqslant 2\left|\int_{\Delta}^{\infty}e^{2\pi i\alpha_n x^n}\,dx\right|.$$

Using the substitute $u=\alpha_n x^n$ and routine considerations (c. f. estimate of U_3 in Lemma 1) we see that the integral at the right-hand side of the above relation can be estimated by the quantity

$$\frac{\sqrt{2}}{n}\,\alpha_n^{-\frac{1}{n}}(\alpha_n\Delta^n)^{-1+\frac{1}{n}}<\frac{1}{n}\sqrt{2}\,c^{-n+1}P^{-1},$$

i.e.

$$|R|<\frac{1}{n}2\sqrt{2}\,c^{-n+1}P^{-1}.$$

Besides we have

$$\int_{-\infty}^{\infty}e^{2\pi i\alpha_n x^n}dx=\alpha_n^{-\frac{1}{n}}\int_{-\infty}^{\infty}e^{2\pi i u^n}\,du=\alpha_n^{-\frac{1}{n}}\left(\int_{-\infty}^{\infty}\cos 2\pi u^n du+i\int_{-\infty}^{\infty}\sin 2\pi u^n du\right);$$

$$\int_{-\infty}^{\infty}\cos 2\pi u^n du=\frac{2\cos\frac{\pi}{2n}}{\sqrt[n]{2\pi}}\,\Gamma\left(1+\frac{1}{n}\right)=2\,æ(n);$$

$$\left|\int_{-\infty}^{\infty} e^{2\pi i \alpha x^n} dx\right| \geqslant 2\,æ(n)\alpha_n^{-\frac{1}{n}}.$$

The above estimates imply the validity of the following one for $|J|$:

$$|J| \geqslant 2\,æ(n)\alpha_n^{-\frac{1}{n}} - 12en^3c^{-\frac{1}{n-1}}P^{-1} - \frac{4\pi c^2 c_1}{1-cc_1}P^{-1} - \frac{1}{n}2\sqrt{2}\,c^{-n+1}P^{-1}$$

$$\geqslant P^{-1}\left(æ(n) - 12en^3c^{-\frac{1}{n-1}} - \frac{4\pi c^2 c_1}{1-cc_1} - \frac{2}{c^{n-1}}\right).$$

Furthermore,

$$æ(n) = \frac{\cos\frac{\pi}{2n}}{\sqrt[n]{2\pi}}\,\Gamma\left(1+\frac{1}{n}\right) > \frac{1}{8};$$

$$c = (600\,n)^{3n}, \qquad c_1 = (600\,n)^{-9n},$$

$$æ(n) - 12en^3c^{-\frac{1}{n-1}} - \frac{4\pi c^2 c_1}{1-cc_1} - 2c^{-n+1} > \frac{1}{16}.$$

Thus

$$|J| > \frac{1}{16}P^{-1}.$$

If we use that estimate in the relation (5), then we see that

$$\theta > \sum_{m=n}^{\infty} P_m P_m^n c_1^{1+2+\cdots+n-1} P_m^{1+2+\cdots+n-1}\left(\frac{1}{16}P_m^{-1}\right)^{2k}$$

$$= c_1^{\frac{n^2-n}{2}}\,16^{-2k}\sum_{m=n}^{\infty}(600\,n)^{15m\left(\frac{n^2+n}{2}+1-2k\right)}.$$

This implies that the integral θ diverges for

$$2k \leqslant \frac{n^2+n}{2} + 1.$$

This completes the proof of the theorem.

As an analogy of θ, the following integral θ_1 can be also considered

$$\theta_1 = \int_{-\infty}^{\infty}\cdots\int_{-\infty}^{\infty}\left|\int_0^1 e^{2\pi i(\alpha_n x^n + \alpha_m x^m + \cdots + \alpha_r x^r)}dx\right|^{2k} d\alpha_n\,d\alpha_m\cdots d\alpha_r.$$

The integral θ_1 corresponds to the system of Diophantine equations similar to (1) but such that it contains only equations with respect to the unknown summands of the form $x^n, x^m, \cdots, x_r$, while all the other equations are absent (incomplete system of equations). However, the effect of an essential difference between the exponents of convergence of the integrals θ and θ_1 is observed here. The following assertion is valid.

Theorem 3. *Let the positive integers $r, \cdots, m, n$ satisfy the relations*

$$1 \leqslant r < \cdots < m < n,$$

$$r + \cdots + m + n < \frac{n^2+n}{2}.$$

Then the integral θ_1 converges for

$$2k > n + m + \cdots + r$$

and diverges for

$$2k \leqslant n + m + \cdots + r.$$

For the proof of that theorem, c.f. [5] – [7].

Exponential integrals I_r can be often encountered in mathematical analysis, probability theory and mathematical statistics, in theoretical physics. Most problems require estimates of absolute values of such integrals from above. As we have just shown those estimates are also required by the analytic number theory.

Very recently some new important applications of trigonometric integrals were found by K. I. Oskolkov. I note only one field of those applications (c.f. [8]).

Let $\gamma_n, \cdots, \gamma_1$ be some fixed real numbers Consider the following partial differential equation in two independent real variables ξ and t (ξ is space, and t-time-coordinate, resp.):

$$\left(\frac{1}{2\pi i}\frac{\partial}{\partial t}\right)\psi = \left(\gamma_n\left(\frac{1}{2\pi i}\frac{\partial}{\partial \xi}\right)^n + \cdots + \gamma_1\frac{1}{2\pi i}\frac{\partial}{\partial \xi}\right)\psi \tag{6}$$

and pose the Cauchy initial value problem for that equation

$$\psi(\xi, t)|_{t=0} = F(\xi), \quad \xi \in (-\infty, \infty). \tag{6a}$$

Here $F(\xi)$ is a certain initial-value function, and an a priori assumption concerning $F(\xi)$ is that it is compactly supported and belongs to L^2.

Equations of the type (6) are called (onedimensional) equations of Schrödinger type. That type includes, for example, the following ones:

$$\frac{\partial \psi}{\partial t} = i\frac{\partial^2 \psi}{\partial \xi^2}, \quad \frac{\partial \psi}{\partial t} = -\frac{\partial^3 \psi}{\partial \xi^3},$$

$$\frac{\partial\psi}{\partial t}=i\frac{\partial^4\psi}{\partial\xi^4}+\frac{\partial^5\psi}{\partial\xi^5}.$$

The first one of the latter is the classical Schrödinger equation for the wave function of a free onedimensional particle of mass $\frac{1}{2}$, and the second one- the degenerate equation of Korteweg- de Vries.

The solution of the Cauchy problem (6), (6a) by Fourier method (we understand the solution in the generalized sense, and the initial condition- in that of the convergence of the function $\psi(\xi,t)$ to $F(\xi)$ as $t\to 0$ in L^2 on compact subsets of the real line) is defined by the formula:

$$\psi(\xi,t)=\int_{-\infty}^{\infty}\hat{F}(x)e^{2\pi i(\xi x+t(\gamma_n x^n+\cdots+\gamma_1 x))}dx,$$

where

$$\hat{F}(x)=\int_{-\infty}^{\infty}F(\xi)e^{-2\pi i\xi x}\,d\xi$$

denotes the Fourier transform of F.

More generally, consider the following exponential integral

$$V(F;\alpha_n,\cdots,\alpha_1)=\int_{-\infty}^{\infty}\hat{F}(x)e^{2\pi i(\alpha_n x^n+\cdots+\alpha_1 x)}dx,$$

as a function of n real variables $\alpha_n,\cdots,\alpha_1$. Then, for each fixed $\gamma_n,\cdots,\gamma_1$, the solution $\psi(\xi,t)$ of the corresponding Cauchy problem (6), (6a) is the trace of the function V on the appropriate two-dimensional hyperplane in the n-dimensional space of the variables $\alpha_n,\cdots,\alpha_1$:

$$\alpha_n=\gamma_n t,\cdots,\alpha_2=\gamma_2 t,\ \alpha_1=\xi+\gamma_1 t,\ (-\infty<\xi<\infty,\ -\infty<t<\infty)$$

Thus, there is an obvious connection between the exponential integrals V and the solutions of Cauchy problems of type (6), (6a). It is also plane, that investigation of local and global properties of the functions V is a more wide problem than that of Ψ. That problem involves the study of classicality conditions of the solutions, stability of the latter with respect to small variations of the coefficients and analysis of the boundary character of the initial condition (6a).

We prove here the following result only, containing nontrivial sufficient condition for pointwise convergence everywhere and uniform boundedness of the integral V (and in particular, that of each generalized solution of the problem (6), (6a) for the corresponding initial functions F).

Theorem 4. *Let $F(\xi)\longrightarrow 0$ ($|\xi|\longrightarrow\infty$), the total variation var F of F be bounded, and for $0<\lambda<\Lambda<\infty$*

$$V_{\lambda,\Lambda}(F;\alpha_n,\cdots,\alpha_1)=\int_{\lambda<|x|<\Lambda}\hat{F}(x)e^{2\pi i(\alpha_n x^n+\cdots+\alpha_1 x)}dx.$$

Then the following estimate is valid:

$$|V_{\lambda,\Lambda}(F;\alpha_n,\cdots,\alpha_1)|\ll n\ \mathrm{var}\ F,$$

the factor in the symbol $\ll$ being an absolute positive constant. Moreover, for $\lambda\longrightarrow 0$, $\Lambda\longrightarrow\infty$, the limit $V(F;\alpha_n,\cdots,\alpha_1)$ exists for all real $\alpha_n,\cdots,\alpha_1$.

Thus, in the case of var $F<\infty$ the improper integral V is convergent in the Cauchy principle value sense (with the two singularities at $x=0$ and $x=\infty$), and Theorem 4 provides a generalization of the well-known Jordan test for pointwise convergence of Fourier integrals.

Proof. Integrating by parts (we make use of the conditions $F(\xi)\longrightarrow 0$ ($|\xi|\longrightarrow\infty$), var $F<\infty$) gives:

$$\hat{F}(x)=\int_{-\infty}^{\infty}F(\xi)e^{-2\pi i\xi x}d\xi=-\frac{F(\xi)e^{-2\pi i\xi x}}{2\pi ix}\Bigg|_{-\infty}^{\infty}$$

$$+\frac{1}{2\pi ix}\int_{-\infty}^{\infty}e^{-2\pi i\xi x}\,dF(\xi)=\frac{1}{2\pi ix}\int_{-\infty}^{\infty}e^{-2\pi i\xi x}dF(\xi),\quad(x\neq 0),$$

$$V_{\lambda,\Lambda}(F;\alpha_n,\cdots,\alpha_1)=\int_{\lambda<|x|<\Lambda}\frac{1}{2\pi ix}\left(\int_{-\infty}^{\infty}e^{-2\pi i\xi x}dF(\xi)\right)$$

$$\times e^{2\pi i(\alpha_n x^n+\cdots+\alpha_1 x)}dx=\int_{-\infty}^{\infty}G_{\lambda,\Lambda}(\alpha_n,\cdots,\alpha_2,\alpha_1-\xi)dF(\xi),$$

where

$$G_{\lambda,\Lambda}(\alpha_n,\cdots,\alpha_1)=\int_{\lambda<|x|<\Lambda}\frac{e^{2\pi i(\alpha_n x^n+\cdots+\alpha_1 x)}}{2\pi ix}\,dx$$

$$= \int_{\lambda}^{\Lambda} \frac{e^{2\pi i f(x)} - e^{2\pi i f(-x)}}{2\pi i x} dx, f(x) = \alpha_n x^n + \cdots + \alpha_1 x.$$

Note, in the latter integral we have passed over to integration over positive x. Thus, to prove Theorem 3 it sufficies to establish that the following estimate is valid

$$| G_{\lambda,\Lambda}(\alpha_n, \cdots, \alpha_1)| \ll n$$

(with an absolute constant in the symbol $\ll$), and that the limit

$$G(\alpha_n, \cdots, \alpha_1) = \lim_{\lambda \to 0, \Lambda \to \infty} G_{\lambda,\Lambda}(\alpha_n, \cdots, \alpha_1) = \text{v.p.} \int_{-\infty}^{\infty} \frac{e^{2\pi i (\alpha_n x^n + \cdots + \alpha_1 x)}}{2\pi i x} dx$$

exists for all real $\alpha_n, \cdots, \alpha_1$. More than that, using the estimate of exponential integrals J (c.f. the above mentioned I.M. Vinogradov's Lemma 2) we prove that after one partial integration, the improper conditionally convergent integral G turns absolutely convergent

$$G(\alpha_n, \cdots, \alpha_1) = \frac{1}{2\pi i} \int_0^{\infty} \frac{1}{x^2} \left(\int_{|y| \leqslant x} e^{2\pi i f(y)} \text{sign } y \, dy \right) dx,$$

and

$$\int_0^{\infty} \frac{1}{x^2} \left| \int_{|y| \leqslant x} e^{2\pi i f(y)} \text{sign } y \, dy \right| dx \ll n.$$

With no loss of generality we may assume that $f \not\equiv 0$. Along with $f(x) = \alpha_n x^n + \cdots + \alpha_1 x \not\equiv 0$, consider another nontrivial polynomial $g(x) = \beta_n x^n + \cdots + \beta_1 x$ with real coefficients $\beta_n, \cdots, \beta_1$, and denote

$$J(f; x) = \int_0^x e^{2\pi i f(y)} dy, \ J(g; x) = \int_0^x e^{2\pi i g(y)} dy, \ (x > 0),$$

$$f_*(x) = |\alpha_n| x^n + \cdots + |\alpha_1| x; \ g_*(x) = |\beta_n| x^n + \cdots + |\beta_1| x,$$

$$h(x) = f(x) - g(x), \ h_*(x) = (f - g)_*(x) = |\alpha_n - \beta_n| x^n + \cdots + |\alpha_1 - \beta_1| x,$$

$$A(f, g) = \text{v.p.} \int_0^{\infty} \frac{e^{2\pi i f(x)} - e^{2\pi i g(x)}}{x} dx = \lim_{\substack{\lambda \to 0 \\ \Lambda \to \infty}} \int_{\lambda}^{\Lambda} \frac{e^{2\pi i f(x)} - e^{2\pi i g(x)}}{x} dx.$$

Integrating by part, we obtain for $0 < \lambda < \Lambda < \infty$:

$$\int_\lambda^\Lambda \frac{e^{2\pi i f(x)} - e^{2\pi i g(x)}}{x} dx = \int_\lambda^\Lambda \frac{(J(f;x)-J(g;x))'}{x} dx$$

$$= \frac{J(f;x)-J(g;x)}{x}\bigg|_\lambda^\Lambda + \int_\lambda^\Lambda \frac{J(f;x)-J(g;x)}{x^2} dx.$$

For the absolute value $x^{-1}|J(f;x)-J(g;x)|$, we use the two trivial estimates:

$$x^{-1}|J(f;x)-J(g;x)| \leqslant 2,$$

$$|J(f;x)-J(g;x)| \leqslant \int_0^x \left|\sin \pi h(y)\right| dy$$

$$\leqslant \pi \int_0^x h_*(y)\,dy \leqslant \pi x h_*(y), \quad (x \geqslant 0)$$

and the non-trivial one, which is a corollary from that due to I. M. Vinogradov (c.f. Lemma 2):

$$|J(f,x)| \leqslant 32\sqrt{2}\, x f_*^{-\frac{1}{n}}(x), \quad |J(f,x)| \leqslant 32\sqrt{2}\, g_*^{-\frac{1}{n}}(x).$$

Indeed, by Lemma 2 we have

$$|J(f,1)| \leqslant \min(1, 32u^{-\frac{1}{n}}), \quad u = \max_{1\leqslant v\leqslant n} |\alpha_v|.$$

As $f_*(1) = |\alpha_n| + \cdots + |\alpha_1| \leqslant nu$, we have, $u^{-\frac{1}{n}} \leqslant n^{\frac{1}{n}} f_*^{-\frac{1}{n}}(1) \leqslant \sqrt{2}\, f_*^{-\frac{1}{n}}(1)$, and thus the above estimates of $J(f;x)$, $J(g;x)$ immediately follow after the linear change of variables $y' = yx^{-1}$.

Summing up, we have

$$x^{-1}|J(f;x)-J(g;x)| \ll \min(1, h_*(x), f_*^{-\frac{1}{n}}(x) + g_*^{-\frac{1}{n}}(x)), \quad (x>0)$$

and in particular

$$\left| \frac{J(f;x)-J(g;x)}{x} \bigg|_\lambda^\Lambda \right| \ll \min(1, h_*(\lambda) + f_*^{-\frac{1}{n}}(\Lambda) + g_*^{-\frac{1}{n}}(\Lambda)),$$

$$\left| \int_\lambda^\Lambda \frac{J(f,x)-J(g,x)}{x^2} \right| dx \leqslant \int_0^\infty \frac{|J(f,x)-J(g,x)|}{x^2} dx,$$

$$\ll \int_0^\infty \frac{\min(h_*(x),\ f_*^{-\frac{1}{n}}(x)+g_*^{-\frac{1}{n}}(x))}{x}\,dx$$

$$\leqslant \inf_{y>0}\left(\int_0^y \frac{h_*(x)}{x}\,dx+\int_y^\infty \frac{f_*^{-\frac{1}{n}}(x)}{x}\,dx+\int_y^\infty \frac{g_*^{-\frac{1}{n}}(x)}{x}\,dx\right).$$

Moreover, for each $y>0$

$$\int_0^y \frac{h_*(x)}{x}\,dx\leqslant h_*(y);\int_y^\infty \frac{f_*^{-\frac{1}{n}}(x)}{x}\,dx \ll nf_*^{-\frac{1}{n}}(y);\int_y^\infty \frac{g_*^{-\frac{1}{n}}(x)}{x}\,dx\ll ng_*^{-\frac{1}{n}}(y).$$

The first of the above estimates is obvious, since $h_*(x)$ is a polynomial with nonnegative coefficients. To prove the remaining two, we subdivide the domain $[y, \infty)$ of integration into the segments $\Delta_j=[ye^j, ye^{j+1})$, $j=0, 1, \cdots$. Then, using the fact that $f_*(x)$ is a polynomial with nonnegative coefficients, we see that

$$f_*(x)\geqslant e^j f_*(y) \text{ for } x\in\Delta_j;$$

$$\int_{\Delta_j} \frac{f_*^{-\frac{1}{n}}(x)}{x}\,dx\leqslant e^{-\frac{j}{n}} f_*^{-\frac{1}{n}}(y)\int_{\Delta_j}\frac{dx}{x}=e^{-\frac{j}{n}} f_*^{-\frac{1}{n}}(y);$$

$$\int_y^\infty \frac{f_*^{-\frac{1}{n}}(x)}{x}\,dx\leqslant f_*^{-\frac{1}{n}}(y)\sum_{j=0}^\infty e^{-\frac{j}{n}}=\frac{f_*^{-\frac{1}{n}}(y)}{1-e^{-1/n}}\ll nf_*^{-\frac{1}{n}}(y).$$

Thus

$$\left|\int_\lambda^\Lambda \frac{e^{2\pi i f(x)}-e^{2\pi i g(x)}}{x}\,dx-\int_\lambda^\Lambda \frac{J(f,x)-J(g,x)}{x^2}\,dx\right|$$

$$\ll \min(1, h_*(\lambda)+f_*^{-\frac{1}{n}}(\Lambda)+g_*^{-\frac{1}{n}}(\Lambda))\longrightarrow 0,\ (\lambda\longrightarrow +0,\ \Lambda\longrightarrow +\infty)$$

(note that $f_*(0)=g_*(0)=h_*(0)=0$, and that $f_*(x)\longrightarrow\infty$, $g_*(x)\longrightarrow\infty$ for $x\longrightarrow\infty$ due to the condition that $f(x)\not\equiv 0$, $g(x)\not\equiv 0$, and

$$\int_0^\infty \frac{|J(f;x)-J(g;x)|}{x^2}\,dx\ll nB(f,g),$$

where

$$B(f,g)=\inf_{y>0}(h_*(y)+f_*^{-\frac{1}{n}}(y)+g_*^{-\frac{1}{n}}(y)).$$

Thus

$$\int_{\lambda}^{\Lambda} \frac{e^{2\pi i f(x)}-e^{2\pi i g(x)}}{x}\, dx \ll 1+nB(f,g),$$

and the integral $A(f, g)$ turns absolutely convergent after integration by parts:

$$\text{v.p.} \int_0^{\infty} \frac{e^{2\pi i f(x)}-e^{2\pi i g(x)}}{x^2}\, dx = \int_0^{\infty} \frac{J(f;x)-J(g;x)}{x^2}\, dx$$

In purticular, let $g(x)=f(-x)$. Since $g_*(y)\equiv f_*(y)$, $h_*(y)\leqslant 2 f_*(y)=2g_*(y)$, $f_*(0)=g_*(0)=0$ and $f_*(y)$ monotonically increases to $+\infty$ as $y\to\infty$, there exists such a point $y_*>0$, that $f_*(y_*)=g_*(y_*)=1$, and it is plain that $h_*(y_*)\leqslant 2$ at that very point. This proves that in the case $g(x)=f(-x)$ we have $B(f, g)\leqslant 4\ll 1$, which completes the proof. Remark. The trigonometric series of the type

$$V(F;\alpha_n,\cdots,\alpha_1)=\sum_x \hat{F}(x)e^{2\pi i(\alpha_n x^n+\cdots+\alpha_1 x)}$$

are discrete analogues of the above exponential integrals. Here, the summation is taken over all integers x, positive and negative and $\hat{F}(x)$ stands for Fourier coefficients of some (periodical with period, say, 1) function $F(\xi)$:

$$\hat{F}(x)=\int_0^1 F(\xi)e^{-2\pi i x\xi}d\xi,\quad (x=0,\pm 1,\cdots).$$

The following analogue of theorem 3 is valid for such series.

Theorem 5. *Let $F(\xi)$ be a periodic, with period 1, function of the real variable ξ, and let the total variation* var$(F;[0,1))$ *of F over the period be bounded. Denote*

$$V_N(F;\alpha_n,\cdots,\alpha_1)=\sum_{|x|\leqslant N} \hat{F}(x)e^{2\pi i(\alpha_n x^n+\cdots+\alpha_1 x)},\quad N=1,2,\cdots$$

the symmetrical partial sums of the V-series of F. Then the following estimate is valid:

$$|V_N(F;\alpha_n,\cdots,\alpha_1)|\ll_n \operatorname{var} F+\sup|F|,$$

where the factor in the symbol $\ll_n$ depends only on n. Moreover, as $N\to\infty$, the limit of the quantities V_N exists for all real $\alpha_n,\cdots,\alpha_1$.

This assertion means, that the V-series of each function F of bounded variation is everywhere convergent in the sense of symmetrical summation, i. e. in the Cauchy principle value sense. Thus, Theorem 5 asserts the validity for the V-series of a complete analogue of the well-known classical Dirichlet-Jordan test in the theory of ordinary trigonometric series (which correspond to the case $n=1$). Plainly, the convergence is conditional, and pointwise only, which happens each time when the initial function has jumps somewhere on the period. But, like in the case of V-integrals, after partial summation. (symmetric with respect to $x=0$), the series turns into an absolutely convergent one:

$$V(F;\alpha_n,\cdots,\alpha_1)=\hat{F}(0)+\sum_{x=1}^{\infty}\frac{1}{x(x+1)}\sum_{0<|y|\leqslant x}\hat{F}(y)e^{2\pi i(\alpha_n y^n+\cdots+\alpha_1 y)}|y|,$$

and

$$\sum_{x=1}^{\infty}\frac{1}{x(x+1)}\left|\sum_{0<|y|\leqslant x}\hat{F}(y)e^{2\pi i(\alpha_n y^n+\cdots+\alpha_1 y)}|y|\right| \ll_n \text{var } F.$$

As in the case of integrals, the main point of the proof is to establish the uniform boundedness of the following sequence:

$$H_N(\alpha_n,\cdots,\alpha_1)=\sum_{1\leqslant|x|\leqslant N}\frac{e^{2\pi i(\alpha_n x^n+\cdots+\alpha_1 x)}}{2\pi i x}.$$

This property was proved by G.I. Arkhipov and K.I. Oskolkov in their joint paper [9], and the background of the proof in [9] is I. M. Vinogradov's method of exponential sums. (I. M. Vinogradov's famous estimates of exponential integrals and sums are crucial for consideration of integrals and series of the above type, which generalize the classical Fourier integrals and series for the case of real polynomials of n-th degree, $n\geqslant 2$, in the complex exponent. For that reason, K. I. Oskolkov called them Vinogradov integrals, and resp., series of the initial function F, or shortly, V-integrals and V-series. Moreover, due to the formal identity

$$V(F;0,\cdots,0,\xi)=F(\xi), \quad (-\infty<\xi<\infty)$$

the values of V-integrals and those sums of V-series as functions of the real variables α_n, $\cdots$, α_1 were called Vinogradov continuation of F, or shortly, its' V-continuation).

Note, that boundedness of the sequence $\{H_N\}$ was used in [9] to establish the exact growth order of Lebesgue constants for functions with polynomial spectra (c. f. also [10], [11]). That is, applications to problems of the "pure" theory of trigonometric series were given only. However, like Theorem 3, Theorem 4 is obviously applicable to investigation of the properties of the solutions of the Cauchy initial value problem for Schrödinger equations (6), (6a) in the cases, when the initial function $F(\xi)$ is periodic (c. f. [8,9]). While the one-dimensional exponential integrals has been studied thoroughly well, the theory of multiple exponential integrals, which is much more rich and important for applications, has not been suffucuently developed as, yet.

For example, consider an analogue of the above Cauchy problem for the Schrödinger-type equation (6), (6a) when the one-dimensional space coordinate is substituted by the multidimensional vector $\xi=(\xi_r,\cdots,\xi_1)$. Then, and especially when the space differential operator at the right hand side of (6) contains mixed differentiation, the Fourier method of solution immediately leades to essentially multiple exponential integrals. I note the paper [5] and the monographs [6], [7] which contain elaboration of that theory.

In connection with the theory of multiple exponential integrals I formulate here the following open problem. Compute the exponent of convergence $2k_0$ of the integral θ_2,

$$\theta_2=\int_{-\infty}^{\infty}\cdots\int_{-\infty}^{\infty}\left|\int_0^1\cdots\int_0^1 e^{2\pi i f(x_1,\cdots,x_r)}dx_1\cdots dx_r\right|^{2k}\overline{d\alpha},$$

where

$$f(x_1,\cdots,x_r)=\sum_{t_1=0}^{n_1}\cdots\sum_{t_r=0}^{n_r}\alpha(t_1,\cdots,t_r)x_1^{t_1}\cdots x_r^{t_r},\quad \alpha(0,\cdots,0)=0$$

Note also, that in [7] the following estimate from above is obtained for the quantity k_0:

$$k_0\leqslant 0.5\,nm,$$

where

$$n=\max(n_1,\cdots,n_r),\quad m=(n_1+1)\cdots(n_r+1)-1.$$

§2. The Zeros of Riemann's Zeta-Function on Short Segments of the Critical Line. A Brief Survey of Results

Riemann's zeta-function $\zeta(s)$, $s=\sigma+it$, $i^2=-1$, is defined for $\operatorname{Re} s=\sigma>1$ by the formula

$$\zeta(s)=\sum_{n=1}^{\infty}\frac{1}{n^s}. \tag{7}$$

The functional equation for the $\theta(x)$,

$$\theta(x)=\sum_{n=-\infty}^{\infty} e^{-\pi x n^2}, x>0,$$

namely,

$$\theta(x^{-1})=\sqrt{x}\,\theta(x),$$

implies the relation

$$\pi^{-\frac{s}{2}}\Gamma\left(\frac{s}{2}\right)\zeta(s)=\frac{1}{s(s-1)}+\int_1^{\infty}\left(x^{-\frac{s+1}{2}}+x^{\frac{s}{2}-1}\right)\theta_1(x)dx, \tag{8}$$

where

$$\theta_1(x)=\frac{1}{2}\theta(x)-\frac{1}{2}=\sum_{n=1}^{\infty}e^{-\pi x n^2},$$

$\Gamma(s)$ is Euler's gamma-function. Since for $x\geqslant 1$ the inequalities

$$0<\theta_1(x)<e^{-x}$$

are valid, the improper integral at the right-hand side of (8) converges for every s. i.e. the function $\zeta(s)$ is extended by the formula (8) onto the whole of s-plane.

The right-hand side of (8) remains unchanged if we substitute s for $1-s$, i.e.

$$\pi^{-\frac{s}{2}}\Gamma\left(\frac{s}{2}\right)\zeta(s)=\pi^{-\frac{1-s}{2}}\Gamma\left(\frac{1-s}{2}\right)\zeta(1-s). \tag{9}$$

The relation (9) is called functional equation of $\zeta(s)$.

Along with (9) the following equality, called Euler's identity, is also valid:

$$\zeta(s)=\prod_p\left(1-\frac{1}{p^s}\right)^{-1}. \tag{10}$$

The right-hand side of (10) contains the pruduct over all prime numbers p. The identity (10) enables reduction of various problems of the theory of primes to those concerning $\zeta(s)$, and in particular, to the problem of location of the zeros of $\zeta(s)$.

For example, one of such problems is that of asymptotic behaviour, as $x\to+\infty$, of the function $\pi(x)$, i.e. the number of the primes which do not exceed a given x.

Identity (9) implies that the numbers $s=-2,-4,\cdots,-2n,\cdots$ are zeros of $\xi(s)$. Those zeros are called "trivial". Besides trivial ones, $\zeta(s)$ possesses infinitely many complex-valued zeros, which are situated in the strip $0\leqslant \operatorname{Re} s\leqslant 1$. The complex-valued zeros of $\zeta(s)$ are situated symmetrically with respect to the lines $\operatorname{Re} s=1/2$ and $\operatorname{Im} s=0$, which is also a consequence from (9).

Riemann cojectured (Riemann's Conjecture — R. C.) that all of the complex-valued zeros of $\zeta(s)$ are situated on the line $\mathrm{Re}s = 1/2$, and the latter was called "the critical line".

R.C. implies that

$$R(x) = \pi(x) - \int_2^x \frac{du}{\log u} = O(\sqrt{x} \log x). \tag{11}$$

Conversely, (11) and even a slightly weaker relation

$$R(x) = O(x^{\frac{1}{2}+\varepsilon}), \ \varepsilon > 0, \tag{12}$$

would imply the validity of R.C. Thus, R.C. and (12) are equivalent assertions.

It is well-known that

$$R(x) = \Omega_{\pm}(\sqrt{x} \frac{\log\log\log x}{\log x}),$$

i.e. R.C. implies the correct order of the estimate of $R(x)$.

R.C. has not been proved as yet (1988).

There is an extensive branch of investigation of the zeros of $\zeta(s)$, which are situated on the critical line. Note that the zeros of $\zeta(s)$ on the critical line are those real of the function $\zeta(1/2 + it)$.

For the sake of convenience, we introduce the following two functions $N_0(T)$ and $N(T)$:

$N_0(T)$ – the number of the zeros of $\zeta(s)$ for $0 < t \leqslant T$;
$N(T)$ – the number of the zeros of $\zeta(s)$ for $0 < t$ Im $s \leqslant T, 0 \leqslant \sigma = \mathrm{Re}s \leqslant 1$.

Plainly, $N_0(T) \leqslant N(T)$, and the R.C. is equivalent to the equality $N_0(T) = N(T)$.

G.H.Hardy [14] proved in 1914, that

$$N_0(T) \to \infty \quad \text{for } T \to \infty.$$

E.Landau wrote in connection with that result: "Among the most essential achievements of modern mathematics there is Mr. G. H. Hardy's paper "On the zeros of the function $\zeta(s)$ of Riemann". [15].

In 1918 and 1921 G.H.Hardy and J.E.Littlwood [16], [17] proved essentialy more, mamely

1. $$N_0(T+H) - N_0(T) > 0, \tag{13}$$

for $$H = T^{0.25+\varepsilon}, \ T \geqslant T_0(\varepsilon) > 0;$$

2. $$N_0(T+H)-N_0(T)\geqslant cH, \tag{14}$$

for $$H=T^{0.5+\varepsilon},\ c=c(\varepsilon)>0,\ T\geqslant T_0(\varepsilon)>0.$$

These results were sources for the following two problems: Prove (13) for possibly small H and prove (14) for possibly small H and possibly large right-hand side. In the order words, the first of the above problems consists in the proof that there is at least one zero of the function $\zeta(\frac{1}{2}+it)$ on a short segment, and the second one – in the possibly most exact estimate of the number of such zeros from below.

It was A. Selberg [18], who proved in 1942 that

$$N_0(T+H)-N_0(T)\geqslant c(N(T+H)-N(T)), \tag{15}$$

where

$$H=T^a,\ a>\frac{1}{2},\ c=c(a)>0,\ T\geqslant T_0(a)>0,$$

and expressed the cojecture that (15) is valid also for $a<\frac{1}{2}$ ("A. Selberg's conjecture").

N. Levinson [19] proved in 1974 that for $H=T(\log T)^{-10}$, the constants c in the inequality (15) is not less that $\frac{1}{3}$.

Y. Möser [20], [21] proved in 1976 and 1980 that (13) is valid for $H=T^{1/6+\varepsilon}$ and (14) for $H=T^{5/12+\varepsilon}$.

The author [22], [23] proved in 1981 and 1983 that (13) is valid for $H=T^{5/32}\log T$ and (14) for $H=T^{27/82}\log^2 T$.

In 1983 A. Ivic [24] proved (13) for $H=T^{\theta+\varepsilon}$,

$$\theta=0.15594583\cdots<\frac{5}{32}=0.15625.$$

That refinement of my result was obtained using application of the method of exponential pairs at the corresponding place of my proof (in [22] I used the estimate of exponential sum by Corput's method).

In 1984 the author [25] proved the validity of A. Selberg's conjecture with $a=\frac{27}{82}=\frac{1}{3}-\frac{1}{246}$. It is plain that the number $a=27/82$ in this result can be slightly refined using the method of exponential paris.

Finally, in [26] the author proved that (15) is valid for arbitrary small $a>0$ for "almost all T". More exactly, that assertion sounds as follows: "the relation (15) is true for each $a>0$ for all T from the segmant $Y<T<2Y$, $Y\geqslant Y_0(a)>0$, with the possible exception of a certain subset E, whose measure does not exceed $Y^{1-0.5a''}$.

It would be interesting to prove an analogous theorem for $H=\exp(\log Y)^{\alpha}$, $0<\alpha<1$, or $H=(\log Y)^{\beta}$, where $\beta>0$ is a constant.

The paper of A.Selberg [18] quoted above contains also the proof of the assertion:

Let $N(\sigma,T)$ denote the number of the zeros of $\zeta(s)$ having the form $s=\sigma_1+it$, $\sigma_1>\sigma$, $0<t\leqslant T$; then for $H=T^a$, $a>1/2$, $\frac{1}{2}<\sigma\leqslant 1$, uniformly with respect to σ the following estimate holds true:

$$N(\sigma,T+H)-N(\sigma,T)=O\left(\frac{H}{\sigma-0.5}\right) \tag{16}$$

Along with the cojecture concerning the zeros of $\zeta(s)$ on short segments of the critical line, that paper contains also the conjecture that (15) is valid for some $a<1/2$ too. The latter was proved by the author [27] for $a>27/82$. The following assertion is the corollary from (16).

Let $\Phi(t)\to+\infty$ for $t\to+\infty$; then the domain

$$\left|\sigma-\frac{1}{2}\right|<\frac{\Phi(t)}{\log t},\quad T<t\leqslant T+H,$$

contains all the complex valued zeros of $\zeta(s)$ with possible exception of not more than

$$O\left(\frac{H\log T}{\Phi(T)}\right);$$

in the other words, the above domain contains

$$\left(N(T+H)-N(T)\right)\left(1+O\left(\frac{1}{\Phi(T)}\right)\right)$$

of the zeros of the function $\zeta(s)$.

Let us briefly sketch the proof of the results and what has led to the progress.

1. If in (8) we substitute

$$s=\frac{1}{2}+it,$$

we obtain:

$$\xi(t)=\pi^{-\frac{it}{2}}\Gamma\left(\frac{1}{4}+\frac{it}{2}\right)\zeta\left(\frac{1}{2}+it\right)$$

$$=\pi^{-\frac{it}{2}}\Gamma\left(\frac{1}{4}-\frac{it}{2}\right)\zeta\left(\frac{1}{2}-it\right)=\overline{\xi(t)},$$

that is $\xi(t)$ takes on real values for real t. An analogous property is also possessed by the function $Z(t)$,

$$Z(t)=\frac{\xi(t)}{\left|\Gamma\left(\frac{1}{4}+\frac{it}{2}\right)\right|}=e^{i\theta(t)}\zeta\left(\frac{1}{2}+it\right),$$

which I call Hardy's function. The zeros $Z(t)$ are those of the function $\zeta(\frac{1}{2}+it)$. If

$$\int_T^{T+H}|Z(t)|\,dt\neq\left|\int_T^{T+H}Z(t)\,dt\right|,$$

then $Z(t)$ for $T<t<T+H$ changes it's sign, i. e. $\zeta(\frac{1}{2}+it)=0$ for some $t\in(T,$ $T+H)$.

This is the background of the proofs of results of type (13).

2. To prove (14), the following speculations are also applied. Define the set $E\subset(T,T+H)$ in such a way that if $t\in E$, if

$$\int_t^{t+h}|Z(u)|\,du\neq\left|\int_t^{t+h}Z(u)\,du\right|.$$

Here $h=A\log^{-1}T$, $A>0$ is a constant. The following sequence of relations is valid:

$$\int_E dt\left(\int_t^{t+h}|Z(u)|\,du\right)\geqslant\int_E dt\left(\int_t^{t+h}|Z(u)|\,du-\left|\int_t^{t+h}Z(u)du\right|\right)$$

$$=\int_T^{T+H}dt\left(\int_t^{t+h}|Z(u)|\,du-\left|\int_t^{t+h}Z(u)\,du\right|\right);$$

$$I_1=\int_T^{T+H}dt\left(\int_t^{t+h}|Z(u)|\,du\right)$$

$$<\int_E dt\left(\int_t^{t+h}|Z(u)|\,du\right)+\int_T^{T+H}dt\left|\int_t^{t+h}Z(u)du\right|$$

$$\leqslant\sqrt{\mu(E)I_2}+\sqrt{HI_3},\tag{17}$$

where

$$I_2=\int_T^{T+H}dt\left(\int_t^{t+h}|Z(u)|\,du\right)^2,$$

$$I_3=\int_T^{T+H} dt\left(\int_t^{t+h} Z(u)du\right)^2.$$

Estimates of I_1 from below and of I_2, I_3 from above, imply the following estimate of $\mu(E)$:

$$\mu(E)>cH,\ c>0$$

From here on, (14) follows in an easy manner.

3. To prove (17), Cauchy's inequality is applied:

$$\left(\int_T^{T+H} f(t)dt\right)^2 \leqslant H\int_T^{T+H} f^2(t)dt$$

However, that inequality is close to an exact one, only if $f(t)$ is close to 1. Thus instead of $Z(t)$, the function $G(t)$ is considered,

$$G(t)=\xi(t)\left|\sum_{n\leqslant X}\alpha(n)n^{it}\right|^2,$$

where the choice of $\alpha(n)$ is done optimally.

4. When estimating the integrals I_2 and I_3, (17). gives birth to integrals of the type.

$$\int_{T_1}^{T_2}\left|\sum_{n\leqslant N}\beta(n)n^{it}\right|^2 dt=(T_2-T_1)\sum_{n\leqslant N}|\beta(n)|^2$$

$$-i\sum_{n\neq m\leqslant N}\beta(n)\beta(m)\frac{\left(\frac{n}{m}\right)^{iT_2}-\left(\frac{n}{m}\right)^{iT_1}}{\log\frac{n}{m}}.$$

The second, double sum over n and m admits nontrivial estimates, using the fact that its summands "oscilliate" due to the factors

$$\left(\frac{n}{m}\right)^{iT}=\cos T\log\frac{n}{m}+i\sin T\log\frac{n}{m}.$$

Clearly, the latter is possible not for all $\beta(n)$.

5. The technics of investigation has been considerably simplified as a whole

I note one more new effect. The real zeros of $Z(t)$ are those of $\zeta(s)$ on the critical line. In 1981, the author [22] proved the following theorem:

For each $k\geqslant 0$, $t\geqslant T_0(k)>0$,

$$H\geqslant T^{\frac{1}{6k+6}}\log T,$$

the segment $(T,T+H)$ contains a zero of odd order, of the function $Z^{(k)}(t)$.

The above means that $Z(t)$ does "essentially oscillate".

A theorem on the zeros of $Z^{(k)}(t)$, uniform with respect to the parameters k and t was proved by my pupil, Miss A. A. Lavrik (in print):

There exists such a constant T_0 that for

$$T \geqslant T_0, \quad 0 \leqslant k \leqslant \frac{1}{6} \log T,$$

$$H \geqslant \max\left(g\pi \log T, T^{\frac{1}{6k+6}} \log^{\frac{2}{k+1}} T\right),$$

the segment $(T, T+H)$ contains a zero of odd order of the function $Z^{(k)}(t)$.

That implies in particular, that for

$$T \geqslant T_0, \quad \frac{\log T}{6 \log \log T} \leqslant k \leqslant \frac{\log T}{6},$$

the segment

$$(T, T + g\pi \log T)$$

contains a zero of odd order of the function $Z^{(k)}(t)$.

Again, using the method of exponential pairs, A. Ivic [28] made a refinement of my theorem concerning zeros of $Z^{(k)}(t)$ and proved that $Z^{(k)}(t)$ possesses a zero of odd order on the interval $(T, T+H)$, where

$$T \geqslant T_0(k), \quad H = T^{a(k)} \log T,$$

$$a(k) = \begin{cases} \min\left(\frac{3}{37}, \mu\left(\frac{1}{2}\right) + \varepsilon\right), \text{ if } k = 1, \\ \min\left(\frac{27}{164k+168}, \frac{1}{k}\mu\left(\frac{1}{2}\right) + \varepsilon\right), \text{ if } k \geqslant 2. \end{cases}$$

The above theorems on the zeros of the function $\zeta(s)$ on the critical line and those on the real zeros of $Z^{(k)}(t)$ implicitely confirm once again, that the R. C. seems to be true.

§ 3. A. Selberg's Conjecture

Below we prove a theorem, the assertion of which in a weaker form was cojectured by A. Selberg [18, p. 5].

Theorem 6. *Let ε be an arbitrary positive number, not exceeding 0.001, $T \geqslant T_0(\varepsilon) > 0$, $H = T^{27/82+\varepsilon}$. Then there exists such a positive constant $c = c(\varepsilon)$ that*

$$N_0(T+H) - N_0(T) \geqslant cH \log T. \tag{18}$$

Since, by the Mangholdt's formula,

$$N(T)=\frac{T}{2\pi}\log\frac{T}{2\pi}-\frac{T}{2\pi}+O(\log T),$$

the relation (18) implies validity of A. Selberg's conjecture with $a=27/82$. To prove the theorem, we need a series of new difinitions and auxiliary statements, connected with them.

First of all, introduce the function $F(t)$, which I call Hardy-Selberg's function (c.f. [25], [26]).

Definition 1. *Hardy-Selberg's function $F(t)$ is defined by the equality*

$$F(t)=Z(t)\left|\varphi\left(\frac{1}{2}+it\right)\right|^2,$$

where

$$\varphi(s)=\varphi_x(s)=\sum_{\nu\leqslant X}\frac{\beta(\nu)}{\nu^s},$$

$$\beta(\nu)=\begin{cases}\alpha(\nu)\left(1-\dfrac{\log\nu}{\log X}\right), & 1\leqslant\nu\leqslant X,\\ 0 & ,\ \nu\geqslant X,\end{cases}$$

and the real numbers $\alpha(\nu)$ are defined by the relation

$$\frac{1}{\sqrt{\zeta(s)}}=\sum_{\nu=1}^{\infty}\frac{\alpha(\nu)}{\nu^s},\quad \operatorname{Re} s>1.$$

Thus, the function $F(t)$ does also depend on a real parameter $X>1$. which has to be optimally chosen according to the problem under consideration. The definition of $F(t)$ implies that $F(t)$ takes on real values for real t, and that the real zeros of $F(t)$ of odd order are the real zeros of odd order of the function $\zeta(\frac{1}{2}+it)$, i. e. the zeros of odd order of the function $\zeta(s)$, situated on the critical line.

Lemma 3. *For $2\leqslant T<t\leqslant T+H$, $H\leqslant\sqrt[3]{T}$, $X\leqslant H^{0.01}$, $P=\sqrt{T/2\pi}$, the following equality is valid*

$$F(t)=F_1(t)+\overline{F_1(t)}+O(\log^{-5}T),$$

where

$$F_1(t)=e^{i\theta_1(t)}\sum_{\lambda<P}\frac{a(\lambda)}{\sqrt{\lambda}}\lambda^{-it},$$

$$\theta_1(t)=t\log P-\frac{T}{2}-\frac{\pi}{8},$$

$$a(\lambda)=\sum_{\frac{n\nu_2}{\nu_1}=\lambda}\frac{\beta(\nu_1)\beta(\nu_2)}{\nu_1},$$

and λ is some positive rational numbers, whose denumerators do not exceed X, and the constant in the O symbol is absolute.

The proof of the Lemma (c.f. [25], [29]) is analogous to the proof of the approximate functional equation of $\zeta(s)$, due to G. H. Hardy and J. E. Littlewood (c.f. [16], [17], [30]).

The estimates of some special exponential sums, which are stated in the following lemma, are the main tools in the proof of the Theorem 6. They also define the result of the Theorem 6.

Lemma 4. *Let ε and ε_1 be arbitrary positive numbers, not exceeding* 0.001, $A\geqslant 10$ *is some fixed number,*

$$P=\sqrt{T/2\pi},\ P_0=\sqrt{Y/2\pi},\ 0<X\leqslant P,\ L=\log P,$$

$$h=AL^{-1},\ 0<H\leqslant\sqrt[3]{Y},\ H_1=H+h.$$

Consider the following three sums:

$$W_0(T)=\sum_{\lambda_1<\lambda_2<P}\frac{a(\lambda_1)a(\lambda_2)}{\sqrt{\lambda_1\lambda_2}}\left(\frac{\lambda_2}{\lambda_1}\right)^{iT}e^{-\left(\frac{H_1}{2}\log\frac{\lambda_2}{\lambda_1}\right)^2};$$

$$W_1(T)=\sum_{\lambda_1<\lambda_2<P_0^{1-\varepsilon}}\frac{a(\lambda_1)a(\lambda_2)}{\sqrt{\lambda_1\lambda_2}}\left(\frac{\lambda_2}{\lambda_1}\right)^{iT}B(\lambda_1)\overline{B(\lambda_2)}\,e^{-\left(\frac{H}{2}\log\frac{\lambda_2}{\lambda_1}\right)^2},$$

where

$$B(\lambda)=\left(\left(\frac{P}{\lambda}\right)^{ih}-1\right)\left(\log\frac{P}{\lambda}\right)^{-1};$$

$$W_2(T)=\sum_{P_0^{1-\varepsilon_1}\leqslant\lambda_1<\lambda_2<P}\frac{a(\lambda_1)a(\lambda_2)}{\sqrt{\lambda_1\lambda_2}}\left(\frac{\lambda_2}{\lambda_1}\right)^{iT}e^{-\left(\frac{H_1}{2}\log\frac{\lambda_2}{\lambda_1}\right)^2}.$$

If $X=Y^{0.01\varepsilon}$, $H=Y^{27/82+\varepsilon}$, then for the sums $W_j(T)$, $j=0,1,2$, the following estimates are valid:

$$W_0\ll Y^{-\varepsilon},$$

$$W_1\ll(\varepsilon_1^{-2}L^{-2}+\varepsilon_1^{-1}AL^{-2})Y^{-\varepsilon},$$

$$W_2\ll Y^{-\varepsilon},$$

where the constants in the symbol $\ll$ are absolute.

The proof of that lemma is contained in [25] (c.f. also [29]).

Difinition 2. *Let $\beta(v)$ be the real numbers from the Definition 1. Then we call the following sum*

$$S(\theta)=\sum_{v_1,v_2,v_3,v_4}\left(\frac{q}{v_1v_3}\right)^{1-\theta}\frac{\beta(v_1)\beta(v_2)\beta(v_3)\beta(v_4)}{v_2v_4},$$

Selberg's sum, where $q=(v_1v_4,v_2v_3)$, $0\leqslant\theta\leqslant 0.5$, and the summation is taken over all positive integers v_1,v_2,v_3,v_4.

Since $\beta(v)=0$ for $v\geqslant X$, we may assume in Definition 2 that $1\leqslant v_1,v_2,v_3,v_4<X$.

The sum $S(\theta)$ was introduced and computed by A. Selberg in his paper [18]

Lemma 5. *For $0\leqslant\theta\leqslant 0.5$, the following estimate is valid*

$$S(\theta)\ll\frac{X^{2\theta}}{\log X},$$

the constant in the symbol $\ll$ being absolute one.

For the proof of the lemma see [18] (c.f. also [29]).

Proof of Theorem 6. Fix an $X=T^{0.01\varepsilon}$ and consider the Hardy-Selberg's function $F(t)$. By the Lemma 3 we have

$$F(t)=F_1(t)+\overline{F_1(t)}+O(\log^{-5}T),$$

where

$$F_1(t)=e^{i\theta_1(t)}\sum_{\lambda<P}\frac{a(\lambda)}{\sqrt{\lambda}}\lambda^{-it},$$

$$\theta_1(t)=t\log P-\frac{T}{2}-\frac{\pi}{8},\quad P=\sqrt{\frac{T}{2\pi}},$$

$$a(\lambda)=\sum_{\frac{hv_2}{v_1}=\lambda}\frac{\beta(v_1)\beta(v_2)}{v_1},\quad T\leqslant t\leqslant T+H,$$

and λ is positive rational numbers whose denumerators do not exceed X. With no loss of generality, we may take such a T. that

$$\frac{T}{2}+\frac{\pi}{8}=2\pi K,$$

where K is some integer. Then

$$F_1(t)=e^{it\log P}\sum_{\lambda<P}\frac{a(\lambda)}{\sqrt{\lambda}}\lambda^{-it}.$$

Take $h=A\log^{-1}T$ where $A\geqslant 10$ is some constant, whose value will be exactly defined somewhat below, and denote E the subset of the segment $(T, T+H)$ for which the following inequality is valid:

$$\int_t^{t+h} |F(u)|\, du > \left| \int_t^{t+h} F(u) du \right|, t \in E.$$

Then, as it was done in § 2, we see that

$$I_1 \leqslant \sqrt{\mu(E) I_2} + \sqrt{HI_3}, \tag{$*$}$$

where

$$I_1 = \int_T^{T+H} \int_t^{t+h} |F(u)|\, du\, dt,$$

$$I_2 = \int_T^{T+H} \left(\int_t^{t+h} |F(u)|\, du \right)^2 dt,$$

$$I_3 = \int_T^{T+H} \left(\int_t^{t+h} F(u)\, du \right)^2 dt.$$

We estimate the integral I_1 from below. First of all we subsequently obtain the following relations:

$$I_1 \geqslant h \int_{T+h}^{T+H} |F(u)|\, du \geqslant h \left| \int_{T+h}^{T+H} \zeta(\frac{1}{2} + it) \varphi^2 (\frac{1}{2} + it) dt \right|.$$

Let Γ denote the rectangle $s = \frac{1}{2} + it, s = 2 + it, T + h \leqslant t \leqslant T + H, s = \sigma + i(T+h)$, $s = \sigma + i(T+H), \frac{1}{2} \leqslant \sigma \leqslant 2$. Then the equality

$$\int_\Gamma \zeta(s) \varphi^2(s) ds = 0$$

is valid, equivalent to the following one:

$$\int_{T+h}^{T+H} \zeta(\frac{1}{2} + it) \varphi^2 (\frac{1}{2} + it) dt = \int_{T+h}^{T+H} \zeta(2 + it) \varphi^2 (2 + it) dt$$

$$+ i \int_{1/2}^{2} \zeta(\sigma + i(T+H))\, \varphi^2(\sigma + i(T+H)) d\sigma$$

$$- i \int_{1/2}^{2} \zeta(\sigma + i(T+h)) \varphi^2 (\sigma + i(T+h)) d\sigma. \tag{19}$$

The definition of $\varphi(s)$ implies that

$$\beta(1)=1, |\beta(\nu)| \leqslant 1, \nu \geqslant 1.$$

Thus, for Re $s>1$

$$\zeta(s)\varphi^2(s)=\sum_{n=1}^{\infty} \sum_{\nu_1, \nu_2<X} \frac{\beta(\nu_1)\beta(\nu_2)}{(n\nu_1 \nu_2)^s} = \sum_{n=1}^{\infty} \frac{b(n)}{n^s} = 1+\sum_{n=2}^{\infty} \frac{b(n)}{n^s},$$

where $|b(n)| \leqslant \tau_3(n)$. Hence,

$$\int_{T+h}^{T+H} \zeta(2+it)\varphi^2(2+it)dt = H-h+\sum_{n=2}^{\infty} \frac{b(n)}{n^2} \int_{T+h}^{T+H} n^{-it}dt = H-h+O(1). \tag{20}$$

Furthermore, making use of the estimates

$$\zeta(\sigma+it)=O(t^{1/6}\log t), \quad \varphi(\sigma+it)=O(\sqrt{X}),$$

$$\sigma \geqslant \frac{1}{2}, \quad t \geqslant T > T_0 > 0,$$

we obtain ($T_1 = T+H, T+h$):

$$\int_{1/2}^{2} \zeta(\sigma+iT_1)\varphi^2(\sigma+iT_1)d\sigma = O(T^{1/6}X\log T). \tag{21}$$

Relations (19)—(21) imply the desired estimate of I_1 from below:

$$I_1 \geqslant hH-h^2+O(T^{1/6}X\log T).$$

We estimate the integral I_2 from above. Trivially,

$$I_2 \ll h^2 \int_{T}^{T+H_1} |F(t)|^2 dt \ll h^2(J+HL^{-10}), \tag{22}$$

where

$$J=\int_{T}^{T+H} |F_1(t)|^2 dt, H_1 = H+h, L=\log T.$$

We use the following formula:

$$\int_{-\infty}^{\infty} e^{-t^2-i\alpha t}dt = \sqrt{\pi}\, e^{-\left(\frac{\alpha}{2}\right)^2}. \tag{23}$$

We obtain

$$J \ll \int_{-\infty}^{\infty} e^{-(t/H_1)^2} \left| \sum_{\lambda<P} \frac{a(\lambda)}{\sqrt{\lambda}} \lambda^{iT} \lambda^{it} \right|^2 dt$$

$$= \sum_{\lambda_1, \lambda_2 < P} \frac{a(\lambda_1)a(\lambda_2)}{\sqrt{\lambda_1\lambda_2}} \left(\frac{\lambda_1}{\lambda_2}\right)^{iT} \int_{-\infty}^{\infty} e^{-\left(\frac{t}{H_1}\right)^2 + it\log\frac{\lambda_1}{\lambda_2}} dt$$

$$= \sqrt{\pi}\, H_1 \sum_{\lambda_1, \lambda_2 < P} \frac{a(\lambda_1)a(\lambda_2)}{\sqrt{\lambda_1\,\lambda_2}} \left(\frac{\lambda_1}{\lambda_2}\right)^{iT} e^{-\left(\frac{H_1}{2}\log\frac{\lambda_1}{\lambda_2}\right)^2}.$$

Representing the latter double sum as a sum of two summands, one of which corresponds to $\lambda_1 = \lambda_2$, we get

$$J \ll H_1 \left(\sum\nolimits_0 + W_0\right), \tag{24}$$

where

$$\sum\nolimits_0 = \sum_{\lambda < P} \frac{a^2(\lambda)}{\lambda},$$

$$W_0 = \left| \sum_{\lambda_1 < \lambda_2 < P} \frac{a(\lambda_1)a(\lambda_2)}{\sqrt{\lambda_1\lambda_2}} \left(\frac{\lambda_2}{\lambda_1}\right)^{iT} e^{-\left(\frac{H_1}{2}\log\frac{\lambda_2}{\lambda_1}\right)^2} \right|$$

To estimate $\sum_0$ consider, for a real number θ, $\frac{1}{4} \leqslant \theta < 1/2$ the sum $\sum(\theta)$,

$$\sum(\theta) = \sum_{\lambda < Z} \frac{a^2(\lambda)}{\lambda^{2\theta}}, \tag{25}$$

and compute it. Assume $\sqrt{P} \leqslant Z \leqslant P$. Then, using definitions of the numbers $a(\lambda)$ and λ, we subsequently obtain:

$$\sum(\theta) = \sum_{\lambda < Z} \frac{1}{\lambda^{2\theta}} \sum_{\frac{n_1 v_1}{v_2} = \frac{n_2 v_3}{v_4} = \lambda} \frac{\beta(v_1)\beta(v_2)\beta(v_3)\beta(v_4)}{v_2 v_4}$$

$$= \sum_{v_1, v_2, v_3, v_4} \frac{\beta(v_1)\beta(v_2)\beta(v_3)\beta(v_4)}{v_1^{\theta} v_2^{1-\theta} v_3^{\theta} v_4^{1-\theta}} \sum_{\frac{n_1 v_1}{v_2} = \frac{n_2 v_3}{v_4} < Z} \frac{1}{n_1^{\theta} n_2^{\theta}}.$$

Denote q the largest common divisor of $v_1\, v_4$ and $v_2\, v_3$ that is $q = (v_1 v_4, v_2 v_3)$. Then

$$v_1 v_4 = aq, \qquad v_2 v_3 = bq, \qquad (a, b) = 1.$$

Hence, the equality

$$n_1 v_1 v_4 = n_2 v_2 v_3$$

implies that

$$a n_1 = b n_2, \qquad n_1 = bm, \qquad n_2 = am.$$

Hence

$$\sum_{\frac{n_1 v_1}{v_2} = \frac{n_2 v_3}{v_4} < Z} \frac{1}{n_1^\theta n_2^\theta} = \frac{1}{a^\theta b^\theta} \sum_{m < \frac{Z v_2}{v_1 b}} \frac{1}{m^{2\theta}} .$$

We apply the Euler's summation formula to the latter sum, and obtain

$$\sum_{\frac{1}{2} < m < \frac{Z v_2}{v_1 b}} \frac{1}{m^{2\theta}} = \int_{1/2}^{Z v_2 (v, b)^{-1}} x^{-2\theta} dx + O\left(\frac{v_1^{2\theta} b^{2\theta}}{v_2^{2\theta} Z^{2\theta}} \right)$$

$$+ 2\theta \int_{1/2}^{Z v_2 (v, b)^{-1}} \left(\frac{1}{2} - \{x\} \right) x^{-1-2\theta} \, dx = \frac{Z^{1-2\theta}}{1-2\theta} \frac{v_2^{1-2\theta}}{v_1^{1-2\theta} b^{1-2\theta}} + C(\theta)$$

$$+ O(v_1^{2\theta} b^{2\theta} v_2^{-2\theta} Z^{-2\theta}),$$

where

$$C(\theta) = \frac{2^{2\theta-1}}{-1+2\theta} + 2\theta \int_{1/2}^{\infty} \left(\frac{1}{2} - \{x\} \right) x^{-1-2\theta} dx.$$

Thus, if we recall the definition of Selberg's sum $S(\theta)$, we get the following formula for $\Sigma(\theta)$:

$$\sum(\theta) = \sum_{v_1, v_2, v_3, v_4} \frac{\beta(v_1)\beta(v_2)\beta(v_3)\beta(v_4)}{v_1^\theta v_2^{1-\theta} v_3^\theta v_4^{1-\theta}} \cdot \frac{1}{a^\theta b^\theta}$$

$$\times \frac{Z^{1-2\theta}}{1-2\theta} \cdot \frac{v_2^{1-2\theta}}{v_1^{1-2\theta} b^{1-2\theta}} + C(\theta) + O\left(\frac{v_1^{2\theta} b^{2\theta}}{v_2^{2\theta} Z^{2\theta}} \right)$$

$$= \frac{Z^{1-2\theta}}{1-2\theta} S(0) + C(\theta) S(1-2\theta) + O(Z^{-2\theta} X^{2+2\theta}).$$

Let in (25)

$$Z = P, \qquad O = \frac{1}{2} - \frac{1}{2L} ;$$

then we obtain

$$\sum(\theta) = \sum_{\lambda < P} \frac{a^2(\lambda)}{\lambda} \lambda^{\frac{1}{L}} > e^{-1} \sum_{\lambda < P} \frac{a^2(\lambda)}{\lambda} = e^{-1} \sum\nolimits_0 .$$

Now we apply Lemma 5 on the estimate of $S(\theta)$, and have:

$$\sum(\theta)=\frac{P^{1-2\theta}}{1-2\theta}S(0)+C(\theta)S(1-2\theta)+O(P^{-2\theta}X^{2+2\theta})=O\left(\frac{\log P}{\log X}\right);$$

$$\sum_0=O\left(\frac{\log P}{\log X}\right). \tag{26}$$

To estimate W_0, we apply Lemma 4, letting $Y=T$ in it. Then we get

$$W_0=O(T^{-\varepsilon}) \tag{27}$$

Thus (22), (24), (26), (27) imply the following estimate of the integral I_2:

$$I_2\ll h^2H_1\left(\frac{\log P}{\log X}+T^{-\varepsilon}\right)\leqslant c(\varepsilon)h^2H_1. \tag{28}$$

Next we estimate the integral I_3 from above. As in the case of I_2, we arrive at the inequality

$$I_3\ll J+Hh^2L^{-10},$$

where

$$J=\int_T^{T+H}\left|\int_t^{t+h}F_1(u)du\right|^2dt,$$

$$F_1(u)=e^{iu\log P}\sum_{\lambda<P}\frac{a(\lambda)}{\sqrt{\lambda}}\lambda^{-iu}.$$

Let ε_1 be a positive number, not exceeding 0.001, whose explicite value will be defined later on. If we subdivide in $F_1(u)$ the summation over λ into two parts: $\lambda<P^{1-\varepsilon_1}$ and $P^{1-\varepsilon_1}\leqslant\lambda<P$, we obtain the relation

$$J\leqslant J_1+J_2, \tag{29}$$

where

$$J_1=\int_T^{T+H}\left|\int_t^{t+h}\sum_{\lambda<P^{1-\varepsilon_1}}\frac{a(\lambda)}{\sqrt{\lambda}}\left(\frac{P}{\lambda}\right)^{iu}du\right|^2dt,$$

$$J_2=\int_T^{T+H}\left|\int_t^{t+h}\sum_{P^{1-\varepsilon_1}\leqslant\lambda<P}\frac{a(\lambda)}{\sqrt{\lambda}}\left(\frac{P}{\lambda}\right)^{iu}du\right|^2dt.$$

We estimate the integral J_1 after integration over u. Then, applying (23), we obtain:

$$J_1=\int_T^{T+H}\left|\sum_{\lambda<P^{1-\varepsilon_1}}\frac{a(\lambda)}{\sqrt{\lambda}}\left(\frac{P}{\lambda}\right)^{it}\frac{\left(\frac{P}{\lambda}\right)^{ih}-1}{\log\frac{P}{\lambda}}\right|^2dt$$

$$\ll\int_{-\infty}^{\infty}e^{-\left(\frac{t}{H}\right)^2}\left|\sum_{\lambda<P^{1-\varepsilon_1}}\frac{a(\lambda)}{\sqrt{\lambda}}\left(\frac{P}{\lambda}\right)^{i(t+T)}\frac{\left(\frac{P}{\lambda}\right)^{ih}-1}{\log\frac{P}{\lambda}}\right|^2dt$$

$$\ll H\left(\sum\nolimits_1 + W_1\right), \tag{30}$$

where

$$\sum\nolimits_1 = \sum_{\lambda < P^{1-\varepsilon_1}} \frac{a^2(\lambda)}{\lambda \log^2 \frac{P}{\lambda}},$$

$$W_1 = \left| \sum_{\lambda_1 < \lambda_2 < P^{1-\varepsilon_1}} \frac{a(\lambda_1)a(\lambda_2)}{\sqrt{\lambda_1\lambda_2}} \left(\frac{\lambda_2}{\lambda_1}\right)^{iT} B(\lambda_1)\overline{B(\lambda_2)} e^{-\left(\frac{H}{2}\log\frac{\lambda_2}{\lambda_1}\right)^2} \right|.$$

If we use the estimate (26) for Σ_1, we obtain:

$$\sum\nolimits_1 \ll \frac{1}{\varepsilon_1^2 \log^2 P} \sum_{\lambda < P} \frac{a^2(\lambda)}{\lambda} \ll \frac{1}{\varepsilon_1^2 \log P \log X}.$$

To estimate W_1, we apply Lemma 4; then we get

$$W_1 \ll (A\varepsilon_1^{-1}L^{-2} + \varepsilon_1^{-2}L^{-2})T^{-\varepsilon}.$$

Hence,

$$J_1 \ll H(\varepsilon_1^{-2}\log^{-1}P\log^{-1}X + A\varepsilon_1^{-1}L^{-2}T^{-\varepsilon} + \varepsilon_1^{-2}L^{-2}T^{-\varepsilon}).$$

We estimate the integral J_2 like it was done above with the integral I_2

$$J_2 \ll h^2 \int_T^{T+H_1} \left| \sum_{P^{1-\varepsilon_1} \leqslant \lambda < P} \frac{a(\lambda)}{\sqrt{\lambda}} \lambda^{it} \right|^2 dt \ll h^2 H_1\left(\sum\nolimits_2 + W_2\right),$$

where

$$\sum\nolimits_2 = \sum_{P^{1-\varepsilon_1} \leqslant \lambda < P} \frac{a^2(\lambda)}{\lambda},$$

$$W_2 = \left| \sum_{P^{1-\varepsilon_1} \leqslant \lambda_1 < \lambda_2 < P} \frac{a(\lambda_1)a(\lambda_2)}{\sqrt{\lambda_1\lambda_2}} \left(\frac{\lambda_2}{\lambda_1}\right)^{iT} e^{-\left(\frac{H}{2}\log\frac{\lambda_2}{\lambda_1}\right)^2} \right|.$$

Let

$$\theta = \frac{1}{2} - \frac{1}{2\log P};$$

then an application of Lemma 5 on the estimate of $S(\theta)$, and of the formula (26) give:

$$\sum\nolimits_2 \ll \sum_{P^{1-\varepsilon_1} \leqslant \lambda < P} \frac{a^2(\lambda)}{\lambda^{2\theta}} = \frac{P^{1-2\theta} - P^{(1-\varepsilon_1)(1-2\theta)}}{1-2\theta} S(0) + O(P^{-1+\varepsilon_1}X^3) \ll \frac{\varepsilon_1 \log P}{\log X}.$$

To estimate W_2, we make use of Lemma 4; we obtain:

$$W_2 \ll T^{-1}$$

Hence,

$$J_2 \ll h^2 H_1 (\varepsilon_1 \log P \log^{-1} X + T^{-\varepsilon});$$

$$J \ll J_1 + J_2;$$

$$I_3 \ll J + h^2 HL^{-10} \leqslant c_1 H_1 (\varepsilon_1^{-2} \log^{-1} P \log^{-1} X + A\varepsilon_1^{-1} L^{-2} T^{-\varepsilon}$$
$$+ \varepsilon_1^{-2} L^{-2} T^{-\varepsilon} + \varepsilon_1 h^2 \log P \log^{-1} X + h^2 T^{-\varepsilon} + h^2 L^{-10})$$

The quantities X and h were defined in the following way,

$$X = T^{0.01\varepsilon}, \quad h = AL^{-1};$$

and thus

$$I_3 \leqslant c_2 Hh^2 (1 + AL^{-1} T^{-0.25})(A^{-2}\varepsilon_1^{-2}\varepsilon^{-1} + A^{-1}\varepsilon_1^{-1} T^{-\varepsilon} + A^{-2}\varepsilon_1^{-2} T^{-\varepsilon} + \varepsilon_1 \varepsilon^{-1} + T^{-\varepsilon}).$$

where $c_2 > 0$ is an absolute constant. Now take

$$A = ((32c_2 + 32)\varepsilon^{-1})^{1.5}, \quad \varepsilon_1 = (32c_2 + 32)^{-1}\varepsilon,$$

and $T_0 = T_0(\varepsilon) > 0$ such that for $T \geqslant T_0$ the inequality

$$I_3 \leqslant \frac{1}{4} Hh^2.$$

be valid. Thus, the estimate of I_1 from below and those from above for I_2 and I_3 and the formula $(*)$ give us:

$$\sqrt{\mu(E) I_2} \geqslant I_1 - \sqrt{HI_3} \geqslant \frac{1}{2} hH - h^2 + O(T^{1/6} X \log T) \geqslant \frac{1}{3} hH;$$

$$\mu(E) \geqslant c_3 H; \qquad c_3 = c_3(\varepsilon) > 0.$$

Subdivide the interval $(T, T+H)$ into intervals of the form $(mh, mh+h)$, $m = [Th^{-1}], [Th^{-1}]+1, \cdots, [(T+H)h^{-1}]$. Then there are at least $[c_3 Hh^{-1}] - 2$ of these intervals which contain a point t from the set E. If an interval $(mh, mh+h)$ contains a point t from E, then in the interval $(t, t+h)$ and consequently also in the interval $(mh, mh+2h)$ there is at least one zero of odd order of the function $\zeta(\frac{1}{2} + it)$. Consequently, the number of the zeros of odd order of the function $\zeta(\frac{1}{2} + it)$ on the interval $(T, T+H)$ is not less than

$$\frac{1}{2}([c_3 Hh^{-1}] - 2) \geqslant c_4 H \log T, \quad c_4 > 0, \qquad \text{Q.E.D.}$$

§4. Distribution of the Zeros of Riemann's Zeta-Function on the Critical Line

As it was mentioned above, the background of the proof of A. Selberg's conjecture is formed by the estimate of an exponential sum of a special type —— the Lemma 4, it also defines the bound of the quantity $H=T^a, a>\frac{27}{82}$.

It is clear that "in the mean" the sums $W_j(T)$, $j=0,1,2$, can be estimated much more exactly. It is this consideration which enables to prove the two following assertions.

Theorem 7. *Let* $\varepsilon>0$ *be an arbitrarily small fixed number* $Y\geqslant Y_0(\varepsilon)>0$, $H=Y^\varepsilon$, $0.5\,Y\leqslant T\leqslant Y$. *Consider the relation*

$$N_0(T+H)-N_0(T)\geqslant cH\log T, \tag{31}$$

where $c=c(\varepsilon)>0$ *is a certain constant which depends on* ε *only and let* E_1 *stand for the set of those* T *from the segment* $0.5\,Y\leqslant T\leqslant Y$ *for which* (31) *does not hold. Then for the measure* $\mu(E_1)$ *of that set the following estimate is valid*:

$$\mu(E_1)\ll Y^{1-0.5\varepsilon}.$$

Theorem 8. *Let* $\varepsilon>0$ *be an arbitrary small fixed number* $Y\geqslant Y_0(\varepsilon)>0$, $H=Y^\varepsilon$, $K=[YH^{-1}]$. *For* $k=K+1, K+2,\cdots, 2K-1, 2K$ consider the intervals $(kH, kH+H)$. *Then each of the latter intervals, with a possible exception of not more than* $k^{1-0.5\varepsilon}$ *contains* $\geqslant c_1H\log Y$ *of the zeros of odd order of the function* $\zeta(1/2+it)$, $c_1=c_1(\varepsilon)>0$.

The two following lemmas and their corollaries are in the background of the proof of the theorems.

Lemma 6. *Keeping the assumptions and notations of Lemma* 4 *the following inequality is valid*:

$$\int_{0.5Y}^{Y}(W_0^2(T)+W_1^2(T)+W_2^2(T))dT\ll(\varepsilon_1^{-4}+\varepsilon_1^{-2}A^2)YH^{-1}X^{12}L^7.$$

Corollary. *Let* δ *be an arbitrary positive number, not exceeding* 1, E_1 *be the set of such* T, $0.5\,Y\leqslant T\leqslant Y$, *that*

$$W_0^2(T)+W_1^2(T)+W_2^2(T)\geqslant(\varepsilon_1^{-4}+\varepsilon_1^{-2}A^2)Y^{1-\delta}H^{-1}X^{12}L^7.$$

Then for the measure $\mu(E_1)$ *of that set the following estimate holds true*:

$$\mu(E_1)\ll Y^\delta.$$

Lemma 7. *Keeping the assumptions and notations of Lemma* 4, $K=[YH^{-1}]$, *the following inequality is valid*

$$\sum_{k=K+1}^{2K}(W_0^2(kH)+W_1^2(kH)+W_2^2(kH))\ll(\varepsilon_1^{-4}+\varepsilon_1^{-2}A^2)KH^{-1}X^{12}L^8.$$

Corollary. *Let δ be an arbitrary positive number less than or equal to* 1, E_2 *be the set of those* k, $K<k\leqslant 2K$, *for which*

$$W_0^2(kH)+W_1^2(kH)+W_2^2(kH)\geqslant(\varepsilon_1^{-4}+\varepsilon_1^{-2}A^2)K^{1-\delta}H^{-1}X^{12}L^8.$$

Then for the number of the elements $\mu(E_2)$ *of that set the following estimate is valid*:

$$\mu(E_2)\ll K^{\delta}.$$

Proof of Theorem 7. We follow the lines of the proof of Theorem 6. Let $X=H^{0.01}$, $L=\log Y$, $h=AL^{-1}$ and define the function $F(t)$ of Hardy-Selberg. Furthermore, put i in the Corollary to the Lemma 6, $\delta=1-0.5\varepsilon$, and consider the numbers T from the segments $0.5Y\leqslant T\leqslant Y$, which do not belong to the set E_1. For those T the following estimate holds true:

$$W_0^2(T)+W_1^2(T)+W_2^2(T)<(\varepsilon_1^{-4}+\varepsilon_1^{-2}A^2)H^{-0.5}X^{12}L^7 \tag{32}$$

Delete from the above set of T the ones, for which the following inequality is valid:

$$\left|\int_{0.5}^{2}\zeta(\sigma+i(T+H))\varphi^2(\sigma+i(T+H))d\sigma\right|$$

$$+\left|\int_{0.5}^{2}\zeta(\sigma+i(T+h))\varphi^2(\sigma+i(T+h))d\sigma\right|>HL^{-1}.$$

The measure of the set of thus deleted T is of the order

$$O(YH^{-2}L^{14})\ll Y^{1-\varepsilon}.$$

Below we consider the remaining T only, belonging to the segment $0.5\ Y\leqslant T\leqslant Y$ the measure of the deleted T is of the order $\ll Y^{1-0.5\varepsilon}$. Denote E the subset of the segment $(T,T+H)$, for which the following inequality is valid:

$$\int_t^{t+h}|F(u)|\,du>\left|\int_t^{t+h}F(u)du\right|,\quad t\in E$$

As in Theorem 6, we arrive at the inequality

$$I_1\leqslant\sqrt{\mu(E)I_2}+\sqrt{HI_3}\ , \tag{33}$$

where

$$I_1=\int_T^{T+H}\int_t^{t+h}|F(u)|\,du\,dt,$$

$$I_2=\int_T^{T+H}\left(\int_t^{t+h}|F(u)|\,du\right)^2dt.$$

$$I_3=\int_T^{T+H}\left|\int_t^{t+h}F(u)du\right|^2 dt.$$

Like in § 3, we get the following estimate from below for I_1:

$$I_1\geqslant hH-h^2+O(hHL^{-1}).$$

As for I_2, we proved in §3 that

$$I_2\ll h^2H_1(\textstyle\sum_0+W_0+L^{-10});\qquad H_1=H+h,$$

where

$$\sum_0=\sum_{\lambda<P}\frac{a^2(\lambda)}{\lambda},$$

$$W_0=\left|\sum_{\lambda_1<\lambda<P}\frac{a(\lambda_1)a(\lambda_2)}{\sqrt{\lambda_1\lambda_2}}\left(\frac{\lambda_2}{\lambda_1}\right)^{iT}e^{-\left(\frac{H_1}{2}\log\frac{\lambda_2}{\lambda_1}\right)^2}\right|.$$

The sum $\sum_0$ was estimated in §3, using A. Selberg's lemma:

$$\sum_0\ll\frac{\log P}{\log X}.$$

For the sum W_0 the following estimate is valid, which is a corollary from Corollary:

$$W_0\ll\sqrt{\varepsilon_1^{-4}+\varepsilon_1^{-2}A^2}\;H^{-0.25}X^6L^{3.5}.$$

Thus we obtain

$$I_2\ll h^2H(1+AL^{-1}H^{-1})(\varepsilon_1^{-1}+(\varepsilon_1^{-2}+\varepsilon_1^{-1}A)H^{-0.15}L^{3.5}).$$

Using §3, I_3 can be estimated as follows:

$$I_3\ll J_1+J_2+Hh^2L^{-10},\tag{34}$$

where

$$J_1=\int_T^{T+H}\left|\int_t^{t+h}\sum_{\lambda<P_0^{1-\varepsilon}}\frac{a(\lambda)}{\sqrt{\lambda}}\left(\frac{P}{\lambda}\right)^{iu}du\right|^2dt,$$

$$J_2=\int_T^{T+H}\left|\int_t^{t+h}\sum_{P_0^{1-\varepsilon_1}<\lambda<P}\frac{a(\lambda)}{\sqrt{\lambda}}\left(\frac{P}{\lambda}\right)^{iu}du\right|^2dt.$$

For the integral J_1 we have

$$J_1\ll H(\textstyle\sum_1+W_1),$$

where

$$\sum_1 = \sum_{\lambda < P_0^{1-\varepsilon}} \frac{a^2(\lambda)}{\lambda \log^2 \frac{P}{\lambda}},$$

$$W_1 = \left| \sum_{\lambda_1 < \lambda_2 < P_0^{1-\varepsilon_1}} \frac{a(\lambda_1)a(\lambda_2)}{\sqrt{\lambda_1\lambda_2}} \left(\frac{\lambda_2}{\lambda_1}\right)^{iT} \cdot B(\lambda_1)\overline{(B(\lambda_2)} e^{-\left(\frac{H}{2}\log\frac{2\lambda}{\lambda_1}\right)^2} \right|.$$

The sum $\sum_1$ has been estimated above in §3:

$$\sum_1 \ll \varepsilon_1^{-2} \log^{-1} P_0 \log^{-1} X. \tag{35}$$

For the sum W_1, the following estimate is valid,

$$W_1 \leqslant \sqrt{\varepsilon_1^{-4} + \varepsilon_1^{-2} A^2}\; H^{-0.25} X^6 L^{3.5}. \tag{36}$$

An analogous estimate holds for the integral J_2:

$$J_2 \ll H_1 h^2 \left(\sum_2 + W_2\right), \tag{37}$$

where

$$\sum_{-2} = \sum_{P_0^{1-\varepsilon} \leqslant \lambda < P} \frac{a^2(\lambda)}{\lambda},$$

$$W_2 = \left| \sum_{P_0^{1-\varepsilon_1} \leqslant \lambda_1 < \lambda_2 < P} \frac{a(\lambda_1)a(\lambda_2)}{\sqrt{\lambda_1\lambda_2}} \left(\frac{\lambda_2}{\lambda_1}\right)^{iT} e^{-\left(\frac{H_1}{2}\log\frac{\lambda_2}{\lambda_1}\right)^2} \right|.$$

We have estimated the sum $\sum_2$ in §3:

$$\sum_2 \ll \varepsilon_1 \log P \log^{-1} X. \tag{38}$$

For the sum W_2, the following estimate is valid, c.f. (36)

$$W_2 \ll \sqrt{\varepsilon_1^{-4} + \varepsilon_1^{-2} A^2}\; H^{-0.25} X^6 L^{3.5} \tag{39}$$

Thus, (34)—(39) imply

$$I_3 \leqslant cHh^2(1 + AL^{-1}H^{-1})(L^{-10} + h^{-2}\varepsilon_1^{-2} \log^{-1} Y \log^{-1} X$$
$$+ \text{æ} h^{-2} + \varepsilon_1 \log P \log^{-1} X + \text{æ}),$$

where

$$\text{æ} = \sqrt{\varepsilon_1^{-4} + \varepsilon_1^{-2} A^2}\; H^{-0.25} X^6 L^{3.5},$$

$c>10$ is an absolute constant. Now, define A and ε_1 by the relations:

$$A=(1600\,c\,\varepsilon^{-1})^{1.5},\qquad \varepsilon_1=(1600\,c)^{-1}\,\varepsilon\,;$$

then the following inequality would be valid:

$$(ch^{-2}\varepsilon_1^{-2}\log^{-1}Y\log^{-1}X+\varepsilon_1\log P\log^{-1}X)\leqslant\frac{1}{8}\,.$$

Furthermore, consider $Y_0(\varepsilon)>0$ so large that for $Y\geqslant Y_0(\varepsilon)$ the inequality

$$(cL^{-10}+æh^{-2}+æ)\leqslant\frac{1}{8}\,.$$

is valid. Thus for $Y\geqslant Y_0(\varepsilon)$ we obtain

$$I_3\leqslant\frac{1}{4}\,Hh^2. \tag{40}$$

Using (33), (40) and the estimates of I_2, we arrive at the inequality

$$\mu(E)>c_1(\varepsilon)H,\qquad c_1(\varepsilon)>0.$$

From here the assertion of Theorem 7 follows immediately. The proof of Theorem 8 repeats, with abvious modifications, that of Theorem 7.

I express my sincere gratitude to Professor K. I. Oskolkov for his assistance in the preparation of this paper and for its translation into English.

References

[1] L.K.Hua, On Tarry's problem, *Quart J. Math*, Oxford Ser. 9(1938), 315 — 320.

[2] L.K.Hua, On the number of solutions of Tarry's Problem, *Acta Sci. Sinica* 1(1952), 1 — 76.

[3] L.K.Hua. On Tarry's problem, Third All-Union Math Congr. (Moscow, 1956), **Vol. IV**. *Izdat Akad. Nauk. SSSR*, Moscow, 1959, 140 — 143(Russian).

[4] Arhibov, G. I., Karatsuba, A. A, Cubarekov, V. N., Proof of the convergence of singular integral of the Tarry problem, *Dokl. AH SSSR*, 1979, **T. 248**, No.2, 268 — 272.

[5] Arhibov, G. I., Karatsuba, A. A., Cubarekov, V. N., Trigonometrical integrals, *Izv. AH SSSR, Ser. Math*, 1979, **T.43**, No.5, 971 — 1003.

[6] Arhibov, G. I., Karatsuba, A. A., Cubarekov, V.N., Multiple Trigonometrical Sums, *Tryd. MIAH*, 1980, **T.151**.

[7] Arhibov, G. I., Karatsuba, A. A., Cubarekov, V. N., Theory of multiple Trigonometrical Sums, Moscow, "**Science**", 1987.

[8] Osckolkov, K. I., I. M. Vinogradov's series and integrals and their applications, *Tryd, MIAH*, 1989.

[9] Osckolkov, K. I., On properties of a class of I. M. Vinogradov's series, *Dokl. AH SSSR*, **T.300**, 1988, 803 — 807.

[10] Arhibov, G. I., Ockolkov, K. I., On a class of Trigonometrical sum and its applications, *Math. sb.*, **T.134**, 10, 1987, 147 — 157.

[11] Ockolkov, K.I., On Spectral uniform convergence, *Dokl. AH SSSR*, **T.288**, No.1, 1986, 54 — 58.

[12] Ockolkov, K.I., On continuous function with polynomial spectrum, See "Investigations on the theory of approximations of functions", *BFAH*, SSSR, 1987, 187 — 200.

[13] Riemann, G. F. B., Selected works, M, OGIZ, 1948.

[14] Hardy, G. H., Sur les zeros de la fonction $\zeta(s)$ de Riemann, *Compt. Rend. Acad. Sci.*, 1914, v. **158**, 1012 — 1014.

[15] Landau, E., Über die Hardysche Eutdeckung unendlich vieler Nullstellen der Zeta-function mit reelen Teil 1/2, *Math. Ann.*, 1915, b.**76**, 212 — 243.

[16] Hardy, G. H., Littlewood, J. E., Contributions to the theory of Riemann zata-function and the theory of distribution of primes, *Acta Math.*, 1918, v.**41**, 119 — 196.

[17] Hardy, G. H., Littlewood, J. E., The zeros of Riemann's zeta-function and the critical line, *Math. Zs.*, 1921, v.**10**, 283 — 317.

[18] Selberg, A., On the zeros of Riemann's zeta-function, *Skr. Norske Vid. Akad. Oslo*, 1942, V.**10**.

[19] Levinson, N., More than one third of the zeros of Riemann's zeta-function are on $\sigma=1/2$, *Adv. in Math.*, 1974, v.**13**, 383 — 436.

[20] Mozer, Ja., On a theorem of Hardy-Littlewood in the theory of Riemann zeta function, *Acta Arith*, **31**, 1976, 45 — 51.

[21] Mozer, Ja., improved theorems of Hardy-Littlewood on the density of zeros of the function $\zeta\left(\frac{1}{2}+it\right)$, *Acta Math*, Univ. Comen. Britislava, 42 — 43, 1983, 41 — 50.

[22] Karatsuba, A. A., On distance between consecutive zeros of Riemann zeta function lying on the critical line, *Dokl. AH SSSR*, **272**, 6, 1983, 1312 — 1314.

[23] Karatsuba, A. A., On zeros of Riemann zeta function on an interval of the critical line, *Dokl. AH SSSR*, **272**, 1983, 1312 — 1314.

[24] Ivic, A., Topics in recent zeta-function theory, *Publ. Math. d'Orsay*, Universite Paris-Sud, Orsay, 1983.

[25] Karatsuba, A. A., On zeros of the function $\zeta(s)$ on certain interval of the critical line, *Izv. AH SSSR. Ser. Mat.* **48**, 1984, 569 — 584.

[26] Karatsuba, A. A., On distribution of zeros of $\zeta\left(\frac{1}{2}+it\right)$, *Izv. AH SSSR, Ser. Mat.* **48**, 1984, 1214 — 1224.

[27] Karatsuba, A. A., On zeros of the function $\zeta(s)$ on the neighbourhood of the critical line, *Izv. AH SSSR, Ser. Mat.* **49**, 1985, 326 — 333.

[28] Ivic, A., On a problem connected with zeros of $\zeta(s)$ on the critical line, *Mh. Math.* 1987, **104**, 17 — 27.

[29] Karatsuba. A. A., Riemann zeta function and its zeros, *YMH*, **40**, 1985, 19 — 70.

[30] Titchmarsh, E. K., Theory of Riemann zeta function, *M.*: *EL*, 1953.

[31] Karatsuba, A. A., On real zeros of the function $\zeta\left(\frac{1}{2}+it\right)$, *YMH*, **40**, 1985, 171 — 172.

SMALL PRIME SOLUTIONS OF A PAIR OF LINEAR EQUATIONS IN FIVE VARIABLES

Ming-Chit Liu and Kai-Man Tsang
Department of Mathematics, University of Hong Kong
Pokfulam Road, Hong Kong

1. Introduction and Main Results

In a recent paper [8] the problem of obtaining a precise bound for the smallest solution in primes p_1, p_2, p_3 of the linear equation

$$a_1p_1 + a_2p_2 + a_3p_3 = b \tag{1.1}$$

was solved completely. The problem was first raised by Baker in connection with his well-known work [1] on the solubility of certain diophantine inequalities involving primes. Our results in [8] include, as particular cases, the well-known Vinogradov's three prime theorem and Linnik's theorem on the least prime in an arithmetic progression.

In the present paper we shall refine the useful techniques we developed in [8] to prove in Theorem 1 below that, subject to some natural conditions on the coefficients, the equations

$$\sum_{j=1}^{5} a_{\lambda j}p_j = b_\lambda, \quad \lambda = 1, 2 \tag{1.2}$$

can be solved simultaneously in primes $p_1, \cdots, p_5$ which are bounded by some power of the coefficients $a_{\lambda j}$ and b_λ. Our Theorem 1 is indeed an extension of Baker's work in [1] from one equation to two equations and the bound given in (1.6) below is the best possible bound for its small prime solutions. The problem on the solubility of equations (1.2) was discussed by Hua in [5, §31] and [6, Chapter 12], in connection with generalization of the Vinogradov theorem and conjectures concerning primes represented by polynomials with integral coefficients. The solubility of n simultaneous linear equations in $2n+1$ prime variables like (1.2) was first demonstrated by Wu [10] some thirty years ago (for earlier work in this context, see van der Corput [2]). Our studies here are apparently the first concerning sharp *bounds* for small solutions. The method used in our proof of Theorem 1 can certainly be adapted to the general case of n linear equations in $2n+1$ prime variables.

For $a_{\lambda j}$, b_λ in the equations (1.2) we define

$$\Delta_{ij} := \begin{vmatrix} a_{1i} & a_{1j} \\ a_{2i} & a_{2j} \end{vmatrix} \quad \text{and} \quad \Delta_{ib} := \begin{vmatrix} a_{1i} & b_1 \\ a_{2i} & b_2 \end{vmatrix} \quad \text{for} \quad 1 \le i,\ j \le 5\,.$$

The determinants Δ_{ij} play the same role as the a's in (1.1). Accordingly, we assume that

$$\Delta_{ij} \neq 0 \quad \text{for all} \quad 1 \le i < j \le 5 \quad \text{and} \quad \gcd(\Delta_{ij})_{1 \le i < j \le 5} = 1\,. \tag{1.3}$$

There are two other natural conditions for the solubility of (1.2), namely, the solubility of the corresponding congruences

$$\sum_{j=1}^{5} a_{\lambda j} \ell_j \equiv b_\lambda \pmod{p}, \quad \lambda = 1,2 \tag{1.4}$$

with $1 \le \ell_1, \cdots, \ell_5 \le p-1$ for each prime p, and the solubility of

$$\sum_{j=1}^{5} a_{\lambda j} y_j = b_\lambda, \qquad \lambda = 1,2$$

in positive real numbers $y_1, \cdots, y_5$. However, in view of (1.2), the latter condition here is not strong enough, since the positive solutions y_j may be too small, say $< 2 \le p$. So for simplicity, we assume the slightly stronger condition that the pair of linear equations,

$$\sum_{j=1}^{5} a_{\lambda j} y_j = 0, \qquad \lambda = 1,2 \tag{1.5}$$

is soluble in positive real y_j. These conditions (1.4) and (1.5) are necessary in the sense that they ensure the positiveness of the singular series and singular integral occurred in the main term of our estimates. They are parallel to the conditions of congruent solubility and positive solubility in Hua's work on Tarry's problem [6, Chapter 11]. We shall give in §7 concise and practical criteria for checking conditions (1.4), (1.5) in terms of the Δ_{ij}'s.

We shall prove

THEOREM 1. *Subject to* (1.3) *and the solubility of* (1.4) *and* (1.5), *the pair of equations in* (1.2) *is soluble in primes* $p_1, \cdots, p_5$ *satisfying*

$$\max\{p_1, \cdots, p_5\} \le (|b_1| + |b_2| + 1) B^A \tag{1.6}$$

where $B := \max_{\lambda=1,2,j=1,\cdots,5} \{2, |a_{\lambda j}|\}$ *and* $A > 0$ *is an effective absolute constant.*

The bound we obtain in (1.6), apart from the exact value of A, is sharp. Indeed, for any positive even integer a the coefficients of the pair of equations

$$ap_1 + p_2 + p_3 + 2p_4 - 3p_5 = 1, \quad ap_1 - p_2 + 3p_3 + p_4 + 2p_5 = 1$$

satisfy the assumptions in (1.3), (1.4) and (1.5). The latter two conditions can be easily checked with the help of our Propositions 7.1 and 7.2 in §7. Any solution $p_1, \cdots, p_5$ clearly satisfies

$$\max\{p_1, \cdots, p_5\} \ge p_2 \ge 2a + 11\,.$$

Hence, the constant A in (1.6) is not less than 1.

The definitive bound, which we obtain in (1.6), is a consequence of an important result (see (2.7) below) on the distribution of the zeros of L-functions near the line $\sigma = 1$ due to Gallagher [7], and we use (2.7) to prove our fundamental lemma (Lemma 2.1 below). By this we obtain, in Lemma 6.1, a good estimate for the minor term M_2. Another novel idea in our proof is the treatment of the interlocking relation of the terms in M_1 and M_3, by which we can overcome the difficulty arising from the possible existence of the Siegel zeros. This is done in Lemmas 5.2, 5.3 and 5.4.

2. The Fundamental Lemma

Throughout this paper, p, with or without subscripts, always denotes a prime number. $c_1, c_2, \cdots$ etc. are effective positive constants and $\delta > 0$ is a sufficiently small positive number depending only on the c_j's. Let

$$N \geq (|b_1| + |b_2| + 1)B^{1/\delta^2} , \tag{2.1}$$

$$Q := N^\delta, \; L := NQ^{-1/90}, \; T := Q^{c_1} \quad \text{and} \quad \tau := N^{-1}T^{1/4} , \tag{2.2}$$

where B is given in Theorem 1 and c_1 is large. Then

$$B \leq Q^\delta . \tag{2.3}$$

We use $\chi (\mathrm{mod}\ q)$ and $\chi_0 (\mathrm{mod}\ q)$ to denote a Dirichlet character and the principal character modulo q respectively. It is known that [3, §14] there exists a small c_2 such that there is at most one primitive character $\tilde{\chi}$ to a modulus $\tilde{r} \leq T$, for which the L-function $L(s, \tilde{\chi})$ has a zero in the region

$$\sigma > 1 - (c_2/\log T), \quad |t| \leq T ,$$

and if there is such an exceptional $\tilde{\chi}$, the corresponding exceptional zero $\tilde{\beta}$ is real, simple and unique and $\tilde{\chi}$ is quadratic. Furthermore, for some c_3, the $\tilde{\beta}$ satisfies

$$c_3/(\tilde{r}^{1/2} \log^2 \tilde{r}) < 1 - \tilde{\beta} < c_2/\log T . \tag{2.4}$$

For any $q \leq T$ and $\chi (\mathrm{mod}\ q)$, let

$$\sideset{}{'}\sum_{|\gamma| \leq T} \tag{2.5}$$

denote the summation over all zeros $\rho = \beta + i\gamma$ of $L(s, \chi)$ lying inside the region $|\gamma| \leq T$, $1/2 \leq \beta \leq 1 - (c_2/\log T)$. So $\tilde{\beta}$ is excluded from the sum (2.5). Let

$$\Omega := \begin{cases} (1 - \tilde{\beta}) \log T & \text{if } \tilde{\beta} \text{ exists,} \\ 1 & \text{otherwise.} \end{cases} \tag{2.6}$$

LEMMA 2.1. *If $y \geq \sqrt{N}$ then there is a c_4 such that*

$$\sum_{q\le T}\ \sum_{\chi(\text{mod } q)}^{*}\ \sum_{|\gamma|\le T}{}' y^{\beta-1} \ll \Omega^5 \exp(-c_4/\delta)$$

where $\sum^*$ *is the summation taken over all primitive characters* $\chi(\text{mod } q)$.

PROOF: This is Lemma 2.1 in [8]. The main ingredient in the proof is the following result due to Gallagher [7, Theorem 6], namely,

$$\sum_{q\le T}\ \sum_{\chi(\text{mod } q)}^{*} N_\chi(\alpha,T) \ll T^{c_5(1-\alpha)} \tag{2.7}$$

where $1/2 \le \alpha \le 1$ and $N_\chi(\alpha,T)$ denotes the number of zeros $\beta+i\gamma$ of $L(s,\chi)$ in the region $\alpha\le\beta<1$, $|\gamma|\le T$. Using different ideas, Selberg [9] proved the quantitative result: *for any* $T\ge 1$, $Q\ge 1$, $\alpha>1/2$ *and* $\varepsilon>0$, *there exists a constant* $C(\varepsilon)$, *depending only on* ε, *such that*

$$\sum_{q\le Q}\ \sum_{\chi(\text{mod } q)}^{*} N_\chi(\alpha,T) \le C(\varepsilon)(T^{3+\varepsilon}Q^{5+\varepsilon})^{1-\alpha}\ .$$

3. Minor Arcs

For any real y, we write $e(y)$ for $e^{i2\pi y}$ and $e_q(y)$ for $e(y/q)$. Define

$$\begin{cases} S(y) := \displaystyle\sum_{L<n\le N} \Lambda(n)e(ny), \quad S_\chi(y) := \sum_{L<n\le N} \Lambda(n)\chi(n)e(ny)\ , \\ I(y) := \displaystyle\int_L^N e(xy)dx, \quad \tilde{I}(y) := \int_L^N x^{\tilde{\beta}-1}e(xy)dx \quad \text{and} \\ I_\chi(y) := \displaystyle\int_L^N e(xy)\sum_{|\gamma|\le T}{}' x^{\rho-1}dx\ , \end{cases} \tag{3.1}$$

where the symbol $\sum'_{|\gamma|\le T}$ is defined in (2.5) and $\Lambda(n)$ is the von Mangoldt function. For integers h_1, h_2, q satisfying $q\le Q$ and

$$1\le h_1,\ h_2\le q, \qquad (h_1,h_2,q)=1\ , \tag{3.2}$$

define

$$\mathcal{M}(h_1,h_2,q) := \{(x_1,x_2)\in \mathbb{R}^2 : |x_\lambda - h_\lambda q^{-1}| \le \tau q^{-1} \quad \text{for} \quad \lambda=1,2\}\ ,$$
$$\mathcal{M} := \bigcup \mathcal{M}(h_1,h_2,q) \quad \text{and} \quad \mathcal{M}' := [\tau,1+\tau]^2 \setminus \mathcal{M}\ .$$

The above union is taken over all $q\le Q$ and all h_1, h_2 satisfying (3.2). In view of (2.2), it is not difficult to see that the squares $\mathcal{M}(h_1,h_2,q)$ are disjoint and all lie inside the square $[\tau,1+\tau]^2$. Let us put

$$\underline{x}_b := b_1x_1+b_2x_2, \quad \underline{x}_j := a_{1j}x_1+a_{2j}x_2 \quad \text{for} \quad j=1,\cdots,5\ ,$$

and

$$I(\underline{b}) := \sum \Lambda(n_1) \cdots \Lambda(n_5) , \tag{3.3}$$

where the summation $\sum$ is over all integers n_j satisfying $L < n_1, \cdots, n_5 \leq N$ and $\sum_{j=1}^{5} a_{\lambda j} n_j = b_\lambda$ for $\lambda = 1, 2$. Then

$$\begin{aligned} I(\underline{b}) &= \int_{\tau}^{1+\tau}\int_{\tau}^{1+\tau} e(-\underline{x}_b) \prod_{j=1}^{5} S(\underline{x}_j) dx_1 dx_2 \\ &= \Big(\iint_{\mathcal{M}} + \iint_{\mathcal{M}'}\Big) e(-\underline{x}_b) \prod_{j=1}^{5} S(\underline{x}_j) dx_1 dx_2 := I_1(\underline{b}) + I_2(\underline{b}) , \end{aligned} \tag{3.4}$$

say. We shall show in Lemma 6.2 that $I_1(\underline{b}) \gg N^3 Q^{-1/6}$ and in Lemma 3.2 below that $I_2(\underline{b}) \ll N^3 Q^{-1/5}$. In view of (3.4), we then have $I(\underline{b}) \gg N^3 Q^{-1/6}$. That is, the system of equations (1.2) has at least $N^3 Q^{-1/6} \log^{-5} N$ solutions in primes $p_1, \cdots, p_5$ satisfying $\max p_j \leq N$, since the main contribution to the $\sum$ in (3.3) comes from those terms with the n_j's equal to primes. Taking $A = \delta^{-2}$, we deduce from (2.1) our Theorem 1.

LEMMA 3.1. *If* $(x_1, x_2) \in \mathcal{M}'$ *and* $1 \leq m < n \leq 5$ *then*

$$\min\{|S(\underline{x}_m)|, |S(\underline{x}_n)|\} \ll N B^{1/2} Q^{-1/4} \log^4 N .$$

PROOF: Write $P = 2B^2 \tau^{-1} Q^{1/2}$. By Dirichlet's theorem on diophantine approximation, there are integers a_j, q_j satisfying $1 \leq q_j \leq P$, $(a_j, q_j) = 1$ and $|a_{1j} x_1 + a_{2j} x_2 - a_j q_j^{-1}| < (q_j P)^{-1}$ for $j = m, n$. Since $\Delta_{mn} \neq 0$, solving the two equations

$$a_{1j} r_1 + a_{2j} r_2 = a_j / q_j, \qquad j = m, n ,$$

we have $r_i = k_i / q$, where $(k_1, k_2, q) = 1$ and q is a positive factor of $q_m q_n \Delta_{mn}$. Thus,

$$1 \leq q \leq |\Delta_{mn}| q_m q_n \leq 2B^2 q_m q_n . \tag{3.5}$$

Let $t_j = a_{1j} x_1 + a_{2j} x_2 - a_j q_j^{-1}$ for $j = m, n$. Then $|t_j| < (q_j P)^{-1}$ and

$$t_j = a_{1j}(x_1 - k_1 q^{-1}) + a_{2j}(x_2 - k_2 q^{-1}) .$$

From the last two equations, we have

$$x_1 - k_1 q^{-1} = (a_{2n} t_m - a_{2m} t_n)/\Delta_{mn}, \quad x_2 - k_2 q^{-1} = (a_{1m} t_n - a_{1n} t_m)/\Delta_{mn} .$$

This gives

$$|x_1 - k_1 q^{-1}|, \quad |x_2 - k_2 q^{-1}| \leq B P^{-1} (q_m^{-1} + q_n^{-1}) .$$

We claim that $\max\{q_m, q_n\} \geq Q^{1/2} (2B)^{-1}$. Indeed, assuming the contrary, we have

$$BP^{-1}(q_m^{-1} + q_n^{-1}) = B(q_m + q_n)(P q_m q_n)^{-1} < Q^{1/2} (P q_m q_n)^{-1} \leq 2B^2 Q^{1/2} (Pq)^{-1} = \tau / q .$$

The last inequality follows from (3.5). Thus, $|x_1 - k_1 q^{-1}|$, $|x_2 - k_2 q^{-1}| < \tau/q$. This together with $\tau \leq x_1$, $x_2 \leq \tau + 1$ gives $1 \leq k_1$, $k_2 \leq q$. Applying (3.5) again we have

$q < 2B^2(Q^{1/2}/(2B))^2 < Q$. This contradicts that $(x_1, x_2) \in \mathcal{M}'$. So we have

$$Q^{1/2}(2B)^{-1} \le \max\{q_m, q_n\} \le P .$$

Let $q_m \ge Q^{1/2}(2B)^{-1}$, say. Using the Vinogradov lemma (see [3, p.143]) we have

$$S(\underline{x}_m) \ll (NQ^{-1/4}B^{1/2} + N^{4/5} + N^{1/2}P^{1/2}) \log^4 N .$$

Then by (2.2) and (2.3) we obtain the inequality in our lemma.

LEMMA 3.2. *We have* $I_2(\underline{b}) \ll N^3Q^{-1/5}$.

PROOF: By (3.4) and Lemma 3.1 with $m = 1$, $n = 5$ we have

$$\begin{aligned} I_2(\underline{b}) &\ll NB^{1/2}Q^{-1/4}\log^4 N \int_0^1\int_0^1 \left(\prod_{j=2}^{5}|S(\underline{x}_j)| + \prod_{j=1}^{4}|S(\underline{x}_j)|\right)dx_1dx_2 \\ &\ll N^3B^{1/2}Q^{-1/4}\log^6 N , \end{aligned}$$

since for $j \ne \ell$

$$\begin{aligned} &\int_0^1\int_0^1 |S(\underline{x}_j)|^2|S(\underline{x}_\ell)|^2 dx_1dx_2 \\ &= \sum_{L<n_1,\cdots,n_4\le N} \Lambda(n_1)\cdots\Lambda(n_4)\prod_{\lambda=1}^{2}\int_0^1 e(x_\lambda(a_{\lambda j}(n_1-n_2)+a_{\lambda\ell}(n_3-n_4)))dx_\lambda \\ &= \left(\sum_{L<n\le N}\Lambda(n)^2\right)^2 \ll (N\log N)^2 . \end{aligned}$$

In view of (2.3) our lemma follows.

4. Some Lemmas for Singular Series and Singular Integral

For any character $\chi(\text{mod } q)$ let

$$C_\chi(m) := \sum_{\ell=1}^{q}\chi(\ell)e_q(m\ell) \quad\text{and}\quad C_q(m) := C_{\chi_0}(m) . \tag{4.1}$$

Write

$$\underline{h}_b := b_1h_1 + b_2h_2 \quad\text{and}\quad \underline{h}_j := a_{1j}h_1 + a_{2j}h_2 \quad\text{for } j = 1,\cdots,5 .$$

For any positive integer q, we shall use $\sum_{(q)}$ to denote the summation over all 5-tuples $(\ell_1,\cdots,\ell_5)$ satisfying the conditions

$$1 \le \ell_j \le q,\ (q,\ell_j) = 1,\ \sum_{j=1}^{5}a_{\lambda j}\ell_j \equiv b_\lambda \ (\text{mod } q) \quad\text{for } \lambda = 1,2 . \tag{4.2}$$

Define

(4.3) $$N(q) := \sum_{(q)} 1$$

and

(4.4) $$A(q) := \phi(q)^{-5} \sum_{\underline{h}}{}' e_q(-\underline{h}_b) \prod_{j=1}^{5} C_q(\underline{h}_j) \ .$$

The sum $\sum_{\underline{h}}{}'$ is over all h_1, h_2 satisfying (3.2). Since $\gcd(\Delta_{ij})_{1\le i<j\le 5} = 1$ we see that, for any prime p, $N(p) \le \phi(p)^3$. Our assumption on the solubility of (1.4) is equivalent to

(4.5) $$N(p) \ge 1 \ .$$

Further, let

(4.6) $$s(p) := 1 + A(p) \ .$$

LEMMA 4.1. (1) *Both $A(q)$ and $N(q)$ are multiplicative functions of q.*
(2) *For any integer $k \ge 2$, $A(p^k) = 0$.*
(3) $s(p) = p^2\phi(p)^{-5}N(p)$. *Hence $s(p) > 0$ for any p.*
(4) *For any positive integer k, $N(p^k) = p^{3(k-1)}N(p)$.*
(5) $q^2\phi(q)^{-5}N(q) = \prod_{p|q} s(p)$.

PROOF: (1) This can be proved by an argument similar to that of Lemma 4.1 in [8]. See also [6, p.100].
(2) It is known that (a special case of (4.15) below) if $\mu(m)$ and $\phi(m)$ are the Möbius function and Euler function respectively, then

(4.7) $$C_q(m) = \mu(q/(m,q))\phi(q)/\phi(q/(m,q)) \ .$$

From this we see that if $k \ge 2$ and if $\prod_{j=1}^{5} C_{p^k}(\underline{h}_j) \ne 0$ then p must divide all the five $\underline{h}_j = a_{1j}h_1 + a_{2j}h_2$ $(j = 1, \cdots, 5)$. Then p divides all Δ_{ij} for $1 \le i < j \le 5$ since the sum $\sum_{\underline{h}}{}'$ in (4.4) is taken over all h_1, h_2 with $(h_1, h_2, p) = 1$. But then this violates (1.3). Thus, $A(p^k) = 0$ for $k \ge 2$.
(3) In (4.4) with $q = p$, the omitted term in the sum $\sum_{\underline{h}}{}'$ corresponds to $h_1 = h_2 = p$. So, by (4.6) and (4.3)

$$s(p) = \phi(p)^{-5} \sum_{1\le h_1,h_2\le p} e_p(-\underline{h}_b) \prod_{j=1}^{5} C_p(\underline{h}_j) = \phi(p)^{-5}\sum_{(p)} p^2 = \phi(p)^{-5}p^2N(p) \ .$$

That $s(p) > 0$ follows from this and (4.5).
(4) In view of (1.3), let $p \nmid \Delta_{12}$, say. The pair of congruences

$$\sum_{j=1}^{5} a_{\lambda j}\ell_j \equiv b_\lambda \pmod{p^k}, \quad \lambda = 1,2$$

is equivalent to

$$(4.8) \qquad \Delta_{12}\ell_1 + \sum_{j=3}^{5} \Delta_{j2}\ell_j \equiv \Delta_{b2},\ \Delta_{12}\ell_2 + \sum_{j=3}^{5} \Delta_{1j}\ell_j \equiv \Delta_{1b} \pmod{p^k}.$$

It follows from (4.3) and (4.8) that $N(p^k)$ is equal to the number of triples (ℓ_3, ℓ_4, ℓ_5) satisfying

$$(4.9) \quad 1 \le \ell_j \le p^k,\ p \nmid \ell_j \text{ for } j = 3,4,5 \quad \text{and} \quad \sum_{j=3}^{5} \Delta_{\lambda j}\ell_j \not\equiv \Delta_{\lambda b} \pmod{p},\ \lambda = 1,2.$$

On writing $\ell_j = s_j + t_j p$ for $j = 3,4,5$, these conditions become

$$1 \le s_j \le p-1,\ 0 \le t_j \le p^{k-1}-1,\ j = 3,4,5 \quad \text{and} \quad \sum_{j=3}^{5} \Delta_{\lambda j}s_j \not\equiv \Delta_{\lambda b} \pmod{p},\ \lambda = 1,2.$$

We see that the number of triples (s_3, s_4, s_5) satisfying these conditions is precisely $N(p)$. Thus, $N(p^k) = p^{3(k-1)}N(p)$ as desired.
(5) This follows readily from parts (1), (3) and (4).

We partition the set of all primes into

$$(4.10) \qquad P_G := \{p : p \nmid \Delta_{ij} \text{ for any } 1 \le i < j \le 5\} \quad \text{and} \quad P_B := \{p : p \notin P_G\}.$$

Note that 2 is always in P_B. We shall see in our estimations below that those $p \in P_G$ are "good" primes.

LEMMA 4.2. *(1) For any p, $-7p^{-2} < A(p) \le 6p^{-1}$.*
(2) If $p \in P_G$ then $|A(p)| \le 16p^{-3}$.
(3) $\prod_p (1 + |A(p)|) \ll \prod_p (1 + A(p)) \ll \log^6(\log N)$.
(4) $\prod_p s(p)$ converges absolutely and $\prod_p s(p) > c_6 > 0$.
(5) For any $y \ge 1$, $\sum_{q \ge y} |A(q)| \ll y^{-2}B^{21}\log^{16}(y+1)$.

PROOF: (1) By (4.6) and Lemma 4.1(3), for any p

$$(4.11) \qquad A(p) = p^2\phi(p)^{-5}N(p) - 1.$$

Using the trivial bound $\phi(p)^3$ for $N(p)$, the upper bound for $A(p)$ follows immediately. For the lower bound of $A(p)$ we shall prove that

$$(4.12) \qquad N(p) \ge \phi(p)^2(p-3) \quad \text{for} \quad p \ge 5.$$

Then by (4.11) we have

$$A(p) > -7p^{-2} \qquad \text{for any} \quad p \geq 2 \,,$$

since $A(2) = 3$ and, by (4.5), $A(3) \geq 3^2\phi(3)^{-5} - 1$. In view of (1.3), let $p \not| \Delta_{12}$, say. Same as in the proof of Lemma 4.1(4), $N(p)$ is the number of triples (ℓ_3, ℓ_4, ℓ_5) in the cube $[1, p-1]^3$ satisfying none of the following two conditions:

$$(\alpha): \quad \Delta_{1b} \equiv \sum_{j=3}^{5} \Delta_{1j}\ell_j \pmod p, \qquad (\beta): \quad \Delta_{2b} \equiv \sum_{j=3}^{5} \Delta_{2j}\ell_j \pmod p \,.$$

So, by a simple inclusion-exclusion argument,

$$N(p) = (p-1)^3 - M(\alpha) - M(\beta) + M(\alpha, \beta) \,, \tag{4.13}$$

where $M(\alpha)$ (respectively $M(\beta)$ and $M(\alpha,\beta)$) denotes the number of triples (ℓ_3, ℓ_4, ℓ_5) in the cube $[1, p-1]^3$ satisfying condition (α) (respectively (β) and both (α), (β)). Consider the condition (α). If $p|(\Delta_{13}, \Delta_{14}, \Delta_{15})$, in view of (4.5), (4.9), we have $p \not| \Delta_{1b}$ and hence $M(\alpha) = 0$. If $p \not| (\Delta_{13}, \Delta_{14}, \Delta_{15})$, then condition (α) admits at most $\phi(p)^2$ solutions (ℓ_3, ℓ_4, ℓ_5). In any case $M(\alpha) \leq \phi(p)^2$ and likewise, $M(\beta) \leq \phi(p)^2$. Hence, from (4.13) we deduce that $N(p) \geq (p-1)^3 - 2\phi(p)^2$ which is just (4.12).

(2) Let $p \in P_G$. For each pair h_1, h_2 with $(h_1, h_2, p) = 1$, p can divide at most one of the five $\underline{h}_j = a_{1j}h_1 + a_{2j}h_2$, $j = 1, \cdots, 5$, since $p|(\underline{h}_i, \underline{h}_j)$ implies $p|\Delta_{ij}$. Furthermore, for each $j = 1, \cdots, 5$, as p cannot divide both a_{1j} and a_{2j}, there are exactly $p-1$ pairs of h_1, h_2 satisfying $\underline{h}_j \equiv 0 \pmod p$ and $(h_1, h_2, p) = 1$. Hence, in view of the fact that (see (4.7))

$$C_p(m) = \begin{cases} -1 & \text{if } p \not| m \,, \\ \phi(p) & \text{if } p|m \,, \end{cases}$$

we deduce from (4.4),

$$\begin{aligned} |A(p)| &\leq \phi(p)^{-5} {\sum_{\underline{h}}}' \prod_{j=1}^{5} |C_p(\underline{h}_j)| \leq \phi(p)^{-5}\{(p^2-1) - 5(p-1) + 5(p-1)\phi(p)\} \\ &= \phi(p)^{-4}(6p-9) < 16p^{-3} \end{aligned}$$

as $2 \notin P_G$. This proves part(2). Parts (3), (4) and (5) can be deduced from part (1) and Lemma 4.1 by a similar argument as in Lemma 4.4 in [8].

For $j = 1, \cdots, 5$ let $\chi_j \pmod{r_j}$ be primitive characters. Denote the least common multiple of the r_j's by $[r_1, \cdots, r_5]$ and, when q is divisible by $[r_1, \cdots, r_5]$, define

$$Z(q) = Z(q;\ \chi_1, \chi_2, \cdots, \chi_5) := {\sum_{\underline{h}}}' e_q(-\underline{h}_b) \prod_{j=1}^{5} C_{\chi_j\chi_0}(\underline{h}_j) \,, \tag{4.14}$$

where χ_0 is modulo q and $\sum_{\underline{h}}'$ is over all h_1, h_2 satisfying (3.2).

LEMMA 4.3. *Let* $r = [r_1, \cdots, r_5]$.

(1) *We have* $Z(r) = r^2 \sum_{(r)} \prod_{j=1}^{5} \chi_j(\ell_j)$ *where the summation* $\sum_{(r)}$ *is defined in* (4.2).

(2) *Let $r|q$ and $q = q'q''$ such that $(r, q'') = 1$ and every prime factor of q' divides r. Then $Z(q) = Z(q')\phi(q'')^5 A(q'')$ and $Z(q') = 0$ if $q' > r$.*
(3) $\sum_{q \le Q,\ r|q} \phi(q)^{-5} Z(q) \ll \prod_p s(p)$.

PROOF: (1) From (4.1) we have

$$\sum_{1 \le h_1, h_2 \le r} e_r(-\underline{h}_b) \prod_{j=1}^{5} C_{\chi_j \chi_0}(\underline{h}_j) = r^2 \sum_{(r)} \prod_{j=1}^{5} \chi_j(\ell_j) .$$

Comparing this with (4.14) for $q = r$, we see that part (1) follows if the two summations $\sum'_{\underline{h}}$ and $\sum_{1 \le h_1, h_2 \le r}$ are the same. It is known that [4, p.450] for any integer m, if $\chi \pmod q$ is induced by a primitive character $\chi^* \pmod{r^*}$ and $q_1 = q/(q, m)$, then

$$C_\chi(m) = \begin{cases} \overline{\chi}^*(m/(q,m))\phi(q)\phi(q_1)^{-1}\mu(q_1/r^*)\chi^*(q_1/r^*)C_{\chi^*}(1), & \text{if } r^* | q_1 , \\ 0, & \text{if } r^* \nmid q_1 . \end{cases} \tag{4.15}$$

Suppose $(h_1, h_2, r) \ne 1$ and $p|(h_1, h_2, r)$. As $r = [r_1, \cdots, r_5]$, we may let $p \nmid r/r_1$, say. Then $r_1 \nmid r/(r, a_{11}h_1 + a_{21}h_2)$. By (4.15), $C_{\chi_1\chi_0}(\underline{h}_1) = 0$. Thus $\prod_{j=1}^{5} C_{\chi_j\chi_0}(\underline{h}_j) = 0$ whenever $(h_1, h_2, r) \ne 1$. This shows that the condition $(h_1, h_2, r) = 1$ in the summation $\sum'_{\underline{h}}$ is superfluous and hence part (1) follows.
Parts (2) and (3) can be proved by similar arguments as in Lemmas 4.5 and 4.6 in [8]. In the proof of part (3) we need parts (1), (2), Lemma 4.1(1), (5) and Lemma 4.2(3).

LEMMA 4.4. *For any complex numbers ρ_j $(j = 1, \cdots, 5)$ satisfying $0 < Re\, \rho_j \le 1$, we have*

$$\iint_{\mathbb{R}^2} \left(\prod_{j=1}^{5} \int_L^N x^{\rho_j - 1} e(\underline{\eta}_j x) dx \right) e(-\underline{\eta}_b) d\eta_1 d\eta_2$$
$$= N^3 |\Delta_{45}|^{-1} \int_D \prod_{j=1}^{5} (N x_j)^{\rho_j - 1} dx_1 dx_2 dx_3 , \tag{4.16}$$

where

$$x_4 = f_4(x_1, x_2, x_3) := \Delta_{45}^{-1}(\Delta_{51}x_1 + \Delta_{52}x_2 + \Delta_{53}x_3 + \Delta_{b5}N^{-1}) ,$$
$$x_5 = f_5(x_1, x_2, x_3) := \Delta_{54}^{-1}(\Delta_{41}x_1 + \Delta_{42}x_2 + \Delta_{43}x_3 + \Delta_{b4}N^{-1})$$

and

$$D := \{(x_1, x_2, x_3) : LN^{-1} \le x_1, x_2, x_3, f_4(x_1, x_2, x_3), f_5(x_1, x_2, x_3) \le 1\} . \tag{4.17}$$

Furthermore,

$$|\Delta_{45}|^{-1} \int_D dx_1 dx_2 dx_3 \gg B^{-9} . \tag{4.18}$$

PROOF: Equality (4.16) can be proved by a similar argument as in Lemma 4.7 in [8]. For (4.18) we may assume, WLOG, that $\Delta_{45} > 0$. The conditions $LN^{-1} \le f_4, f_5 \le 1$ in the domain $\mathcal{D}$ are equivalent to the inequalities

$$\Delta_{45}LN^{-1} - \Delta_{b5}N^{-1} \le \Delta_{51}x_1 + \Delta_{52}x_2 + \Delta_{53}x_3 \le \Delta_{45} - \Delta_{b5}N^{-1} ,$$
$$\Delta_{54} - \Delta_{b4}N^{-1} \le \Delta_{41}x_1 + \Delta_{42}x_2 + \Delta_{43}x_3 \le \Delta_{54}LN^{-1} - \Delta_{b4}N^{-1} .$$

Dropping the terms $\Delta_{b5}N^{-1}$ and $\Delta_{b4}N^{-1}$ from these inequalities causes an error in the integral (4.18) that is $\ll (|\Delta_{b4}| + |\Delta_{b5}|)N^{-1} \le (|b_1| + |b_2|)BN^{-1}$. In view of (2.1) this is negligible compared with the lower bound in (4.18). So, we estimate the volume of the solid in $\mathbb{R}^3$ defined by:

$$\text{(4.19)} \qquad \begin{cases} LN^{-1} \le x_1, x_2, x_3 \le 1 , \\ \Delta_{45}LN^{-1} \le \Delta_{51}x_1 + \Delta_{52}x_2 + \Delta_{53}x_3 \le \Delta_{45} , \\ \Delta_{54} \le \Delta_{41}x_1 + \Delta_{42}x_2 + \Delta_{43}x_3 \le \Delta_{54}LN^{-1} . \end{cases}$$

In view of condition (1.5), by relabelling the subscripts of the coefficients $a_{\lambda j}$ and suitably scaling the variables y_j, we see that the system $\sum_{j=1}^{5} a_{\lambda j}y_j = 0$, $\lambda = 1,2$ has a solution with $1/2 = y_4 \ge y_5 \ge y_1, y_2, y_3 > 0$. From the first equation of this system (use the second equation if $a_{14} = 0$), we see that

$$1/2 \le |a_{14}y_4| \le |a_{15}|y_5 + |a_{11}|y_1 + |a_{12}|y_2 + |a_{13}|y_3 \le 4By_5 .$$

Thus, $y_5 \ge (8B)^{-1}$. Let $x_j = y_j + \theta_j$ for $j = 1,2,3$. Then

$$(\Delta_{51}x_1 + \Delta_{52}x_2 + \Delta_{53}x_3) = \frac{1}{2}\Delta_{45} + (\Delta_{51}\theta_1 + \Delta_{52}\theta_2 + \Delta_{53}\theta_3) ,$$
$$(\Delta_{41}x_1 + \Delta_{42}x_2 + \Delta_{43}x_3) = y_5\Delta_{54} + (\Delta_{41}\theta_1 + \Delta_{42}\theta_2 + \Delta_{43}\theta_3) .$$

For any θ_1, θ_2, $\theta_3 \in [LN^{-1}, \Delta_{45}(96B^3)^{-1}]$, in view of (2.2) and (2.3), we verify that (4.19) is satisfied. Thus, the solid defined by (4.19) has a volume $\gg (\Delta_{45}B^{-3})^3$. This proves Lemma 4.4.

5. Major Arcs

For any $(x_1, x_2) \in \mathcal{M}(h_1, h_2, q)$, we write $\eta_\lambda := x_\lambda - h_\lambda q^{-1}$ for $\lambda = 1,2$, and

$$\underline{\eta}_b := b_1\eta_1 + b_2\eta_2, \quad \underline{\eta}_j := a_{1j}\eta_1 + a_{2j}\eta_2 \qquad \text{for } j = 1, \cdots, 5 .$$

Further, we let

$$\text{(5.1)} \qquad G_j(\underline{h}, q, \underline{\eta}) := \sum_{\chi (\operatorname{mod} q)} C_{\overline{\chi}}(\underline{h}_j) I_\chi(\underline{\eta}_j) ,$$

$$\text{(5.2)} \qquad H_j(\underline{h}, q, \underline{\eta}) := C_q(\underline{h}_j) I(\underline{\eta}_j) - \delta_q C_{\tilde{\chi}\chi_0}(\underline{h}_j) \tilde{I}(\underline{\eta}_j) - G_j(\underline{h}, q, \underline{\eta}) ,$$

for $j = 1, 2, \cdots, 5$, where $\delta_q = 1$ if $\tilde{r}|q$ and $\delta_q = 0$ otherwise. Following similar arguments as used in §3 of [8], we can simplify the expression for $I_1(b)$ given in (3.4) and obtain

$$(5.3)\quad I_1(\underline{b}) = \sum_{q \le Q} \phi(q)^{-5} \sum_{\underline{h}}{}' e_q(-\underline{h}_b) \iint_{\mathbb{R}^2} e(-\underline{\eta}_b) \prod_{j=1}^{5} H_j(\underline{h}, q, \underline{\eta}) d\eta_1 d\eta_2 + O(N^3 T^{-1/9}) .$$

Note that the Fundamental Lemma 2.1 plays an essential role in the simplification mentioned above. Totally there are 243 terms in the product $\prod_{j=1}^{5} H_j(\underline{h}, q, \underline{\eta})$. We classify them into three categories: (T_1): the term $\prod_{j=1}^{5} C_q(\underline{h}_j) I(\underline{\eta}_j)$; (T_2): 211 terms, each has at least one $G_j(\underline{h}, q, \underline{\eta})$ as factor; (T_3): the remaining 31 terms.

For $i = 1, 2, 3$ define

$$(5.4)\quad M_i := \sum_{q \le Q} \phi(q)^{-5} \sum_{\underline{h}}{}' e_q(-\underline{h}_b) \iint_{\mathbb{R}^2} e(-\underline{\eta}_b) \,\{\text{sum of all the terms in } (T_i)\} d\eta_1 d\eta_2 .$$

Note that each M_i is real. In view of (5.3) we have

$$(5.5)\qquad I_1(\underline{b}) = M_1 + M_2 + M_3 + O(N^3 T^{-1/9}) .$$

Put

$$(5.6)\qquad M_0 := N^3 |\Delta_{45}|^{-1} \Big(\prod_p s(p) \Big) \int_D dx_1 dx_2 dx_3 .$$

LEMMA 5.1. *We have* $M_1 = M_0 + O(N^3 Q^{-1})$.

PROOF: From (5.4), (3.1), Lemma 4.4 with $\rho_j = 1$ and (4.4) we have

$$\begin{aligned} M_1 &= \sum_{q \le Q} \phi(q)^{-5} \sum_{\underline{h}}{}' e_q(-\underline{h}_b) \prod_{j=1}^{5} C_q(\underline{h}_j) \iint_{\mathbb{R}^2} \Big(\prod_{j=1}^{5} \int_L^N e(\underline{\eta}_j x) dx \Big) e(-\underline{\eta}_b) d\eta_1 d\eta_2 \\ &= N^3 |\Delta_{45}|^{-1} \Big(\int_D dx_1 dx_2 dx_3 \Big) \sum_{q=1}^{\infty} A(q) + O(N^3 Q^{-2} B^{21} \log^{16} Q) . \end{aligned}$$

The last O-term is obtained by Lemma 4.2(5) and the fact that $\int_D dx_1 dx_2 dx_3 \le 1$. The above main term is M_0 since by Lemma 4.1(1), (2) and (4.6) we have

$$\sum_{q=1}^{\infty} A(q) = \prod_p (1 + A(p)) = \prod_p s(p) .$$

This proves Lemma 5.1.

For distinct $m_1, m_2, \cdots$ taken from the set $\{1, 2, \cdots, 5\}$ we define

$$\mathcal{G}(m_1, m_2, \cdots) := \sum_{(\tilde{r})} \tilde{\chi}(\ell_{m_1})\tilde{\chi}(\ell_{m_2}) \cdots \tag{5.7}$$

where $\sum_{(\tilde{r})}$ is defined in (4.2), and

$$\mathcal{P}(m_1, m_2, \cdots) := \int_{\mathcal{D}} (Nx_{m_1})^{\tilde{\beta}-1}(Nx_{m_2})^{\tilde{\beta}-1} \cdots dx_1 dx_2 dx_3 \tag{5.8}$$

where $\mathcal{D}$ is defined in (4.17). In view of (4.3) and (5.7) we have

$$|\mathcal{G}(m_1, m_2, \cdots)| \leq N(\tilde{r}) \leq \phi(\tilde{r})^3 . \tag{5.9}$$

When $\tilde{r}^2 Q^{-1}$ is small the following lemma gives a good lower estimate.

LEMMA 5.2. *If the exceptional zero $\tilde{\beta}$ exists, then*

$$M_1 + M_3 \geq \Omega^5 M_0 - O(N^3 Q^{-1} \tilde{r}^2) ,$$

where Ω and M_0 are defined in (2.6) and (5.6) respectively.

PROOF: In view of (5.1), (5.2) the 31 terms in (T_3) can be grouped to the following five types of product

$$W_m := (-1)^m \delta_q \prod_{j=1}^{m} C_{\tilde{\chi}\chi_0}(\underline{h}_j)\tilde{I}(\underline{\eta}_j) \prod_{j=m+1}^{5} C_q(\underline{h}_j) I(\underline{\eta}_j)$$

where $m = 1, \cdots, 5$. As usual, an empty product is taken as one. Then, in view of (5.4), M_3 is a sum of 31 terms of the following five types

$$M_{3,m} := \sum_{q \leq Q} \phi(q)^{-5} \sum_{\underline{h}}{}' e_q(-\underline{h}_b) \iint_{\mathbb{R}^2} W_m e(-\underline{\eta}_b) d\eta_1 d\eta_2, \quad m = 1, \cdots, 5 .$$

Substituting from the above,

$$M_{3,m} = (-1)^m \Big\{ \sum_{q \leq Q} \delta_q \phi(q)^{-5} \sum_{\underline{h}}{}' e_q(-\underline{h}_b) \prod_{j=1}^{m} C_{\tilde{\chi}\chi_0}(\underline{h}_j) \prod_{j=m+1}^{5} C_q(\underline{h}_j) \Big\} \times$$
$$\times \Big\{ \iint_{\mathbb{R}^2} e(-\underline{\eta}_b) \prod_{j=1}^{m} \tilde{I}(\underline{\eta}_j) \prod_{j=m+1}^{5} I(\underline{\eta}_j) d\eta_1 d\eta_2 \Big\}$$

$$=: (-1)^m X_m Y_m , \text{ say} . \tag{5.10}$$

Now by (3.1), Lemma 4.4 with $\rho_j = \tilde{\beta}$ or 1 and (5.8) we have

$$Y_m = N^3 |\Delta_{45}|^{-1} \int_{\mathcal{D}} \prod_{j=1}^{m} (Nx_j)^{\tilde{\beta}-1} dx_1 dx_2 dx_3 = N^3 |\Delta_{45}|^{-1} \mathcal{P}(1, \cdots, m) . \tag{5.11}$$

Next, by (5.10), (4.14) with $r_1 = \cdots = r_m = \tilde{r}$, $r_{m+1} = \cdots = r_5 = 1$, and Lemma 4.3(2), (1), i.e. $Z(\tilde{r}) = \tilde{r}^2 \mathcal{G}(1,\cdots,m)$, we have

$$X_m = \sum_{q\le Q} \delta_q \phi(q)^{-5} Z(q) = Z(\tilde{r})\phi(\tilde{r})^{-5} \sum_{q\le Q/\tilde{r},\ (q,\tilde{r})=1} A(q)$$

$$= \tilde{r}^2 \mathcal{G}(1,\cdots,m)\phi(\tilde{r})^{-5} \sum_{(q,\tilde{r})=1} A(q) + O(\tilde{r}^2 Q^{-1}) \ . \tag{5.12}$$

The last O-term follows, since by (5.9), Lemma 4.2(5), (3) and Lemma 4.1(5)

$$\tilde{r}^2 \mathcal{G}(1,\cdots,m)\phi(\tilde{r})^{-5} \sum_{q>Q/\tilde{r}} A(q) \ll \tilde{r}^2 N(\tilde{r})\phi(\tilde{r})^{-5}(\tilde{r}/Q)^2 B^{21}\log^{16} Q$$

$$= \Big(\prod_{p|\tilde{r}} s(p)\Big)\tilde{r}^2 Q^{-2} B^{21}\log^{16} Q \ll \log^6(\log Q)\tilde{r}^2 Q^{-2} B^{21} \log^{16} Q \ .$$

Furthermore, by Lemma 4.1(1), (2) and (4.6)

$$\sum_{(q,\tilde{r})=1} A(q) = \prod_{p\nmid \tilde{r}} (1 + A(p)) = \prod_{p\nmid \tilde{r}} s(p) \ . \tag{5.13}$$

Collecting the results in (5.10) – (5.13), and noting that $|\mathcal{P}(1,\cdots,m)| \le 1$, we have

$$M_{3,m} = (-1)^m N^3 |\Delta_{45}|^{-1} \tilde{r}^2 \phi(\tilde{r})^{-5} \Big(\prod_{p\nmid \tilde{r}} s(p)\Big) \mathcal{G}(1,\cdots,m)\mathcal{P}(1,\cdots,m)$$

$$+ O(N^3 \tilde{r}^2 Q^{-1}) \ . \tag{5.14}$$

As given in Lemma 5.1 and (5.6), the main term M_1 has a similar form, namely,

$$M_1 = N^3|\Delta_{45}|^{-1}\Big(\prod_{p\nmid \tilde{r}} s(p)\Big)\tilde{r}^2\phi(\tilde{r})^{-5}\sum_{(\tilde{r})}\int_D dx_1dx_2dx_3 + O(N^3Q^{-1})$$

since by Lemma 4.1(5) and (4.3)

$$\prod_{p|\tilde{r}} s(p) = \tilde{r}^2 \phi(\tilde{r})^{-5} \sum_{(\tilde{r})} 1 \ . \tag{5.15}$$

Now, adding all the contributions of the form $M_{3,m}$ in M_3 we get, by (5.14),

$$M_1 + M_3 = N^3|\Delta_{45}|^{-1}\Big(\prod_{p\nmid \tilde{r}} s(p)\Big)\ \tilde{r}^2\phi(\tilde{r})^{-5}\Big\{\sum_{(\tilde{r})}\int_D dx_1dx_2dx_3 - \sum_{j=1}^{5} \mathcal{G}(j)\mathcal{P}(j)$$

$$+ \sum_{1\le i<j\le 5} \mathcal{G}(i,j)\mathcal{P}(i,j) - \sum_{1\le i<j<k\le 5} \mathcal{G}(i,j,k)\mathcal{P}(i,j,k)$$

$$+ \sum_{1\le i<j<k<\ell\le 5} \mathcal{G}(i,j,k,\ell)\mathcal{P}(i,j,k,\ell)$$

$$- \mathcal{G}(1,2,3,4,5)\mathcal{P}(1,2,3,4,5)\Big\} + O(N^3\tilde{r}^2Q^{-1}) \ . \tag{5.16}$$

By (5.7) and (5.8) the whole expression inside the above braces is equal to

$$\sum_{(\tilde r)}\int_D\prod_{j=1}^{5}(1-\tilde\chi(\ell_j)(Nx_j)^{\tilde\beta-1})dx_1dx_2dx_3 \ge \sum_{(\tilde r)}\int_D\prod_{j=1}^{5}(1-(Nx_j)^{\tilde\beta-1})dx_1dx_2dx_3$$

$$\ge \sum_{(\tilde r)}(1-L^{\tilde\beta-1})^5\int_D dx_1dx_2dx_3 \tag{5.17}$$

since by (4.17) we have $Nx_j \ge L$. Now

$$1-L^{\tilde\beta-1}\ge 1-\exp\{-\frac{1}{2}(1-\tilde\beta)\log N\}\ge(1-\tilde\beta)\log T=\Omega$$

if the δ in (2.2) is small. Thus, by applying (5.17) in (5.16), we get

$$M_1+M_3\ge N^3|\Delta_{45}|^{-1}\Big(\prod_{p\nmid \tilde r}s(p)\Big)(\tilde r^2\phi(\tilde r)^{-5}\sum_{(\tilde r)}1)\Omega^5\int_D dx_1dx_2dx_3-O(N^3\tilde r^2Q^{-1})\,.$$

In view of (5.6), Lemma 5.2 follows from this inequality and (5.15).

The lower bound for M_1+M_3 in Lemma 5.2 is useful only when $\tilde r^2Q^{-1}$ is small. When it is not so small, we need a different estimation (see Lemma 5.4 below).

LEMMA 5.3. *We have* $\tilde r^2\phi(\tilde r)^{-5}\mathcal{G}(m_1,m_2,\cdots)\ll B^{10}\tilde r^{-1/2}(\log\,\log\tilde r)^3$.

PROOF: It is a well-known property of real primitive characters that the modulus $\tilde r$ of $\tilde\chi$ is of the form $\tilde r=\nu_1\cdots\nu_k$ where $\nu_1=2^t$, $t=0$ or 2 or 3 and $\nu_2<\cdots<\nu_k$ are primes. For $j=1,\cdots,k$ let ψ_j be the real primitive character modulo ν_j. Then

$$\tilde\chi(\text{mod }\tilde r)=\prod_{j=1}^{k}\psi_j(\text{mod }\nu_j)\,. \tag{5.18}$$

For distinct $m_1,m_2,\cdots$ taken from the set $\{1,2,\cdots,5\}$ let

$$R_j(m_1,m_2,\cdots):=$$

$$\sum_{1\le h_1,h_2\le\nu_j}e_{\nu_j}(-\underline{h}_b)C_{\psi_j}(\underline{h}_{m_1})C_{\psi_j}(\underline{h}_{m_2})\cdots\prod_{\substack{\ell=1\\ \ell\ne m_1,m_2,\cdots}}^{5}C_{\nu_j}(\underline{h}_\ell)\,. \tag{5.19}$$

For example,

$$R_j(2,3)=\sum_{1\le h_1,h_2\le\nu_j}e_{\nu_j}(-\underline{h}_b)C_{\psi_j}(\underline{h}_2)C_{\psi_j}(\underline{h}_3)C_{\nu_j}(\underline{h}_1)C_{\nu_j}(\underline{h}_4)C_{\nu_j}(\underline{h}_5)\,.$$

First of all, from (5.19), (4.1) and (4.2), for any ν_j, we have $R_j(m_1,m_2,\cdots)=\nu_j^2\sum_{(\nu_j)}\psi_j(\ell_{m_1})\psi_j(\ell_{m_2})\cdots$ so that

$$|R_j(m_1,m_2,\cdots)|\le\nu_j^2N(\nu_j)\le\nu_j^2\phi(\nu_j)^3\,. \tag{5.20}$$

This bound can be improved considerably if $\nu_j\in P_G$. For simplicity, let us drop the subscript j from ν_j, ψ_j and R_j whenever there is no ambiguity. So $\nu(=\nu_j)$ is a prime in P_G and ψ is a primitive character modulo ν. It is well-known that [3, p.65-66], $C_\psi(m) = \psi(m)C_\psi(1)$ and $|C_\psi(1)| = \sqrt{\nu}$. Thus

$$|C_\psi(\underline{h}_{m_i})| = \begin{cases} 0, & \text{if } \nu | \underline{h}_{m_i} , \\ \sqrt{\nu}, & \text{otherwise} . \end{cases}$$

From this, we can give an upper bound to each term in the sum for $R(m_1, m_2, \cdots)$ on the right side of (5.19). Following essentially the same argument used in the proof of Lemma 4.2(2), we have

$$\begin{aligned} |R(m_1,m_2,\cdots)| &\le \sum_{\substack{1\le h_1,h_2\le \nu \\ (h_1,h_2,\nu)=1}} |C_\psi(\underline{h}_{m_1})|\,|C_\psi(\underline{h}_{m_2})|\cdots \prod_{\substack{\ell=1 \\ \ell\ne m_1,m_2,\cdots}}^{5} |C_\nu(\underline{h}_\ell)| \\ &= (\nu^2-1-5(\nu-1))\nu^{n/2} + (5-n)(\nu-1)\phi(\nu)\nu^{n/2} , \end{aligned}$$

where n is the number of $m_1, m_2, \cdots$. A simple computation shows that this bound is $\le \nu^{5/2}(\nu-1)^2$ for $n = 0, 1, \cdots, 5$. Thus

$$|R(m_1,m_2,\cdots)| \le \nu^{5/2}\phi(\nu)^2 . \tag{5.21}$$

Now, by (5.7) and (4.1),

$$\tilde{r}^2\mathcal{G}(m_1,m_2,\cdots) = \sum_{1\le h_1,h_2\le \tilde{r}} e_{\tilde{r}}(-\underline{h}_b)C_{\tilde{\chi}}(\underline{h}_{m_1})C_{\tilde{\chi}}(\underline{h}_{m_2})\cdots \prod_{\substack{\ell=1 \\ \ell\ne m_1,m_2,\cdots}}^{5} C_{\tilde{r}}(\underline{h}_\ell) . \tag{5.22}$$

By a similar argument as in the proof of Lemma 4.1 in [8] (see also [6, p.100]), we can factorize, according to (5.18), the above $C_{\tilde{\chi}}(\underline{h}_{m_i})$, $C_{\tilde{r}}(\underline{h}_\ell)$ as well as the whole sum on the right side of (5.22). More precisely, we have

$$\tilde{r}^2\mathcal{G}(m_1,m_2,\cdots) = \prod_{j=1}^{k} R_j(m_1,m_2,\cdots) .$$

Applying to this product the bound in (5.20) for $\nu_j \in P_B$ and that in (5.21) for $\nu_j \in P_G$, we have

$$\begin{aligned} |\tilde{r}^2\mathcal{G}(m_1,m_2,\cdots)| &\le \prod_{\nu\in P_B} \nu^2\phi(\nu)^3 \prod_{\nu\in P_G} \nu^{5/2}\phi(\nu)^2 \le \tilde{r}^{5/2}\phi(\tilde{r})^2 \prod_{\nu\in P_B} \nu^{1/2} \\ &\ll \tilde{r}^{5/2}\phi(\tilde{r})^2 B^{10} , \end{aligned} \tag{5.23}$$

since by (4.10) $\prod_{\nu\in P_B} \nu \le \prod_{1\le i<j\le 5} |\Delta_{ij}| \le (2B^2)^{10}$. Then in view of the estimate $\tilde{r} \ll \phi(\tilde{r}) \log\log \tilde{r}$, our lemma follows from (5.23) immediately.

We come now to obtain a lower bound for $M_1 + M_3$ which is useful when $\tilde{r}^2Q^{-1}$ is not small.

LEMMA 5.4. *If the exceptional zero $\tilde{\beta}$ exists then*

$$M_1 + M_3 = M_0 + O(N^3Q^{-1} + N^3B^{10}\tilde{r}^{-1/2}\log N) .$$

PROOF: In the proof of Lemma 5.2 we have shown in (5.10) – (5.12) that $M_{3,m} = (-1)^m X_m Y_m$, $Y_m = N^3|\Delta_{45}|^{-1}\mathcal{P}(1,\cdots,m)$ and $X_m = \tilde{r}^2\phi(\tilde{r})^{-5}\mathcal{G}(1,\cdots,m)\sum_{q\le Q/\tilde{r},\ (q,\tilde{r})=1} A(q)$ where $m=1,\cdots,5$. Then (cf. (5.16))

$$\begin{aligned} M_3 =& N^3\tilde{r}^2(|\Delta_{45}|\phi(\tilde{r})^5)^{-1}\Bigg(\sum_{q\le Q/\tilde{r},\ (\tilde{r},q)=1} A(q)\Bigg)\Bigg\{-\sum_{j=1}^{5}\mathcal{G}(j)\mathcal{P}(j) \\ &+\sum_{1\le i<j\le 5}\mathcal{G}(i,j)\mathcal{P}(i,j)\cdots-\mathcal{G}(1,2,3,4,5)\mathcal{P}(1,2,3,4,5)\Bigg\}. \end{aligned} \tag{5.24}$$

By Lemma 4.1(1) and Lemma 4.2(3) we have

$$\Bigg|\sum_{q\le Q/\tilde{r},\ (\tilde{r},q)=1} A(q)\Bigg| \le \prod_p (1+|A(p)|) \ll \log^6(\log N)\ .$$

So, by Lemma 5.3 and (5.24), $M_3 \ll N^3\tilde{r}^{-1/2}B^{10}\log N$. This together with Lemma 5.1 proves Lemma 5.4.

6. Major Arcs – An Application of the Fundamental Lemma

LEMMA 6.1. *We have*

$$M_2 \ll M_0\Omega^5\exp(-c_4/\delta)$$

where M_2, Ω and M_0 are defined in (5.4), (2.6) and (5.6) respectively.

PROOF: In view of (5.2), each of the 211 terms in (T_2) can be represented by one of the following products

$$(-1)^m\prod_{j=1}^{\ell} G_j(\underline{h},q,\underline{\eta})\prod_{j=\ell+1}^{m}\delta_q C_{\tilde{\chi}\chi_0}(\underline{h}_j)\tilde{I}(\underline{\eta}_j)\prod_{j=m+1}^{5} C_q(\underline{h}_j)I(\underline{\eta}_j) \tag{6.1}$$

where $1\le \ell\le m\le 5$. We shall use Lemma 2.1 to estimate the contribution of these products to M_2 defined in (5.4). For clarity we take $\ell=2$, $m=4$ in (6.1) to illustrate our techniques. The same arguments can be applied to other ℓ and m. Accordingly, in view of (5.4), (5.1) and (6.1) for $\ell=2$, $m=4$ we define

$$\begin{aligned} M_2(2,4) :=& \sum_{q\le Q,\ \tilde{r}|q}\phi(q)^{-5}\sum_{\underline{h}}{}' e_q(-\underline{h}_b)C_q(\underline{h}_5)C_{\tilde{\chi}\chi_0}(\underline{h}_4)C_{\tilde{\chi}\chi_0}(\underline{h}_3)\sum_{\chi_2(\mathrm{mod}\,q)}\times \\ &\times\sum_{\chi_1(\mathrm{mod}\,q)} C_{\overline{\chi}_2}(\underline{h}_2)C_{\overline{\chi}_1}(\underline{h}_1)\iint_{\mathbf{R}^2} e(-\underline{\eta}_b)I(\underline{\eta}_5)\tilde{I}(\underline{\eta}_4)\tilde{I}(\underline{\eta}_3)I_{\chi_2}(\underline{\eta}_2)I_{\chi_1}(\underline{\eta}_1)d\eta_1 d\eta_2\ . \end{aligned} \tag{6.2}$$

By (3.1) the double integral in (6.2) is equal to

$$\sum_{|\gamma_1|\le T}{}'\sum_{|\gamma_2|\le T}{}'\iint_{\mathbf{R}^2} e(-\underline{\eta}_b)\prod_{j=1}^{5}\left(\int_L^N x^{\rho_j-1}e(x\underline{\eta}_j)dx\right)d\eta_1 d\eta_2 \tag{6.3}$$

where $\rho_5=1$, $\rho_4=\rho_3=\tilde{\beta}$ and for $j=1,2$, ρ_j are the zeros of $L(s,\chi_j)$ with

Im $\rho_j = \gamma_j$. Applying Lemma 4.4 to the double integral in (6.3), we see that $M_2(2,4)$ in (6.2) becomes

$$N^3|\Delta_{45}|^{-1}\int_D (N^2x_3x_4)^{\tilde\beta-1}\sum_{q\le Q,\ \tilde r|q}\phi(q)^{-5}\sum_{\underline{h}}{}' e_q(-\underline{h}_b)C_q(\underline{h}_5)C_{\tilde\chi\chi_0}(\underline{h}_4)C_{\tilde\chi\chi_0}(\underline{h}_3)\times$$

$$\times\prod_{j=1}^{2}\Big(\sum_{\chi_j(\bmod q)}\ \sum_{|\gamma_j|\le T}{}'(Nx_j)^{\rho_j-1}C_{\overline{\chi}_j}(\underline{h}_j)\Big)dx_1dx_2dx_3\ . \tag{6.4}$$

Note that each character is induced by a unique primitive character and conversely, for each primitive character modulo r and for each q with $r|q$, there is a unique character modulo q induced by that primitive character. Furthermore, if χ^* (mod r^*) is primitive and χ_0 (mod q) is principal with $r^*|q$, then the L-functions $L(s,\chi^*)$ and $L(s,\chi^*\chi_0)$ have the same set of non-trivial zeros with positive real parts. Therefore, the summations $\sum\limits_{q\le Q,\ \tilde r|q}\prod\limits_{j=1}^{2}\sum\limits_{\chi_j\ (\bmod q)}\sum\limits_{|\gamma_j|\le T}{}'$ in (6.4) can be written as

$$\prod_{j=1}^{2}\Big(\sum_{r_j\le Q}\ \sum_{\chi_j(\bmod r_j)}{}^{*}\ \sum_{|\gamma_j|\le T}{}'\Big)\ \sum_{q\le Q,\ [\tilde r,r_1,r_2]|q}$$

where the summations $\sum^*$ are taken over all primitive characters χ_j (mod r_j). By (4.14) and the above rearrangement, (6.4) becomes

$$M_2(2,4) = N^3|\Delta_{45}|^{-1}\int_D (N^2x_3x_4)^{\tilde\beta-1}\Big(\prod_{j=1}^{2}\Big(\sum_{r_j\le Q}\ \sum_{\chi_j(\bmod r_j)}{}^{*}\ \sum_{|\gamma_j|\le T}{}'(Nx_j)^{\rho_j-1}\Big)\times$$

$$\times\sum_{q\le Q,\ r|q}\phi(q)^{-5}Z(q;\chi_1,\chi_2,\tilde\chi,\tilde\chi,1)\Big)dx_1dx_2dx_3\ .$$

Hence, by Lemma 4.3(3), $M_2(2,4)$ is bounded by

$$N^3|\Delta_{45}|^{-1}\Big(\prod_p s(p)\Big)\int_D (N^2x_3x_4)^{\tilde\beta-1}\prod_{j=1}^{2}\Big(\sum_{r_j\le Q}\ \sum_{\chi_j(\bmod r_j)}{}^{*}\ \sum_{|\gamma_j|\le T}{}'(Nx_j)^{\beta_j-1}\Big)dx_1dx_2dx_3$$

where $\beta_j = \mathrm{Re}\ \rho_j$. Finally, applying Lemma 2.1 with $y = Nx_j \ge L > \sqrt{N}$ for $j = 1,2$, we have

$$M_2(2,4) \ll N^3|\Delta_{45}|^{-1}\Big(\prod_p s(p)\Big)(\Omega^5\exp(-c_4/\delta))^2\int_D dx_1dx_2dx_3\ .$$

This together with (5.6) proves Lemma 6.1.

LEMMA 6.2. (1) *If the exceptional zero $\tilde\beta$ does not exist or if $\tilde\beta$ exists and $\tilde r \ge Q^{1/18}$, then $I_1(\underline{b}) \gg N^3B^{-9}$.*

(2) *If $\tilde\beta$ exists and $\tilde r < Q^{1/18}$, then $I_1(\underline{b}) \gg N^3Q^{-1/6}$.*

PROOF: By (5.6), Lemma 4.2(4) and (4.18) we have

$$M_0 \gg N^3 B^{-9} . \tag{6.5}$$

(1) If $\tilde{\beta}$ does not exist, then the δ_q in (5.2) is zero and by (5.4), there is no M_3. Using (5.5), Lemma 5.1, Lemma 6.1 and (6.5) we have

$$I_1(\underline{b}) = M_1 + M_2 + O(N^3 T^{-1/9}) \gg M_0(1 - c_7 \exp(-c_4/\delta)) - O(N^3 Q^{-1}) \gg N^3 B^{-9} .$$

If $\tilde{\beta}$ exists and $\tilde{r} \geq Q^{1/18}$ then by (5.5), Lemma 5.4, Lemma 6.1 and (6.5) we have

$$I_1(\underline{b}) \gg M_0(1 - c_7 \exp(-c_4/\delta)) - O(N^3 Q^{-1} + N^3 B^{10} \tilde{r}^{-1/2} \log N) \gg N^3 B^{-9} ,$$

since by (2.2), $B^{10} \tilde{r}^{-1/2} \log N = O(Q^{-1/37})$.

(2) If $\tilde{r} < Q^{1/18}$, then by (2.6), (2.4) and $T = Q^{c_1}$, we have

$$\Omega^5 \gg Q^{-1/7} \log^{-5} Q .$$

Hence, combining (5.5), Lemma 5.2, Lemma 6.1 and (6.5),

$$\begin{aligned} I_1(\underline{b}) &\geq M_0 \Omega^5 (1 - c_7 \exp(-c_4/\delta)) - O(N^3 Q^{-1} \tilde{r}^2 + N^3 T^{-1/9}) \\ &\gg N^3 B^{-9} Q^{-1/7} \log^{-5} Q \gg N^3 Q^{-1/6} , \end{aligned}$$

since $T^{-1/9}, Q^{-1} \tilde{r}^2 < Q^{-8/9}$. This completes the proof of Lemma 6.2.

In view of the remark given immediately after (3.4), Theorem 1 is proved.

7. Appendix – Criteria for Congruent Solubility and Positive Solubility

Let S denote the set $\{b, 1, 2, 3, 4, 5\}$ and let $N(p)$ be defined by (4.3). We come now to give simple criteria for the solubility of the congruences (1.4) and equations (1.5) in terms of the Δ_{ij}'s. The proofs of these are quite elementary and we leave them to the interested readers.

PROPOSITION 7.1. (1) *Let $p \geq 5$. $N(p) = 0$ if and only if*

(7.1) *there exists i with $1 \leq i \leq 5$ such that for all $j, k \in S \setminus \{i\}$ we have* $\Delta_{jk} \equiv 0 \pmod p$.

(2) *$N(3) = 0$ if and only if either the statement in (7.1) holds for $p = 3$ or there exists a permutation π of S such that*

$$\begin{aligned} &\Delta_{1'5'} \equiv \Delta_{1'b'} \equiv \Delta_{2'5'} \equiv \Delta_{2'b'} \equiv 0 \pmod 3 \quad \textit{and} \\ &\Delta_{1'3'} \Delta_{1'4'} \Delta_{2'3'} \Delta_{2'4'} \equiv -1 \pmod 3 \quad \textit{where} \quad j' = \pi(j) . \end{aligned}$$

PROPOSITION 7.2. *The system (1.5) is soluble in positive real y_j, $j = 1, \cdots, 5$ if and only if there are $i, j, k \in S \setminus \{b\}$ such that Δ_{ij}, Δ_{jk}, Δ_{ki} are all positive.*

References

[1] Baker, A., On some diophantine inequalities involving primes, J. reine angew. Math. **228** (1967), 166-181.

[2] Van der Corput, J.G., Propriéte's additives I, Acta Arith. **3** (1939), 180-234.

[3] Davenport, H., Multiplicative Number Theory, 2nd ed. Springer-Verlag, Graduate Text in Mathematics, Vol.74 (1980).

[4] Hasse, H., Vorlesungen über Zahlentheorie, Grundlehren Math. Wiss. Band 59, Springer-Verlag (1964).

[5] Hua, L.K., Die Abschätzung von Exponentialsummen und ihre Angewendung in der Zahlentheorie, Enzykl. Math. Wiss. Band I, Teil 2, Heft 13, Leipzig (1959) (Chinese ed. Beijing, 1963).

[6] Hua, L.K., Additive Theory of Prime Numbers, Translations of Mathematical Monographs, Vol. 13, Amer. Math. Soc., Providence, R.I. 1965.

[7] Gallagher, P.X., A large sieve density estimate near $\sigma = 1$, Invent. Math. **11** (1970), 329-339.

[8] Liu, M.C. and Tsang, K.M., Small prime solutions of linear equations, Proceedings of the 1987 Laval University International Number Theory Conference, 595-624.

[9] Selberg, A., Remarks on sieves, Proceedings of the 1972 Number Theory Conference, University of Colorado, Boulder, 205-216.

[10] Wu, F., On the solutions of the systems of linear equations with prime variables, Acta Math. Sinica, **7** (1957) 102-122 (in Chinese with English summary).

Hecke Operator and Pellian Equation Conjecture*

—— In Memory of My Teacher Late Professor Hua Loo Keng

Lu Hongwen

Department of Mathematics, University of Science and Technology of China, Hefei, Anhuei, P.R.China

Abstract

Let the prime $p \equiv 3 \pmod 4$. In this note, we prove an identity relating Hecke operator to the conjecture that the least solution $t \neq u\sqrt{p}$ for Pellian equation $x^2 - py^2 = 1$ satisfies $p \nmid u$. We also find that the conjecture is a consequence of a conjecture on the Hecke operator.

I. Introduction and Results

Let the prime $p \equiv 3 \pmod 4$. For the least solution $\varepsilon = t + u\sqrt{p}$ of Pellian equation $x^2 - py^2 = 1$, L.J.Mordell[4] proposed **Conjecture M.** $p \nmid u$.

In 1982, we proved

Theorem 1. $4h(p)u/t \equiv (-1)^{(p-3)/4} E_{(p-3)/4} \pmod p$, *where* $h(p)$ *denotes the class number of the real quadratic field* $Q(\sqrt{p})$ *and* E_n *denotes Euler number.*

Let

$$j_p = \begin{cases} 1, \text{if } p \mid u; \\ p, \text{if } p \nmid u. \end{cases}$$

Let T_N be a Hecke operator, i.e. :

$$(T_N f)(\alpha) = \sum_{\substack{de=N \\ d>0,\, m \pmod d}} f((e\alpha + m)/d),$$

for a positive integer N and a function f of α.

---

* Supported by a grant from National Education Committee of P.R.China

For a real quadratic irrational number β. let

$$\Psi(\beta)=\begin{cases}\sum_{j=1}^{k}(-1)^{j+s}a_j, & \text{if } k \text{ even};\\ 0, & \text{if } k \text{ odd}.\end{cases}$$

Here β has a development of the simple continued fractions

$$\beta=[\hat{a}_0,\cdots,\hat{a}_s,\overline{a_1,\cdots,a_k}]$$

with the basic period $\overline{a_1,\cdots,a_k}$.

The main result of this note is as follows:

Theorem 2. *Let the prime* $p\equiv 3 \pmod 4$ and $p\geqslant 7$ *. Then we have*

$$3((p+2)j_p-1)h(-p)=\sum_{/A}\chi(/A)(T_p\Psi)(\alpha_{/A}),$$

where $h(-p)$ *is the class number of the imaginary quadratic field* $Q(\sqrt{-p})$, T_p *is a Hecke operator*, $/A=[a,b+\sqrt{p}]$ *(with* $p\nmid a>0$*) runs through a complete representatives set for the ideal class group of the real quadratic field* $Q(\sqrt{p})$, $\alpha_{/A}=(b+\sqrt{p})/a$ *, and* $\chi(/A)=(\frac{a}{p})_L$ *is Legendre symbol.*

Corollary 1. *Let the prime* $p\equiv 3 \pmod 4$ *and* $p\geqslant 7$. *If the real quadratic field* $Q(\sqrt{p})$ *has class number one, then we have*

$$3((p+2)j_p-1)h(-p)=(T_p\Psi)(\sqrt{p}).$$

Corollary 2. *Let the prime* $p\equiv 3 \pmod 4$ *and* $p\geqslant 7$. *If the real quadratic field* $Q(\sqrt{p})$ *has class number one and Conjecture M holds, then we have*

$$3(p+2)h(-p)=\Psi(p\sqrt{p}).$$

F. Hirzebruch and D. Zagier[5] proved the following:

Theorem 3. *Let the prime* $p\equiv 3 \pmod 4$ *and* $p\geqslant 7$. *Then we have*

$$3h(-p)=\sum_{/A}\chi(/A)\Psi(\alpha_{/A}).$$

Here the notations have the same notations as in Theorem 2.

From Corollary 1 and Theorem 3, we have

Corollary 3. *Let the prime* $p\equiv 3 \pmod 4$ *and* $p\geqslant 7$. *If the real quadratic field* $Q(\sqrt{p})$ *has class number one, then we have*

$$(T_p\Psi)(\sqrt{p})=((p+2)j_p-1)\Psi(\sqrt{p}).$$

According to Corollary 3 and Conjecture M, we put the following:

Conjecture L. *For a prime* $p \equiv 3 \pmod 4$, we have

$$(T_p\Psi)(\alpha)=\lambda_p\Psi(\alpha), \text{ for } \alpha \in Q(\sqrt{p}\ ),$$

with an integer $\lambda_p \equiv -1 \pmod p$.

From Theorem 3, it is easy to see

Theorem 4. *Conjecture L* $\Longrightarrow$ *Conjecture M.*

II. A Kronecker Limit Formula

Let the prime $p \equiv 3 \pmod 4$. Let

$$\chi_1(n)=\left(\frac{-p}{n}\right),\quad \chi_2(n)=\left(\frac{-4p^2}{n}\right) \text{ and } \chi(n)=\left(\frac{4p}{n}\right)$$

be Kronecker symbols. It is clear that $\chi_2=\chi\chi_1$.

Let

$$L(s,\chi_1)=\sum_{n=1}^{\infty}\chi_1(n)n^{-s} \text{ and } L(s,\chi_2)=\sum_{n=1}^{\infty}\chi_2(n)n^{-s}.$$

Similar to my paper [2], there exists a so-called Kronecker limit formula:

$$\begin{aligned} jL(1,\chi_1)L(1,\chi_2)= &\frac{\pi^2}{12\sqrt{p}} \sum_{/A=[a,b+\sqrt{p}\,]}\left(\frac{a}{p}\right)_L(1-p^{-2})\operatorname{Re} z_{/A,j} \\ &-\frac{1}{4p^2\sqrt{p}}\sum_{/A}\left(\frac{\pi}{2}-\gamma_{/A,j}\right)\sum_{m,n \,(\mathrm{mod}\, p)}\left(\frac{am^2+2bmn+cn^2}{p}\right)_L \\ &-\frac{\pi}{p\sqrt{p}}\sum_{/A}\operatorname{Im}\Big(\sum_{u\in\mathbb{Z},u\geqslant 1} e^{2\pi iuz}/A,j^{/p}\sum_{n\mid u,n\geqslant 1}n^{-1}. \qquad (1) \\ &\sum_{m\,(\mathrm{mod}\, p)}\left(\frac{am^2+2bmn+cn^2}{p}\right)_L\cos\frac{2\pi u}{p}(b/a+m/n)\Big), \end{aligned}$$

where $A=[a,b+\sqrt{p}\ ]$ (with $p\nmid a$) runs through a complete reperesentatives set for the ideal class group of the real quadratic field $Q(\sqrt{p}\ )$, $\left(\frac{*}{p}\right)_L$ is a Legendre symbol, j is a positive integer, $z_{/A,j}=\frac{\sqrt{p}}{a}\left(\frac{\varepsilon^j+i\varepsilon^{-j}}{|\varepsilon^j+i\varepsilon^{-j}|}\right)^2$, $\gamma_{/A,j}=$ are $z_{/A,j}(0<\gamma_{/A,j}<\frac{\pi}{2})$, $\varepsilon=t+u\sqrt{p}$ is the least solution of Pellian equation $x^2-py^2=1$, and a,b and c are the integers satifying $b^2-ac=p$.

We omit the proof.

There is no difficulty showing that, if $p\nmid a$, then

$$\sum_{m,n \,(\mathrm{mod}\ p)} \left(\frac{am^2+2bmn+cn^2}{p}\right)_L = p(p-1)\left(\frac{a}{p}\right)_L, \tag{2}$$

and if $p \mid n$ and $n \mid u$, then

$$\sum_{m\,(\mathrm{mod}\ p)} \left(\frac{am^2+2bmn+cn^2}{p}\right)_L \cos\frac{2\pi u}{p}(b/a+m/n) \tag{3}$$

$$=\left(\frac{a}{p}\right)_L \cos\frac{2\pi ub}{ap} \cdot \begin{cases} p-1, \text{ if } p\mid(u/n); \\ -1, \text{ if } p\nmid(u/n). \end{cases}$$

By (3), for Im $z>0$, we have

$$\sum\nolimits_Z \overset{\text{def}}{=\!=} \sum_{\substack{u\in\mathbb{Z}\\ u\geqslant 1}} e^{2\pi iuz/p} \sum_{\substack{n\mid u\\ n\geqslant 1}} n^{-1} \sum_{m\,(\mathrm{mod}\ p)} \left(\frac{am^2+2bmn+cn^2}{p}\right)_L \cos\frac{2\pi u}{p}\left(\frac{b}{a}+\frac{m}{n}\right)$$

$$=\sum_{u=1}^{\infty} e^{2\pi iuz/p}\Bigg(\sum_{\substack{n\mid u\\ p\mid n}} n^{-1}\left(\frac{a}{p}\right)_L \cos\frac{2\pi ub}{ap} \cdot \begin{cases} p-1, \text{ if } p\mid(u/n) \\ -1, \text{ if } p\nmid(u/n) \end{cases}$$

$$+ \sum_{n\mid u,\, p\nmid n} n^{-1} \sum_{m\,(\mathrm{mod}\ p)} \left(\frac{am^2+2bm+c}{p}\right)_L \cos\frac{2\pi u}{p}(b/a+m)\Bigg)$$

$$=\left(\frac{a}{p}\right)_L \sum_{u=1}^{\infty} \frac{\sigma(u)}{u} \cos\frac{2\pi bup}{a} e^{2\pi iupz} \tag{4}$$

$$-\frac{1}{p}\left(\frac{a}{p}\right)_L \sum_{u=1}^{\infty} \frac{\sigma(u)}{u}\cos\frac{2\pi bu}{a} e^{2\pi iuz} + \sum_{m\,(\mathrm{mod}\ p)}\left(\frac{am^2+2bm+c}{p}\right)_L$$

$$\times\left(\sum_{u=1}^{\infty} \frac{\sigma(u)}{u}\cos\frac{2\pi u}{p}\left(\frac{b}{a}+m\right)e^{2\pi iuz/p} - \frac{1}{p}\sum_{u=1} \frac{\sigma(u)}{u}\cos\frac{2\pi ub}{a} e^{2\pi iuz}\right),$$

where $\sigma(u)$ is the sum of all positive divisors of u.

Let, for Im $z>0$,

$$F(z)=\sum_{u=1}^{\infty} \frac{\sigma(u)}{u} e^{2\pi iuz}. \tag{5}$$

It is well known that

$$F(z)=\frac{\pi iz}{12} - \log\eta(z), \tag{6}$$

where $\eta(z)$ is the Dedekind η-function.

From (4), (5) and (6), we have

$$2\sum_z=\left(\frac{a}{p}\right)_L\left(F(p(z+b/a))+F(p(z-b/a))-\frac{1}{p}F(z+b/a)-\frac{1}{p}F(z-b/a)\right)$$

$$+\sum_{m\,(\mathrm{mod}\,p)}\left(\frac{am^2+2bm+c}{p}\right)_L\Big(F((z+b/a+m)/p)+F((z-b/a-m)/p)$$

$$-\frac{1}{p}F(z+b/a)-\frac{1}{p}F(z-b/a)\Big)$$

$$=p(1-p^{-2})\frac{\pi iz}{6}\left(\frac{a}{p}\right)_L+\left(\frac{a}{p}\right)_L(\log\eta(z+b/a)+\log\eta(z-b/a) \tag{7}$$

$$-\log\eta\,(p(z+b/a))-\log\eta\,(p(z-b/a)))$$

$$-\sum_{m\,(\mathrm{mod}\,p)}\left(\frac{am^2+2bm+c}{p}\right)_L(\log\eta((z+b/a+m)/p)+\log\eta((z-b/a-m)/p)),$$

$$\text{since}\sum_{m\,(\mathrm{mod}\,p)}\left(\frac{am^2+2bm+c}{p}\right)_L=\left(\frac{a}{p}\right)_L(p-1),\ \text{for } p\nmid a.$$

Using (1), (2), and (7), we get

$$jL(1,\chi_1)L(1,\chi_2)=-\frac{(p-1)}{4p\sqrt{p}}\sum_{/A}\chi(/A)\left(\frac{\pi}{2}-\gamma_{/A,j}\right)$$

$$+\frac{\pi}{2p\sqrt{p}}\sum_{/A=[a,b+\sqrt{p}]}\chi(/A)(\mathrm{Im}\,\log\eta(p(z_{/A,j}+b/a))+\mathrm{Im}\,\log\eta(p(z_{/A,j}-b/a))$$

$$-\mathrm{Im}\,\log\eta(z_{/A,j}+b/a)-\mathrm{Im}\,\log\eta(z_{/A,j}-b/a)) \tag{8}$$

$$+\frac{\pi}{2p\sqrt{p}}\sum_{/A=[a,b+\sqrt{p}]}\ \sum_{m\,(\mathrm{mod}\,p)}\left(\frac{am^2+2bm+c}{p}\right)_L(\mathrm{Im}\,\log\eta((z_{/A,j}+b/a+m)/p)$$

$$+\mathrm{Im}\,\log\eta((z_{/A,j}-b/a-m)/p)).$$

III. Modular Substitution

Take $j=j_p$. Then ε^j is the least solution for Pellian equation $x^2-p^3y^2=1$. Let $\varepsilon^j=U+\sqrt{p}\,pV$, where U and V are two positive integers. Then we have

$$z_{/A,j}=\frac{2p^2UV+i\sqrt{p}}{a(U^2+p^3V^2)}\ . \tag{9}$$

By (9), it is easy to show that

$$\gamma_{/A,j}=2\arcsin\frac{U}{\sqrt{U^2+p^3V^2}}-\frac{\pi}{2}. \tag{10}$$

Let $M=\begin{pmatrix}A & B\\ C & D\end{pmatrix}$, where $A=U\pm bpV$, $B=-cpV$, $C=apV$ and $D=U\mp bpV$, then $M\in SL_2(\mathbb{Z})$. Let

$$z=(b+i\sqrt{p}\,)/a. \tag{11}$$

It is easy to see that

$$M<z> \underline{\underline{\text{def}}}\,(Az+B)/(Cz+D)=z_{/A,j}\pm b/a.$$

Let

$$M_{\pm m}=\begin{pmatrix}1 & \pm m\\ 0 & 1\end{pmatrix}M\begin{pmatrix}1 & \mp m\\ 0 & 1\end{pmatrix}=\begin{pmatrix}A\pm mC & -(am^2+2bm+c)pV\\ C & \mp mC+D\end{pmatrix},$$

for any integer m.

It is easy to show that

$$M_{\pm m}\in SL_2(\mathbb{Z})\text{ and } M_{\pm m}<z\pm m>=z_{/A,j}\pm b/a\pm m.$$

Let u and n be two positive integers such that

$$un\mid p\ (\text{so } u=n=1;\text{ or } u=1,n=p;\text{ or } u=p,n=1).$$

Let

$$M_{\pm m,u/n}=\begin{pmatrix}A\pm mC & -(am^2+2bm+c)Vpu/n\\ aVpn/u & \mp mC+D\end{pmatrix}. \tag{12}$$

Thus we have

$$M_{\pm m,u/n}\in SL_2(\mathbb{Z})\text{ and } M_{\pm m,u/n}<(z\pm m)u/n>=(z_{/A,j}\pm b/a\pm m)u/n. \tag{13}$$

For Dedekind η-function and a matrix $\hat{M}=\begin{pmatrix}\hat{A} & \hat{B}\\ \hat{C} & \hat{D}\end{pmatrix}\in SL_2(\mathbb{Z})$ with $\hat{C}\neq 0$ and $|\hat{A}+\hat{D}|>2$, we have (cf. [3])

$$\log\eta(\hat{M}<Z>)=\log\eta(Z)+\frac{\pi i}{2}\Phi(\hat{M})+\frac{1}{2}(\operatorname{sign}\hat{C})^2\log\frac{(\operatorname{sign}\hat{C})(\hat{C}Z+\hat{D})}{i}, \tag{14}$$

where

$$\Phi(M)=3+m_0\Psi\left(\frac{\hat{A}-\hat{D}+\sqrt{(\hat{A}+\hat{D})^2-4}}{2\hat{C}}\right)\operatorname{sign}(\hat{C}(\hat{A}+\hat{D})), \tag{15}$$

and the integer m_0 is defined as follows:
Let

$$\hat{U}=|\hat{A}+\hat{D}|,\quad \hat{V}=g.c.d.(\hat{A}-\hat{D},\hat{B},\hat{C})\text{ and }\Delta=\hat{V}^{-2}(\hat{U}^2-4),$$

where Δ is a squre-free positive integer. Then we have

$$\frac{\hat{U}+\hat{V}\sqrt{\Delta}}{2}=\left(\frac{U_0+V_0\sqrt{\Delta}}{2}\right)^{m_0},$$

where $(U_0+V_0\sqrt{\Delta})/Z$ is the least solution of Pellian equation $x^2-y^2\Delta=4$.

IV. The Proof of Theorem 2

From III and by using (9)—(15) we have not much difficulty to prove

$$\text{Im}\log\eta((z_{/A,j}+b/a+m)u/n)+\text{Im}\log\eta((z_{/A,j}-b/a-m)u/n)=\pi/2$$
$$+\frac{m_0\pi}{12}\left(\Psi\left(\frac{u}{n}\cdot\frac{\sqrt{p}+(am+b)}{a}\right)+\Psi\left(\frac{u}{n}\cdot\frac{\sqrt{p}-(am+b)}{a}\right)\right)$$
$$-\arcsin\frac{U}{\sqrt{U^2+p^3V^2}},\tag{16}$$

where

$$m_0=\begin{cases} j_p, \text{ if } n=u=1; \text{ or } n=p, u=1 \text{ and } p\mid(am+b); \\ 1, \text{ if } n=1, u=p; \text{ or } n=p, u=1 \text{ and } p\nmid(am+b).\end{cases}\tag{17}$$

(8), (9), (16) and (17) derives that

$$j_pL(1,\chi_1)L(1,\chi_2)=\frac{\pi^2}{12p\sqrt{p}}\sum_{A=[a,b+\sqrt{p}]}\left(\left(\frac{a}{p}\right)_L\Psi\left(p\frac{\sqrt{p}+b}{a}\right)\right.$$
$$\left.+\sum_{m\,(\text{mod}\,p)}\left(\frac{am^2+2bm+c}{p}\right)_L\Psi\left(\frac{(\sqrt{p}+b)/a+m}{p}\right)\right)$$
$$-\frac{j_p\pi^2}{12p\sqrt{p}}\sum_{/A}\chi(/A)\Psi(/A),\tag{18}$$

because of the facts

$$p\mid(am+b)\Longleftrightarrow p\mid(am^2+2bm+c), \text{ for } p\nmid a,$$

and

$$\Psi\left(\frac{\sqrt{x}-y}{T}\right)=\Psi\left(\frac{\sqrt{x}+y}{T}\right),$$

for any squre-free positive integer x, integer y and squre-free integer T.

Using Theorem 3, the classical class number formula and (18), we get

$$3(p+2)j_p h(-p)=\sum_{/A}\chi(/A)(T_p\Psi)(\alpha_{/A})-\sum_{\substack{A=[a,b+\sqrt{p}]\\ p\mid(am+b)}}\chi(/A)\Psi\left(\frac{\alpha_{/A}+m}{p}\right). \tag{19}$$

For $p \mid (am+b)$, let $am+b=np$, $n\in\mathbb{Z}$. Then we have

$$a(am^2+2bm+c)=(n^2p-1)p,$$

and deduce that

$$n^2p-1=ad, \text{ and } am^2+2bm+c=dp \text{ with } d\in\mathbb{Z}, \tag{20}$$

and

$$nb-1=a(d-mn). \tag{21}$$

Put

$$A=-n,\ B=mn-d,\ C=a \text{ and } D=b, \tag{22}$$

then we have

$$AD-BC=-1 \text{ and } \frac{A(\sqrt{p}-b)/a+B}{C(\sqrt{p}-b)/a+D}=\frac{-(am+b)+\sqrt{D}}{ap},$$

by using (20), (21) and (22).
Therefore we get, for $p \mid (am+b)$,

$$\Psi\left(\frac{b+\sqrt{p}}{a}\right)=\Psi\left(\frac{-b+\sqrt{p}}{a}\right)=-\Psi\left(\frac{-(am+b)+\sqrt{p}}{ap}\right)$$
$$=-\Psi\left(\frac{(am+b)+\sqrt{p}}{ap}\right). \tag{23}$$

Theorem 2 follows from (19) and (23).

V. The Proof for Other Results

Only Corollary 2 needs to be proved. We have no difficulty to get

$$\Psi\left(\frac{\sqrt{p}+m}{p}\right)=\Psi(p\sqrt{p}),\ \text{for}\ p\nmid m,$$

and to prove Corollary 2.

References

[1] H.W. Lu: Congruences for the class number of quadratic fields, *Abh. Math. Sem. Univ. Hamburg*, **52**(1982)254—258.

[2] H. W. Lu: Kronecker limit formula of real quadratic fields (I), *Scientia Sinica* (A), **27**(1984) 1233—1250.

[3] H.W. Lu: Translation of Dedekind η-function, *Acta. Math. Scientia*, (1988).

[4] L.J. Mordell: On a Pellian equation conjecture, *Acta Arith.*, **6**(1960)137—144.

[5] D.Zagier: A Kronecker limit formula for real quadratic fields, *Math. Ann.*, **213**(1975)153—184.

Lower Bound for Number of B-Twins in Short Intervals*

Luo Wenzhi

Department of Mathematics, Beijing University, People's Republic of China.

1. Introduction

Define B as the set consisting of all the integers expressible as sums of two squares of integer and

$$B_2(x, K, L) = |\{n \leqslant x : n \in B, n+1 \in B, n \equiv L \pmod K\}|.$$

In 1974, Hooley[4] and Indlekofer[6] independently proved

$$B_2(x, 1, 1,) >> \frac{x}{\log x}. \tag{1}$$

Later, Kelly[9] successively applied Hooley's method to the case of short interval showing if $\theta > \frac{5}{6}$, then

$$B_2(x, 1, 1) - B_2(x - x^\theta, 1, 1) >> \frac{x^\theta}{\log x} \tag{2}$$

Recently, Bantle[1] proved that (2) holds even when $\theta > \frac{2}{3}$ and also considered the distribution of the B-twins belonging to short interval in arithmetic progression. In the framework of Iwaniec's linear and half-dimensional sieves, he employed the Bombieri Theorem in short interval due to Iwaniec-Huxley and Ricci, in connection with the classical estimate of Kloosterman sum belonging to A. Weil to yield sharp estimation for the corresponding remainder terms which involve exponential sums.

In this paper, we establish a new version of Bombieri Theorem in short intervals which, when combined with Bantle's argument, lead to the following theorem:

Theorem. *Let* $A > 0$, $1 \leqslant K \leqslant (\log x)^A$, $(k, 2L(L+1)) = 1$. *Then there exists a constant c such that*

$$B_2(x, k, L,) - B_2(x - x^\theta, k, L) \geqslant \frac{c}{k} \prod_{\substack{p|k \\ p \equiv 3 \pmod 4}} \frac{p}{p-2} \cdot \frac{x^\theta}{\log x}, \tag{3}$$

provided $x > x_0(A, \theta)$, $1 \geqslant \theta > 0.84375$.

---

* The project is supported by National Natural Science Foundation of China.

Remarks : Our result constitutes an improvement upon Bantle's work which asserts (3) holds for $1 \geqslant \theta \geqslant 0.8521$. Our proof is motivated by an ingenious idea in Professor Pan Chengdong's new meanvalue theorem.

The author wish to express his deep gratitude to professor Pan Chengbiao for suggestion and encouragement.

2. Notation

p — *prime*;

P — the set consisted of all primes;

$\mathscr{P}$ — a subset of P;

$\mathscr{P}(z)$ — the set $\{p \in \mathscr{P} : p < z\}$, sometimes denote $\prod_{\substack{p \in \mathscr{P} \\ p < z}} p$;

$\mathscr{P}_3^{(q)}(z)$ — the set $\{p < z : p \equiv 3 \pmod 4,\ p \nmid q\}$;

$\mathscr{P}_3^{(q)}$ — the set $\mathscr{P}_3^{(q)}(\infty)$;

$\mathscr{P}_3$ — the set $\mathscr{P}_3^{(1)}(\infty)$;

(a, b) — the greatest common divisor of integers a and b;

$\mu(n)$ — Mobius function;

$\tau(n)$ — divisor function;

$\varphi(n)$ — Euler function;

$\Lambda(n)$ — Mangoldt function;

$e(x)$ — $e^{2\pi i x}$;

lix — $\int_2^x \frac{dt}{\log t}$, $x \geqslant 2$;

$[x]$ — the greatest integer not exceeding x;

$\psi(x)$ — $x - [x] - \frac{1}{2}$;

$\sum_{v \bmod d}$ — the summation is over representives of complete residue system mod$d$;

$\sum_{V \bmod d}^{*}$ — the summation is over representives of reduced residue system mod$d$;

ε, η — sufficienty small positive numbers. They may take different values at each occurence;

B, C — absolute constants. They may take different values at each occurence.

3. Sieve Result

Let $\mathscr{A}$ be a finite sequence of integers, $\mathscr{P}$ a prime set $\subseteq P$, denote sifting function as

$$s(\mathscr{A}, \mathscr{P}, z) = \sum_{\substack{a \in \mathscr{A} \\ (a, \mathscr{P}(z))=1}} 1$$

For $d | \mathscr{P}(z)$, define

$$|\mathscr{A}_d| = |\{a \in \mathscr{A}; a \equiv 0 \pmod d\}|,$$

$\omega(d)$ is a multiplicative function defined for $d | \mathscr{P}(x)$, $x > 1$.

Set $r(\mathscr{A}, d) = |\mathscr{A}_d| - \frac{(w l d)}{d} x$, $v(z) \prod_{p<z} (1 - \frac{w(p)}{p})$, $z \geqslant 2$. For convenience, we also define $w(p)=0$, if $p \notin \mathscr{P}$.

The following lemma is required:

Lemma. *Assume $z \geqslant 2$, $D \geqslant 2$, $S = \frac{\log D}{\log Z}$, and $w(p)$ satisfies the following conditions:*

(Ω) *There exist A_0, A_1, such that*

$$W(p) \leqslant A_0, \quad 0 \leqslant \frac{W(p)}{p} \leqslant 1 - \frac{1}{A_1}$$

($\Omega_2(\frac{1}{z}, L)$) *There exist A_2,* and *$L \geqslant 2$, such that for $2 \leqslant w \leqslant z$,*

$$-L \leqslant \sum_{w \leqslant p \leqslant z} \frac{w(p)}{p} \log p - \frac{1}{z} \log \frac{z}{w} \leqslant A_2,$$

then there exists a constant c depending on A_0, A_1, A_2, such that

$$S(\mathscr{A}, \mathscr{P}, z) \leqslant XV(z) \left\{ F_{\frac{1}{2}}(s) + \frac{C}{(\log D)^{0.21}} \right\} + \sum_{\substack{d<D \\ d | \mathscr{P}(z)}} |r(\mathscr{A}, d)|,$$

where $F(s)$ satisfies

$$\begin{cases} s^{\frac{1}{2}} F_{\frac{1}{2}}(s) = 2\sqrt{\frac{e^r}{\pi}}, \ 0 < s \leqslant 2, \ s^{\frac{1}{z}} f_{\frac{1}{z}}(s) = 0, \ 0 < s \leqslant 1. \\ (s^{\frac{1}{z}} F_{\frac{1}{z}}(s))' = \frac{1}{z} s^{\frac{1}{z}} f_{\frac{1}{z}}(s-1), \ s > 2, \ (s^{\frac{1}{z}} f_{\frac{1}{z}}(s))' = \frac{1}{z} s^{-\frac{1}{z}} F_{\frac{1}{z}}(s-1), \ s > 1. \end{cases}$$

For its proof, see [7].

4. Basic Inequality

It is well-known that if $p^a \| n$, $p \equiv 3 \pmod 4$ imply a is a even interger, then $n \in B$, and the contrary is also true. We define

$$\mathscr{A}=\{n(n+1):x-x^{\theta}<n\leqslant x,\ n\equiv L_1(\bmod 16K)\},$$

where x is sufficiently large, $L_1=L\cdot 16\cdot\overline{16}+4\cdot k\cdot\overline{k}$, $(k,2L(L+1))=1$. Here $\overline{16}$, $\overline{K}$ are solutions of the equations $16\cdot\overline{16}\equiv 1\ (\bmod K)$, and $K\cdot\overline{K}\equiv 1\ (\bmod 16)$ respectively. It is easy to see that L_1 is of form $4(4t+1)$, therefore from $n\equiv L_1$ $(\bmod 16K)$ we know n of form $4(4t+1)$; $n+1$ of form $4t+1$; the number of prime factors $\equiv 3\ (\bmod 4)$ of both n and $n+1$ are even (multiplicity being counted). Assume u satisfy $\frac{1}{4}<u<\frac{1}{z}$. If $n(n+1)\in\mathscr{A}$, $(n(n+1),\ \mathscr{P}_3^{(k)}(x^u))=1$, then $n\in B$, $n+1\in B$, unless the following exceptions occur:

1) There exist q, $(q,\mathscr{P}_3)=1$, and p_1, $p_2\in\mathscr{P}_3$, $x^u\leqslant p_1<p_2$, such that $n=q\cdot p_1\cdot p_2$. The number of $n(n+1)\in\mathscr{A}$ satisfying this condition is at most

$$R_1=\sum_{\substack{q\leqslant x^{1-2u}\\ q\equiv 4(16),\ (q,k\mathscr{P}_3^{(k)})=1}}\ \sum_{\substack{x^u\leqslant p_1<\sqrt{\frac{x}{q}}\\ p_1\equiv 3(4)}} s(M_1(qp_1),\ \mathscr{P}_3^{(k)},\ x^u),\tag{1}$$

where $M_1(a)=\left\{ap+1:\ \frac{x-x^{\theta}}{a}<p\leqslant\frac{x}{a},\ ap\equiv L_1\ (\bmod 16K)\right\}$, $4\mid a$.

2) There exist q, $(q,\mathscr{P}_3)=1$, and p_1, $p_2\in\mathscr{P}_3$, $x^u\leqslant p_1<p_2$, such that $n+1=qp_1\cdot p_2$.

The number of $n(n+1)\in\mathscr{A}$ satisfying this condition is at most

$$R_2=\sum_{\substack{q\leqslant x^{1-2u}\\ q\equiv 1(4),\ (q,k\mathscr{P}_3^{(k)})=1}}\ \sum_{\substack{x^u\leqslant p_1<\sqrt{\frac{x}{q}}\\ p_1\equiv 3(4)}} s(M_2(qp_1),\ \mathscr{P}_3^{(k)},\ x^u)+1,\tag{2}$$

where $M_2(a)=\left\{ap-1:\ \frac{x-x^{\theta}}{a}<p\leqslant\frac{x}{a},\ p\equiv\overline{a}\,(L_1+1)(\bmod 16K)\right\}$.

Therefore, in the above notation, we have the inequality

$$B_2(x,K,L)-B_2(x-x^{\theta},K,L)\geqslant S(\mathscr{A},\mathscr{P}_3^{(k)},x^u)-R_1-R_2-1.\tag{3}$$

5. The Estimation of $S(\mathscr{A},\mathscr{P}_3^{(k)},x^u)$

Lemma. *Assume* $\eta>0$, $\theta>\frac{2}{3}$, $\psi=\min\left(1,\ \frac{5}{4}\theta-\frac{1}{6}\right)$, $\frac{1}{4}<u<\frac{1}{z}\psi$. *then there exists a constant* $c\geqslant 1$ *depending at most on* θ, u, A *such that*

$$S(\mathscr{A},\mathscr{P}_3^{(k)},x^u)\geqslant\frac{1}{8\psi}\log\left(\frac{\psi}{\mu}-1\right)\frac{x^{\theta}}{K\log x}\prod_p\frac{p-w(p)}{p-1}(1-c\eta),$$

provided $x \geqslant x_0(\eta, \theta)$, where

$$w(p)=\begin{cases}2 & p\in\mathscr{P}_3^{(k)},\\ 0 & p\notin\mathscr{P}_3^{(k)}.\end{cases}$$

For its proof, see [2].

6. Statement and Proof of the Basic Lemma

The purpose of this section is to prove the following :

Basic Lemma. *Assume* $\frac{19}{24}<\theta\leqslant 1$, $\max\left(\frac{1}{4}, \frac{12}{5}(1-\theta)\right)<u<\frac{\psi}{2}$ $\psi=\min(1, \frac{5}{4}\theta-\frac{1}{6})$. *For any* $A>0$, *there exists a* B *such that*

$$R_1=\sum_{d\leqslant \frac{x^{\theta-\frac{1}{2}}}{\log^{\theta}x}}\max_{(l,d)=1}\max_{z\leqslant x^{\theta}}\max_{\frac{1}{16}x\leqslant y\leqslant x}\left|\sum_{\substack{x^u<a\leqslant x^{1-u}\\(a,d)=1}}g(a)\left(\sum_{\substack{y<an\leqslant y+z\\ an\equiv l(\bmod d)}}\Lambda(n)-\frac{z}{\psi(d)a}\right)\right|$$

$$\ll x^{\theta}\log^{-A}x,$$

where $g(a)$ *is any function satisfying* $|g(a)|\leqslant 1$.

We can infer from the basic lemma by usual method the following:

Corollary. *Assume* $\frac{19}{24}<\theta\leqslant 1$, $\max(\frac{1}{4}, \frac{12}{5}(1-\theta))<u<\frac{\psi}{2}$, $\psi=\min(1, \frac{5}{4}\theta-\frac{1}{6})$. *Then for any* $A>0$, *there exists a* B *such that*

$$R_2=\sum_{d\leqslant x^{\theta-\frac{1}{2}}\log^{-B}x}\max_{(l,d)=1}\max_{z\leqslant x^{\theta}}\max_{\frac{1}{16}x\leqslant y\leqslant x}\left|\sum_{\substack{x^u<a\leqslant x^{1-u}\\(a,d)=1}}g(a)\left(\sum_{\substack{y<ap\leqslant y+z\\ ap\equiv l(d)}}1-\frac{1}{\psi(d)}\int_{\frac{y}{a}}^{\frac{y+z}{a}}\frac{dt}{\log t}\right)\right|$$

$$\ll x^{\theta}\log^{-A}x,$$

where $g(a)$ *is any function satisfying* $|g(a)|\leqslant 1$.

The proof of the basic lemma:

We may think that y only takes the values of half odd integers and z integers. Then,

$$R_1\leqslant\sum_{d\leqslant x^{\theta-\frac{1}{2}}\log^{-B}x}\max_{(l,d)=1}\max_{z\leqslant x^{\theta}}\max_{\frac{1}{16}x\leqslant y\leqslant x}\left|\sum_{\substack{x^u<a\leqslant x^{1-u}\\(a,d)=1}}g(a)\left(\sum_{\substack{y<an\leqslant y+z\\ an\equiv l\ (d)}}\Lambda(n)-\frac{1}{\psi(d)}\sum_{y<an\leqslant y+z}\Lambda(n)\right)\right|$$

$$+\sum_{d\leqslant x^{\theta-\frac{1}{2}}\log^{-B}x}\max_{z\leqslant x^{\theta}}\max_{\frac{1}{16}x\leqslant y\leqslant x}\left|\sum_{\substack{x^u<a\leqslant x^{1-u}\\(a,d)=1}}\frac{g(a)}{\psi(d)}\left(\sum_{y<ax\leqslant y+z}\Lambda(n)-\frac{z}{a}\right)\right|. \tag{1}$$

We first show that the second sum on the right-hand side of (1) $\ll x^{\theta}\log^{-B}x$.

By [11],

$$\sum_{\frac{y}{a}<u\leqslant\frac{y+z}{a}}\Lambda(n)-\frac{z}{a}=-\sum_{|r|\leqslant T}\frac{\left(\frac{y+z}{a}\right)^{\rho}-\left(\frac{y}{a}\right)^{\rho}}{\rho}+O\left(\frac{x\log^2 x}{aT}\right)(\rho=\beta+ir)$$

where $T<<\frac{x}{a}$, ρ's are the non-trivial zeros of $\zeta(s)$. Taking $T=x^{1-\theta}\log^{B_1}x$, B_1 sufficiently large, we easily see

$$\frac{\left(\frac{y+z}{a}\right)^{\rho}-\left(\frac{y}{a}\right)^{\rho}}{\rho}\ll\begin{cases}\frac{z}{a}\left(\frac{y}{a}\right)^{B-1}<<\frac{x^{\theta}}{a}\left(\frac{x}{a}\right)^{B-1}, |r|\leqslant x^{1-\theta},\\ \frac{\left(\frac{y}{a}\right)^{\beta}}{|r|}<<\frac{1}{|r|}\left(\frac{x}{a}\right)^{\beta}, |r|>x^{1-\theta}.\end{cases}$$

Hence, $$\sum_{|r|\leqslant T}\frac{\left(\frac{y+z}{a}\right)^{\rho}-\left(\frac{y}{a}\right)^{\rho}}{\rho}<<\sum_{|r|\leqslant x^{1-\theta}}\frac{x^{\theta}}{a}\left(\frac{x}{a}\right)^{B-1}+\sum_{x^{l-\theta}<|r|\leqslant T}\frac{1}{|r|}\left(\frac{x}{a}\right)^{\beta}$$

$$<<-\left(\frac{x^{\theta}}{a}\right)\int_{\frac{1}{2}}^{1}\left(\frac{x}{a}\right)^{\alpha-1}d\left(\sum_{\substack{|r|\leqslant x^{1-\theta}\\ \beta\geqslant\alpha}}1\right)-\int_{\frac{1}{2}}^{1}\left(\frac{x}{a}\right)^{2}d\left(\sum_{\substack{x^{l-\theta}<|r|\leqslant T\\ \beta\geqslant\alpha}}\frac{1}{|r|}\right).$$

Denote by $N(\alpha, u)$ the number of zeros $\rho=\beta+i\gamma$ of $\zeta(s)$ with $\beta\geqslant\alpha$, $|r|\leqslant u$. The above expression

$$<<\frac{x^{\theta}}{a}N\left(\frac{1}{2}, x^{1-\theta}\right)\left(\frac{x}{a}\right)^{\frac{1}{2}-1}+\frac{x^{\theta}}{a}\int_{\frac{1}{2}}^{1}N(\alpha,x^{1-\theta})\left(\frac{x}{a}\right)^{\alpha-1}\log\left(\frac{x}{a}\right)^{d\alpha}$$

$$+\left(\sum_{\substack{x^{1-\theta}\leqslant|r|\leqslant T\\ \beta\geqslant\frac{1}{2}}}\frac{1}{|r|}\right)\left(\frac{x}{a}\right)^{\frac{1}{2}}+\int_{\frac{1}{2}}^{1}\left(\sum_{\substack{x^{1-B}\leqslant|r|\leqslant T\\ \beta\geqslant\alpha}}\frac{1}{|r|}\right)\left(\frac{x}{a}\right)^{\alpha}\log\left(\frac{x}{a}\right)^{d\alpha}$$

$$<<\log x\cdot\frac{x^{\theta}}{a}\max_{\frac{1}{2}\leqslant\alpha\leqslant1}N(\alpha,x^{1-\theta})\left(\frac{x}{a}\right)^{\alpha-1}+\log x\cdot\max_{\frac{1}{2}\leqslant\alpha\leqslant1}\left(\sum_{\substack{x^{1-\theta}\leqslant|r|\leqslant T\\ \beta\geqslant\alpha}}\frac{1}{|r|}\right)\frac{x}{a}$$

$$<<\log x\cdot\frac{x^{\theta}}{a}\cdot\max_{\frac{1}{2}\leqslant\alpha\leqslant1}N(2, x^{1-\theta})\left(\frac{x}{a}\right)^{\alpha-1}+\log^2 x\cdot\max_{\substack{\frac{1}{2}\leqslant\alpha\leqslant1\\ x^{1-\theta}\leqslant u\leqslant T}}\frac{1}{U}N(\alpha, U)\left(\frac{x}{a}\right)^{2}.$$

(2)

For $N(\alpha, U)$ we have the well- known estimate (see [11], [10], [5])

$$N(\alpha, U)\begin{cases} =0, & \text{if} \quad \alpha\geqslant\alpha_0=1-C(\log x)^{-\frac{4}{5}}, \\ \ll(\log x)^4U^{2(1-\alpha)/2}, & \text{if} \quad \alpha_0\geqslant\alpha\geqslant\frac{5}{6}, \\ \ll U^{\frac{12}{5}(1-\alpha)+\varepsilon}, & \text{if} \quad \frac{5}{6}\geqslant\alpha\geqslant\frac{3}{4}, \\ \ll(\log x)^9U^{\frac{3(1-\alpha)}{2-\alpha}}, & \text{if} \quad \frac{3}{4}\geqslant\alpha\geqslant\frac{1}{2}. \end{cases}$$

Combining this with (2), and noting $u>\frac{12}{5}(1-\theta)$, we obtain

$$\sum_{d\leqslant x^{\theta-\frac{1}{2}}\log^{-B}x}\max_{z\leqslant x^\theta}\max_{\frac{1}{16}x\leqslant y\leqslant x}\left|\sum_{\substack{x^u<a\leqslant x^{1-u}\\(a,d)=1}}\frac{g(a)}{\varphi(d)}\left(\sum_{y<an\leqslant y+Z}\Lambda(n)-\frac{z}{a}\right)\right|.$$

$$\ll x^\theta\log^{-A}x.$$

Next we turn to the first sum on the right-hand side of (1).
If $(a, d)=(L, d)=1$,

$$\sum_{\substack{y<an\leqslant y+z\\an\equiv l\,(d)}}\Lambda(n)=\frac{1}{\varphi(d)}\sum_{y<an\leqslant y+z}\chi_d^0(n)\Lambda(n)+\frac{1}{\varphi(d)}\sum_{\chi_d\neq\chi_d^0}\overline{\chi}(l)\chi(a)\sum_{y<an\leqslant y+z}\chi(n)\Lambda(n)$$

$$=\frac{1}{\varphi(d)}\sum_{y<an\leqslant y+z}\Lambda(n)+\frac{1}{\varphi(d)}\sum_{1<q|d}\sum_{x_q}{}^*\overline{x}(l)x(a)\sum_{\substack{y<an\leqslant y+z\\(n,d)=1}}x(n)\Lambda(n)$$

$$+O\left(\frac{\log d\log y}{\varphi(d)}\right).$$

Hence, the first sum of (1)

$$\leqslant\sum_{d\leqslant x^{\theta-\frac{1}{2}}\log^{-B}x}\frac{1}{\varphi(d)}\sum_{1<q|d}\max_{z\leqslant x^\theta}\max_{\frac{1}{16}x\leqslant y\leqslant x}\sum_{x_q}{}^*\left|\sum_{\substack{x^u<a\leqslant x^{1-u}\\(a,d)=1}}g(a)\chi(a)\sum_{\substack{y<an\leqslant y+z\\(n,d)=1}}\Lambda(n)\chi(n)\right|$$

$$+0(x^\theta\log^{-A}x).\ll\log x\max_{m\leqslant x^{\theta-\frac{1}{2}}\log^{-B}x}\sum_{1<q\leqslant x^{\theta-\frac{1}{2}}\log^{-B}x}\frac{1}{\varphi(q)}\max_{z\leqslant x^\theta}\max_{\frac{1}{16}x\leqslant y\leqslant x}\sum_{x_q}{}^*\Bigg|$$

$$\sum_{\substack{x^u<a\leqslant x^{1-u}\\(a,m)=1}}g(a)\chi(z)\sum_{\substack{y<an\leqslant y+z\\(n,m)=1}}\Lambda(n)x(n)+O(x^\theta\log^{-A}x).$$

we know that

$$\sum_{\substack{y<an\leqslant y+z\\(n,m)=1}}\Lambda(n)\chi(n)=\sum_{y<an\leqslant y+z}\Lambda(n)\chi(n)+O(\log^2x).$$

Let $D_1=\log^{B_2}x$, B_2 sufficiently large, $M(F,U)=\sum_{f\leqslant F}\sum^*_{\chi(\bmod f)}N(\alpha,u,\chi)$ where $N(\alpha,u,\chi)$ denotes the number of zeros ρ of $L(s,\chi)$ with

$$\beta\geqslant\alpha,\ |r|\leqslant u,\ \rho=\beta+i\gamma,\ u\ll x.$$

The following formula holds (see [11])

$$\sum_{\frac{y}{a}<n\leqslant\frac{y+z}{a}}\Lambda(n)\chi(n)=-\sum_{|r|\leqslant T}\frac{\left(\frac{y+z}{a}\right)^{\rho}-\left(\frac{y}{a}\right)^{\rho}}{\rho}+O\left(\frac{x}{a}\ \frac{\log^2 x}{T}\right),$$

where $T\ll\frac{x}{a}$, χ denotes primitive characters. Furthermore, the zero-density theorems imply (see [11], [5], [10])

$$M(F,U)\begin{cases} = & 0 \text{ or } 1 \text{ depending on whether there is an exceptional modulus } f\leqslant F, \quad \text{if} \\ & \alpha\geqslant\alpha_0.\ \alpha_0=1-C\{\max(\log F,(\log x)^{\frac{4}{5}})\}^{-1}, \\ \ll (F^2U)^{\frac{12}{5}(1-\alpha)+\varepsilon}, & \text{if}\quad \frac{3}{4}\leqslant\alpha\leqslant\frac{5}{6}, \\ \ll (F^2U)^{\frac{2(1-\alpha)}{\alpha}}\log^{14}x, & \text{if}\quad \frac{5}{6}\leqslant\alpha\leqslant\alpha_0, \\ \ll (F^2U)^{\frac{3(1-\alpha)}{2-\alpha}}\log^9 x, & \text{if}\quad \frac{1}{2}\leqslant\alpha\leqslant\frac{3}{4}. \end{cases}$$

Combining these with Siegel′s theorem, in a way similiar to the derivation of the estimates for the second sum of (1), we obtain

$$\sum_{1<q\leqslant D_1}\frac{1}{\varphi(q)}\max_{z\leqslant x^{\theta}}\max_{\frac{1}{16}x\leqslant y\leqslant x}\sum_{\chi_q}{}^{*}\left|\sum_{\substack{x^u<a\leqslant x^{1-u}\\(a,m)=1}}g(a)\chi(a)\sum_{\substack{y<an\leqslant y+z\\(n,m)=1}}\Lambda(n)\chi(n)\right|$$

$$\ll x^{\theta}\log^{-A-1}x.$$

thus the proof is reduced to bounding

$$\log x\max_{m\leqslant x^{\theta-\frac{1}{2}}\log^{-B}x}\sum_{D_1<q\leqslant x^{\theta-\frac{1}{2}}\log^{-B}x}\frac{1}{\varphi(q)}\max_{z\leqslant x^{\theta}}\max_{\frac{1}{16}x\leqslant y\leqslant x}\sum_{\chi_q}{}^{*}\left|\sum_{\substack{x^u<a\leqslant x^{1-u}\\(a,m)=1}}g(a)\chi(a)\right.$$

$$\left.\cdot\sum_{\substack{y<an\leqslant y+z\\(n,m)=1}}\Lambda(n)\chi(n)\right|.$$

Let $D_1\leqslant Q\leqslant x^{\theta-\frac{1}{2}}\log^{-B}x$, $Q<Q'\leqslant 2Q$, denote by (q) the range of summation $Q<q\leqslant Q'$. Also assume that $x^u\leqslant E\leqslant x^{1-u}$, $E<E'\leqslant 2E$, denote by (a) the range of

summation $E<a\leqslant E'$. Furthermore we define

$$I_m(Q,E)=\sum_{(q)}\frac{1}{\varphi(q)}\max_{z\leqslant x^\theta}\max_{\frac{1}{16}x\leqslant y\leqslant x}\sum_{x_q}{}^*\left|\sum_{\substack{(a)\\(a,m)=1}}g(a)\chi(a)\sum_{\substack{y<an\leqslant y+z\\(n,m)=1}}d_E^{(m)}(n)\chi(n)\right|.$$

By above discussion, to prove the basic lemma it suffices to show $I_m(Q,E)\ll x\log^{-(A+3)}x$ with the suitable B_2.

If we write
$$g^{(m)}(a)=\begin{cases}g(a), & (m,a)=1,\\ 0\quad, & (m,a)>1,\end{cases}$$
$$d_E^{(m)}(n)=\begin{cases}\Lambda(n), & (n,m)=1,\\ 0\quad, & (n,m)>1,\end{cases}$$
then
$$I_m(Q,E)=\sum_{(q)}\frac{1}{\varphi(q)}\max_{z\leqslant x^\theta}\max_{\frac{1}{16}x\leqslant y\leqslant x}\sum_{x_q}{}^*\left|\sum_{(a)}g^*(a)\chi(a)\sum_{y<an\leqslant y+z}d_E^{(m)}(n)\chi(n)\right|.$$

Set $b=1+\dfrac{1}{\log x}$, $T=x^{10}$, $S=\sigma+it$,

$$d_E^{(m)}(s,\chi)=\sum_{n=1}^{\infty}d_E^{(m)}(n)\chi(n)n^{-s},$$
$$\qquad\qquad\sigma>1.$$
$$g_E^{(m)}(s,\chi)=\sum_{(a)}g^{(m)}(a)\chi(a)a^{-s},$$

By Perron Formula,

$$\sum_{y<an\leqslant y+z}d_E^{(m)}(n)X(n)=\sum_{\frac{y}{a}<n\leqslant\frac{y+z}{a}}d_E^{(m)}(n)X(n)$$
$$=\frac{1}{2\pi i}\int_{(b,T)}d_E^{(m)}(s,\chi)\frac{(y+z)^s-y^s}{a^s}\frac{ds}{s}+O(x^{-3}).$$

Therefore

$$\sum_{(a)}g^{(m)}(a)X(a)\sum_{y<an\leqslant y+z}d_E^{(m)}(n)\chi(n)$$
$$=\frac{1}{2\pi i}\int_{(b,T)}g_E^{(m)}(s,\chi)d_E^{(m)}(s,\chi)\frac{(y+z)^s-y^s}{s}ds+O(x^{-2}).$$

So,

$$I_m(Q,E)\ll\sum_{(q)}\frac{1}{\varphi(q)}\max_{z\ll x^\theta}\max_{\frac{1}{16}x\leqslant y\leqslant x}\sum_{x_q}{}^*\left|\int_{(b,T)}g_E^{(m)}(s,\chi)d_E^{(m)}(s,\chi)\frac{(y+z)^s-y^s}{s}ds\right|.$$
$$+O(x^\theta\log^{-(A+3)}x).$$

Set $M_2=\dfrac{x_0^2}{h}$, $N=2^j$, $J=[4\log^2 x]$, where $h=x^\theta$. If $\operatorname{Re}S=b=1+\dfrac{1}{\log x}$,

then
$$d=d_E^m(s,\chi)=d_1+d_2+O(x^{-3}),$$

where
$$d=\sum_{n\leqslant M_2} d_E^{(m)}(n)\chi(n)n^{-s},$$
$$d=\sum_{M_2<n\leqslant NM_2} d_E^{(m)}(n)\chi(n)n^{-s}.$$

We have

$$\int_{(b,T)} gd\frac{(y+z)^s-y^s}{s}ds=\int_{(b,T)} gd_2\frac{(y+z)^s-y^s}{s}ds+\int_{(\frac{1}{2},T)} gd_1\frac{(y+z)^s-y^s}{s}ds$$
$$+O(x^{-1}\log x),$$

By estimating $(\sigma=\frac{1}{2}$ or $b)\frac{(y+z)^s-y^s}{s}\ll\min(\frac{(x+h)^\sigma}{|t|+2}, hx^{\sigma-1})$ and Schwarz inequality,

$$I_m(Q,E)\ll\frac{\log\log Q}{Q}\left(\sum_{(q)}\sum_{x_q}{}^*\int_{-T}^{T}|d_2(b+it)|^2\cdot\min\left(\frac{(x+h)^b}{|t|+2},hx^{b-1}\right)dt\right)^{\frac{1}{2}}$$
$$\times\left(\sum_{(q)}\sum_{x_q}{}^*|g(b+it)|^2\min\left(\frac{(x+h)^b}{|t|+2},hx^{b-1}\right)dt\right)^{\frac{1}{2}}$$
$$+\frac{\log\log Q}{Q}\left(\sum_{(q)}\sum_{x_q}{}^*\int_{-T}^{T}|d_1(\frac{1}{2}+it)|^2\cdot\min\left(\frac{(x+h)^{\frac{1}{2}}}{|t|+2},hx^{-\frac{1}{2}}\right)dt\right)^{\frac{1}{2}}$$
$$\times\left(\sum_{(q)}\sum_{x_q}{}^*\int_{-T}^{T}\left|g\left(\frac{1}{2}+it\right)\right|^2\min\left(\frac{(x+h)^{\frac{1}{2}}}{|t|+2},hx^{-\frac{1}{2}}\right)dt\right)^{\frac{1}{2}}$$
$$+O(h\log^{-(A+3)}x),$$

and $|d_2|^2\ll\log^2 x\sum_{j=0}^{J-1}|\sum_{2^jM_2<n\leqslant 2^{j+1}M_2} d_E^{(m)}(n)\chi(n)n^{-s}|^2.$

We need the following lemma (see [12]):

Lemma. *Let* $S(s,\chi)=\sum a_n\chi(n)n^{-s}$ *denote a Dirichlet polynomial,* $\frac{1}{2}\leqslant\sigma\leqslant 1+\frac{1}{\log x}$, *then for* $Q\geqslant 1$, $T\geqslant 2$,

$$\sum_{q\leqslant Q}\sum_{x_q}{}^*\int_{-T}^{T}|S(\sigma+it,\chi)|^2\min\left(\frac{(\xi+h)^\sigma}{|t|+2},h\xi^{a-1}\right)dt$$
$$\ll(\xi+h)^\sigma\sum|a_n|^2\left(Q^2\log T+\frac{nh}{h+(\xi+h)^\sigma\xi^{1-\sigma}}\right)n^{-2\sigma}.$$

From this lemma we immediately obtain

$$\log^2 x\cdot\sum_{j=0}^{J-1}\sum_{(q)}\sum_{x_q}{}^{*}\int_{-T}^{T}\left|\sum_{2^jM_2<n\leqslant 2^{j+1}M_2} d_E^{(m)}(n)\chi(n)n^{-(b+it)}\right|^2 \times\min\left(\frac{(x+h)^b}{|t|+2},hx^{b-1}\right)dt$$

$$\ll\log^4 x\sum_{j=0}^{J-1}(x+h)^b\sum_{2^jM_2<n\leqslant 2^{j+1}M_2}\frac{1}{n^{2b}}\left(Q^2\log T+\frac{nh}{h+(x+h)^bx^{1-b}}\right)$$

$$\ll\log^6 x\sum_{j=0}^{J-1}x\sum_{2^jM_2<n\leqslant 2^{j+1}M_2}\frac{1}{n^2}\left(Q^2\log T+\frac{nh}{x}\right)$$

$$\ll\log^6\cdot x\sum_{j=0}^{J-1}x\left(\frac{Q^2\log x}{2^jM_2}+\frac{h}{x}\log^2 x\right)$$

$$\ll\log^8 x\left(\frac{Q^2}{M_2}+\log^2 x\cdot\frac{h}{x}\right)\cdot x$$

$$\ll h\cdot\log^{10}x.$$

$$\sum_q\sum_{x_q}{}^{*}\int_{(b,T)}\left|\sum_{(a)}g^{(m)}(a)x(a)a^{-s}\right|^2\min\left(\frac{(x+h)^b}{|t|+2},hx^{b-1}\right)dt$$

$$\ll x\sum_{(a)}\frac{1}{a^2}\left(Q^2\log x+\frac{ah}{x}\right)$$

$$\ll x\log x\left(\frac{Q^2}{E}+\frac{h}{x}\log x\right),$$

$$\sum_{(q)}\sum_{x_q}{}^{*}\int_{(\frac{1}{2},T)}\left|\sum_{n\leqslant M_2}d_E^{(m)}(n)x(n)n^{-s}\right|^2\min\left(\frac{(x+h)^{\frac{1}{2}}}{|t|+2},hx^{-\frac{1}{2}}\right)dt$$

$$\ll x^{\frac{1}{2}}\sum_{n\leqslant M_2}\frac{\Lambda^2(n)}{n}\left(Q^2\log x+\frac{nh}{x}\right)$$

$$\ll x^{\frac{1}{2}}\log^2 x\left(Q^2\log^2 x+\frac{h}{x}M_2\right)$$

$$\ll x^{\frac{1}{2}}\log^4 x\cdot Q^2,$$

$$\sum_{(q)}\sum_{x_q}{}^{*}\int_{(\frac{1}{2},T)}\left|\sum_{(a)} g^{(m)}(a)x(a)a^{-s}\right|^2 \min\left(\frac{(x+h)^{\frac{1}{2}}}{|t|+2}, hx^{-\frac{1}{2}}\right)dt$$

$$\ll x^{\frac{1}{2}}\sum_{(a)}\frac{1}{a}\left(Q^2\log x+\frac{ah}{x}\right)$$

$$\ll x^{\frac{1}{2}}\left(Q^2\log^2 x+\frac{h}{x}E\right).$$

Combining all the above estimations leads to

$$I_m(Q, E)\ll x^{\theta}\log^{-(A+3)}x$$

and the basic lemma follows.

7. The Estimation of R_1, R_2.

Let $\varepsilon_0=\{a: a=qp_1, q\leqslant x^{1-2u}, q\equiv 4 \pmod{16}, (q, k\,p_3^{(k)})=1, x^u\leqslant p<\sqrt{\frac{x}{q}},$ $p\equiv 3 \pmod 4\}$

$$C=\{ap+1: a\in\varepsilon: \frac{x-x^{\theta}}{a}<p\leqslant\frac{x}{a}, p\equiv\left(\frac{\bar a}{4}\right)\frac{l_1}{4} \pmod{4k}, \text{ then}$$

$$R_1=\sum_{\substack{q\leqslant x^{1-2u} q\equiv 4(16)\\ (q, kP_3^{(k)})=1}}\ \sum_{\substack{x^u\leqslant p_1<\sqrt{\frac{x}{q}}\\ p_1\equiv 3(4)}} s(M_1(qp_1), \mathscr{P}_3^{(k)}, x^u)=s(c, \mathscr{P}_3^{(k)}, x).$$

It is easy to see that

$$|C_d|=\sum_{a\in\varepsilon_0}\left|\left\{p: \frac{x-x^{\theta}}{a}<p\leqslant\frac{x}{a}, p\equiv-\bar a \pmod d, p\equiv\left(\frac{\bar a}{4}\right)\frac{l_1}{4} \pmod{4k}\right\}\right|$$

where $\bar a$, $\left(\frac{\bar a}{4}\right)$ satisfy $a\bar a\equiv 1 \pmod d$ and $\left(\frac{a}{4}\right)\left(\frac{\bar a}{4}\right)\equiv 1 \pmod{4K}$ respectively,

$$\omega_1(p)=\begin{cases}\dfrac{p}{p-1}, & p\in\mathscr{P}_3^{(k)},\\ 0, & p\notin\mathscr{P}_3^{(k)},\end{cases}$$

$$x=\sum_{a\in\varepsilon_0}\frac{1}{\varphi(4k)}\left\{\mathrm{li}\frac{x}{a}-\mathrm{li}\frac{x-x^{\theta}}{a}\right\}=\frac{1}{\varphi(4k)}\left(1+O\left(\frac{1}{\log x}\right)\right)x^{\theta}\sum_{a\in\varepsilon_0}\frac{1}{a\log(\frac{x}{a})},$$

$$r_d=\sum_{x^u<a\leqslant x^{1-u}} g(a)\left(\sum_{\frac{x-x^{\theta}}{4}<\frac{a}{4}p\leqslant\frac{x}{4},\ \frac{a}{4}\cdot p\equiv l^*(4kd)} 1-\frac{1}{\varphi(4kd)}\left\{\mathrm{li}\frac{\frac{x}{4}}{\frac{a}{4}}-\mathrm{li}\frac{\frac{x-x^{\theta}}{4}}{\frac{a}{4}}\right\}\right),$$

where L^* is any integer such that the system of congruence equation

$$p \equiv -\bar{a} \pmod{d},$$

$$p \equiv \left(\frac{\bar{a}}{4}\right)\frac{l_1}{4} \pmod{4K},$$

is equivalent to the equation $\frac{a}{4}p \equiv L^* \pmod{4Kd}$ and $g(a)$ is the characteristic function of ε_0.

From the lemmas stated and proved in the previous sections we deduce

$$R_1 \leqslant xv(x^u)\left\{F_{\frac{1}{2}}\left(\frac{\log x^{\theta-\frac{1}{2}-\varepsilon}}{u\log x}\right)+\frac{c}{(\log x)^{0.21}}+O(x^\theta \log^{-A} x).\right.$$

Because $0<\eta<\dfrac{\theta-\frac{1}{2}-\varepsilon}{u}=\dfrac{\log x^{\theta-\frac{1}{2}-\varepsilon}}{u\log x}\leqslant 2$, we have

$$F_{\frac{1}{2}}\left(\frac{\log x^{\theta-\frac{1}{2}-\varepsilon}}{u\log x}\right)=2\sqrt{\frac{e^r}{\pi}}\sqrt{\frac{u}{\theta-\frac{1}{2}-\varepsilon}},$$

and ([3])

$$v(x^u)=\prod_p\left(1-\frac{\omega(p)}{p}\right)\left(1-\frac{1}{p}\right)^{-\frac{1}{2}}\times\frac{e^{-\frac{1}{2}r}}{u^{\frac{1}{2}}\log^{\frac{1}{2}}x}\left(1+O\left(\frac{1}{\log x}\right)\right),$$

hence

$$R_1 \leqslant \frac{1}{\varphi(k)\sqrt{\pi}}\prod_p\left(1-\frac{w(p)}{p}\right)\left(1-\frac{1}{p}\right)^{-\frac{1}{2}}x^\theta$$

$$\times \sum_{q\leqslant x^{1-2u},\ q\equiv 4(16),\ (q,k\mathscr{S}_3^{(k)})=1}\ \sum_{\substack{x^u\leqslant p_1<\sqrt{\frac{x}{q}}\\ p_1\equiv 3(4)}}\left(1+O\left(\frac{1}{(\log x)^{\frac{1}{5}}}\right)\right)$$

$$\times \frac{\sqrt{2}}{qp_1\log\frac{x}{qp_1}\sqrt{(2\theta-1-\varepsilon)\log x}}.$$

For estimating R, the following lemma is required:

Lemma.

$$\sum_{\substack{q\leqslant x^{1-2u}\\ q\equiv 4\ (16)\\ (q,k\mathscr{S}_3^{(k)})=1}}\ \sum_{\substack{x^u\leqslant p_1<\sqrt{\frac{x}{q}}\\ p_1\equiv 3\ (4)}}\frac{1}{qp_1\log\frac{x}{qp_1}\sqrt{(2\theta-1-\varepsilon)\log x}}$$

$$= \frac{1}{16\sqrt{\pi}} \prod_p \left(1 - \frac{w_2(p)}{p}\right)\left(1 - \frac{1}{p}\right)^{-\frac{1}{2}} \frac{\varphi(k)}{k} \frac{-1}{\log x}$$

$$\times\left(1+O\left(\frac{1}{(\log x)^{\frac{1}{5}}}\right)\right)\int_0^{1-2u} dt \int_u^{\frac{1}{2}-\frac{1}{2}t} \frac{dv}{v(1-v-t)\sqrt{2\theta-1-\varepsilon}\sqrt{t}},$$

Where

$$w_2(p) = \begin{cases} 1, & p \in \mathscr{P}_3^{(k)}, \\ 0, & p \notin \mathscr{P}_3^{(k)}. \end{cases}$$

For its proof, see [2].

We conclude with the aid of the above lemma

$$R_1 \leqslant \frac{\sqrt{2}}{16\pi} \prod_p \frac{p-w(p)}{p-1} \frac{x^\theta}{k\log x}(1+c\eta)$$

$$\cdot \int_0^{1-2u} dt \int_u^{\frac{1}{2}-\frac{1}{2}t} \frac{dv}{v(1-v-t)\sqrt{2\theta-1-\varepsilon}\sqrt{t}},$$

where

$$w(p) = \begin{cases} 2, & p \in \mathscr{P}_3^{(k)}, \\ 0, & p \notin \mathscr{P}_3^{(k)}. \end{cases}$$

Similarly we deduce that

$$R_2 \leqslant \frac{\sqrt{2}}{16\pi} \prod_p \frac{p-w(p)}{p-1} \frac{x^\theta}{k\log x}(1+c\eta)\times\int_0^{1-2u} dt$$

$$\cdot \int_u^{\frac{1}{2}-\frac{1}{2}t} \frac{dv}{v(1-v-t)\sqrt{2\theta-1-\varepsilon}\sqrt{t}}.$$

8. Results

If $\frac{19}{24} < \theta \leqslant 1$, $\max\left(\frac{1}{4}, \frac{12}{5}(1-\theta)\right) < u < \frac{\varphi}{2}$, $\psi = \min\left(1, \frac{5}{4}\theta - \frac{1}{6}\right)$, $1 \leqslant k \leqslant \log^A x$, $(K, 2L(L+1)) = 1$, summing up the discussions in the previous sections we are led to

$$B_2(x, K, L) - B_2(x - x^\theta, K, L)$$

$$\geqslant \frac{1}{8\pi}\prod_p \frac{p-w(p)}{p-1}\ \frac{x^\theta}{k\log x}\ (1-c\eta)\left\{\frac{\pi}{4}\log\left(\frac{\psi}{u}-1\right)-\sqrt{2}\ \mathrm{I}\right\},$$

where $\mathrm{I}=\int_0^{1-2u} dt\int_u^{\frac{1}{2}-\frac{1}{2}t}\frac{dv}{v(1-v-t)\sqrt{2\theta-1-\varepsilon}\ \sqrt{t}}$,

$$w(p)=\begin{cases}2, & p\in\mathscr{P}_3^{(k)},\\ 0, & p\notin\mathscr{P}_3^{(k)}.\end{cases}$$

But

$$I_1=\frac{1}{\sqrt{2\theta-1-\varepsilon}}\int_0^{1-2u}\frac{dt}{\sqrt{t}}\int_u^{\frac{1}{2}-\frac{1}{2}t}\frac{dv}{v(1-v-t)}$$

$$=\frac{1}{\sqrt{2\theta-1-\varepsilon}}\int_0^{1-2u}\frac{1}{\sqrt{t}\ (1-t)}\log\left(\frac{1-t-u}{u}\right)dt.$$

Noting $\log t\leqslant\frac{1}{2t}(t^2-1)$, $t\geqslant 1$, we finally have

$$I_1\leqslant\frac{1}{\sqrt{2\theta-1-\varepsilon}}\int_0^{1-2u}\frac{1}{\sqrt{t}}\ \frac{1}{1-t}\ \frac{\left(\frac{1-t-u}{u}\right)^2-1}{2\left(\frac{1-t-u}{u}\right)}dt$$

$$=\frac{1}{\sqrt{2\theta-1-\varepsilon}\cdot u}\sqrt{1-2u}-\frac{1}{2\sqrt{1-u}\cdot\sqrt{2\theta-1-\varepsilon}}\ln\frac{\sqrt{1-u}+\sqrt{1-2u}}{\sqrt{1-u}-\sqrt{1-2u}}.$$

Taking $u=\frac{3}{8}$, when $\theta>0.84375$, we get

$$\frac{\pi}{\psi}\log\left(\frac{\psi}{u}-1\right)-\sqrt{2I_1}>0.01.$$

And that finishes our proof.

References

[1] G. Bantle, An Asymptotic Formula for B-twins, *Acta Arith.*, **47**(1986), 297—312.

[2] G. Bantle, Untere Abschätzung für die Anzahl der B-Zwillinge auf Kurzen Intervallen, *Acta Arith.*, **46**(1986), 313—329.

[3] H. Halberstam and H.-B. Richert, *Sieve methods*, Academic Press, London, 1974.

[4] C. Hooley, On the intervals between numbers that are sums of two squares III, *J. Reine Angew. Math.*, **267**(1974), 207—218.

[5] M. N. Huxley, Large values of Dirichlet polinomials III, *Acta Arith.*, **26**(1974), 431—440.

[6] K. H. Indlekofer, Scharfe untere Abschätzung für die Anzahlfunktion der B-Zwillinge, *Acta Arith.*, **26**(1974), 207—212.

[7] H. Iwaniec, The Half-dimensional sieve, *Acta Arith.*, **29**(1976), 69—95.

[8] H. Iwaniec and M. N. Huxley, Bombieri's theorem in short intervals, *Mathematika*, **22** (1975), 188—194.

[9] P. J. Kelly, The Number of B-Twins in an interval, Dissertation, Nottingham, 1978.

[10] H. L. Montgomery, Topics in multiplicative number theory, *Lecture notes in Mathematics*, **227**, Springer-Verlag, 1971.

[11] K. Prachar, Primzahlverteilung, Springer-Verlag, 1957.

[12] S. J. Ricci, Mean value theorems for primes in short intervals, *Proc. London Math. Soc.* (3), **37** (1978), 230—242.

Legendre Polynomial over Finite Fields and Factorization of Integers

Kanji Namba

We consider elliptic curves of the form:

$$y^2 = x^3 + ax^2 + bx + c$$

over finite field F_p, and abelean groups associated to the curve. The order of the group suitably parametrized satisfies so-called Gaussian differential equation.

1. As a normal form of elliptic curve, we consider

$$c:\ y^2 = x^3 + qx + r.$$

Take two points (a, b), (c, d) on c, and define their sum (x, y) as the symmetric point to x-axis of the third intersection point of the line passing (a, b), (c, d) and the curve C. Explicitly if $a \neq c$, putting $u = (b-d)/(a-c)$,

$$x = u^2 - (a+c),$$

$$y = -u^3 + (2a+c)u - b,$$

y can be expressed in symmetric form as:

$$y = -u^3 + (ad-bc)/(a-c) - (b+d).$$

If $a = c$, we replace u by derivative at $x = a$, namely by $2yy' = 3x^2 + q$, we put $u = (3a^2 + q)/2b$,

$$x = u^2 - 2a,$$

$$y = -u^3 + 3au - b.$$

Since $b^2 = a^3 + qa + r$, we have

$$x = (3a^2 + q)^2/4b^2 - 2a$$

$$= -(3a^4 + 6qa^2 + 12ra - q^2)/4(a^3 + qa + r).$$

Notice that differential of the numerator equals to -3 times of denominator. This is same as the transition determined by Newton's method applied to

$$y^3 = 3x^4 + 6x^2 + 12rx - q^2.$$

Consider the discriminant $D=4q^3+27r^2$ of $x^3+qx+r=0$. If $D\neq 0$, then the set of the points on C and infinity point forms an abelian group structure. Infinity point playes the role of unit element.

Important points are rationality and transitivity of these operations. In order to compute the order of the group, we consider Legendre symbol:

$$(n/p)=\begin{cases} 1, & \text{if } n \text{ is square mod } p, \\ 0, & \text{if } n=0, \\ -1, & \text{if } n \text{ is not squre mod } p. \end{cases}$$

This is a quadratic character and $1+(n/p)$ is the total number of solutions in finite field $Fp=\{0, 1, \cdots, p-1\}$. And

$$(n/p)=n^{(p-1)/2}$$

mod p is another expression of this.

In order to compute the order of the group associated to elliptic courves, we use notions of finite Fourier analysis. Every element of Fp satisfies the equation

$$x^p-x=0.$$

Consider the residue mod. by this polynomial

$$f(x)=g(x)(x^p-x)+h(x).$$

then we have

$$\sum_{x=0}^{p-1} f(x)=\sum_{x=0}^{p-1} h(x).$$

And for monomials, we have

$$\sum_{x=0}^{p-1} x^k=\begin{cases} 0, \text{ if } 0\leqslant k<p-1, \\ -1, \text{ if } k=p-1. \end{cases}$$

There is essential difference between $k=0$ and $k=p-1$. In the case of $k=0$ it is 0, but if $k=p-1$ then it is -1. By this fact, if the residue is of the form

$$h(x)=c_0+c_1x+\cdots+c_{p-1}x^{p-1},$$

then we have

$$\sum_{x=0}^{p-1} f(x)=-c_{p-1}.$$

General form of expansion is

$$-\sum_{x=0}^{p-1} f(x)x^{p-1-k}=c_k,$$

and therefore

$$f(x)=\sum_{k=0}^{p-1} c_k x^k.$$

This relation is a consequence of

$$\sum_{k=0}^{p-1} x^{p-1-k} y^k = -\delta(x, y).$$

Now, we compute the order of abelian group associated to elliptic curve of the form $y^2=x^3+qr^2+r$. Note that this is not the general normal form $y^2=x^3+qx+r$.

Theorem 1. *The order of abelian group associated to*

$$C: y^2=x^3+qx^2+r$$

over the finite field Fp $(p\neq 2, 3)$ *is of the form*

$$n(p, z)=1+a(p, z)+p,$$

where $z=-27r/4q^3$, *and* $w(z)=\pm a(p, z)$ *satisfies Gaussian differential equation*

$$z(z-1)\frac{d^2w}{dz^2}+(2z-1)\frac{dw}{dz}+\frac{5}{36}w=0,$$

and $w(z)$ *is expressed as polynomial*

$$w(z)=F(1/6, 5/6, 1, z)$$

of degree $n=[p/6]$.

Proof. Put $s=(p-1)/2$. And consider the sum

$$a(p)=\sum_{x=0}^{p-1} (x^3+qx^2+r)^s.$$

Then the order of the group is $1+a(p)+p$. Note that

$$f(x)=(x^3+qx^2+r)^s$$

has degree $3(p-1)/2<2(p-1)$. By previous remark, it is equal to the minus of coefficient of degree $p-1$ term. Since

$$(x^3+qx^2+r)^s=x^{p-1}(x+q+rx^{-2})^s,$$

it is enough to compute the constant term of

$$(x+q+rx^{-2})^s=\sum_n \binom{s}{3n} q^{s-3n}(x+rx^{-2})^{3n}.$$

We take $3n$ for the last term may contain constant term. Now the constant term is

$$\frac{s!}{(3n)!(s-3n)!}\frac{(3n)!}{n!(2n)!}q^{s-3n}r^n$$

$$=q^s\frac{s(s-1)\cdots(s-3n+1)}{n!(2n)!}\left(\frac{r}{q^3}\right)^n.$$

Since $s=(p-1)/2$, we have the coefficient

$$\frac{(p-1)(p-3)\cdots(p-6n+1)}{2^{\wedge}3n\ n!\ (2n)!}=(-1)^n\ \frac{1/2\cdot 3/2\cdots(6n-1)/2}{n!\ (2n)!}.$$

We use usual notation

$$(a)_n=a(a+1)\cdots(a+n-1),$$

then, we have

$$1/2\cdot 3/2\cdots(6n-1)/2=(1/6)_n(3/6)_n(5/6)_n 27^n$$
$$1/2\cdot 2/2\cdots(2n/2)=n!\,(1/2)_n.$$

By $1/2=3/6$, we see that

$$(-1)^n\frac{1/2\cdot 3/2\cdots(6n-1/2)}{n!\ (2n)!}\ (\frac{r}{q^3})=\frac{(1/6)_n(5/6)_n}{(n!)^2}\ (\frac{-27r}{4q^3})^n.$$

This means that, putting $z=-27r/4q^3$,

$$a(p,z)=-q^s\sum_n \frac{(1/6)_n(5/6)_n}{(n!)^2}\ z^n.$$

It is known that 2nd order Fuchsean differential equation with 3 stationary singularities is reduced to Gaussian differential equation of the form

$$z(z-1)\frac{d^2w}{dz^2}+((a+b+1)z-c)\frac{dw}{dz}+abw=0$$

with stationary singularities 0, 1, ∞. There is a formal solution with $w(0)=1$ such that

$$w(z)=\sum_{n=0}^{\infty}\frac{(a)_n(b)_n}{n!(c)}\ z^n.$$

Using Fuchs notation, $w(z)=F(a,b,c,z)$.

This kind of infinite series in general does not make sense in *Fp*. But notice that every prime $p\neq 2$, 3 satisfies $p=\pm 1$ mod 6, so if $n>[p/6]$ then in *Fp*.

$$(1/6)_n(5/6)_n=0,$$

and $F(1/6, 5/6, 1, z)$ is a polynomial solution of degree $[p/6]$. The inequality

$$a(p,z)^2<4p$$

for elliptic curves is proved by H. Hasse.

Conjecture. (M. Sato) *Consider the solutions of*

$$x^2+a(p,z)x+p=0.$$

Arguments distribution of

$$x=\frac{-a(p,z)\pm\sqrt{(a(p,z)^2-4p)}}{2}=\sqrt{p}\exp\left(i\,\theta\,(p,z)\right)$$

is proportional to $\sin^2(\theta)$ *for random p.*

Consider general normal form

$$y^2=x^3+qx+r$$

with the discriminant

$$\begin{vmatrix} 1 & 0 & q & r & \\ & 1 & 0 & q & r \\ 3 & 0 & q & & \\ & 3 & 0 & q & \\ & & 3 & 0 & q \end{vmatrix} = \begin{vmatrix} -2q & -3r & 0 \\ 0 & -2q & -3r \\ 3 & 0 & q \end{vmatrix} = 4q^3+27r^2$$

and the form treated above

$$y^2=x^3+ux^2+v$$

with the discriminant

$$\begin{vmatrix} 1 & u & 0 & v & \\ & 1 & u & 0 & v \\ 3 & 2u & 0 & & \\ & 3 & 2u & 0 & \\ & & 3 & 2u & 0 \end{vmatrix} = -v\begin{vmatrix} -u & 0 & -3v \\ 3 & 2u & 0 \\ & 3 & 2u \end{vmatrix} = v\,(4u^3+27v).$$

Substituting x by $x-u/3$ in the second curve, we have

$$q=-u^3/3 \qquad r=2u^3/27+v.$$

Consider parameters

$$t=-27r^2/4q^3 \qquad z=-27v/4u^3.$$

The values $z=0,1$ corresponds to

$$D=v\,(4u^3+27v)=0.$$

Also we have

$$t=(1-2z)^2$$

and values $z=0,\ 1$ corresponds to $t=1$ and to

$$D=27r^2+4q^3=0.$$

Since $z=1/2\pm\sqrt{}\;t/2$, we have two Gaussian differential equation according to $\pm$ in z:

$$t(t-1)\frac{d^2w}{dt^2}+((2\pm1/2)t-1)\frac{dw}{dt}+\frac{41\pm36}{144}w=0.$$

The primitive residue mod 12 are 1, 5, 7, 11 and 5/144 corresponds to 1,5 and the prime of the form $p=4n+1$, and 77/144 to 7, 11 and the prime of the form $p=4n-1$. In each case the solution

$$F(1/12,\ 5/12,\ 1,\ t)\quad F(7/12,\ 11/12,\ 1,\ t)$$

is a polynomial of order $[p/12]$.

Note in general, Gaussian differential equation

$$z(z-1)\frac{d^2w}{dz^2}+(2z-1)\frac{dw}{dz}-n(n+1)w=0$$

is translated by $x=2z-1$ to Legendre's differential equation

$$(x^2-1)\frac{d^2w}{dx^2}+2x\frac{dw}{dx}-n(n+1)w=0.$$

The polynomial solution $p(n,x)$ such that

$$p(n,1)=1$$

of degree n is called Legendre polynomial.

What obtained in Theorem 1 is

$$a(p,z)=p(n,x),$$

where $n=[p/6]$, $x=2z-1$ and $z=-27r/4q^3$. Any way $a(p,z)$ is the unique solution having value at both stationary singularities 0, 1 and $a(p,0)=1$.

We list here only initial part of values of $p(1/6, x)$:

p	$p(n,x)$
5	1 1 1 1 1
7	−3 −2 −1 0 1 2 3
11	−5 −4 −3 −2 −1 0 1 2 3 4 5
13	−5 −2 4 0 −1 1 6 1 −1 0 4 −2 −5
17	2 5 −6 3 −2 −4 −3 1 8 1 −3 −4 −2 3 −6 5 2
19	−4 5 8 1 −1 −2 −6 2 −1 0 1 −2 6 2 1 −1 −8 −5 4 .

Values $x \neq \pm 1$ for which $p(n,x) = \pm 1$ are specially interesting, and it is known that except for

7, 11, 17, 61, 107, 139, 307, 571

below 1500 have such x. Is there another such prime number? are they finite?

Theorem 2. *The order of the abelian group associated to the elliptic curve of the form*

$$y^2 = x^3 + qx^2 + rx$$

and that of Jacobian normal form

$$y^2 = (1-x^2)(1-kx^2)$$

is of the form

$$n(p,z) = 1 + a(p,z) + p,$$

where $z = 4r/q^2$ *or* $z = 4k/(k+1)^2$ *and*

$$a(p,z) = \pm F(1/4, 3/4, 1, z),$$

which is a polynomial of degree $n = [p/4]$.

Proof. As before we put $s = (p-1)/2$, and compute the coefficient of $x(p-1)$ in

$$(x^3 + qx^2 + rx)^s = x^{(p-1)}(x + q + rx^{-1})^s.$$

The constant term of last part is the sum of

$$\frac{s(s-1)\cdots(s-2n+1)}{n!^2} q^{s-2n} r^n = \left(\frac{q}{p}\right) \frac{(1/4)_n (3/4)_n}{n!^2} \left(\frac{4r}{q^2}\right).$$

So we have for $z = 4r/q^2$,

$$a(p,z) = (q/p) F(1/4, 3/4, 1, z),$$

and for $x = 2z - 1$, $n = [p/4]$,

$$a(p,z) = (q/p) p(n,x).$$

For the discriminant, we have

$$\begin{vmatrix} 1 & q & r & 0 & \\ & 1 & q & r & 0 \\ 3 & 2q & r & & \\ & 3 & 2q & r & \\ & & 3 & 2q & r \end{vmatrix} = r^2 \begin{vmatrix} -2 & -q & \\ & -2 & -q \\ 3 & 2q & r \end{vmatrix} = r^2(4r - q^2).$$

Next case is Jacobian form

$$v^2=(1-x^2)(1-kx^2).$$

In this case, we introduce function corresponds to

$$x=sn(t),\ y=cn(t),\ z=dn(t),$$

which satisfies the condition

$$x^2+y^2=1,$$
$$kx^2+z^2=1,$$

so that we have $v^2=y^2z^2=(1-x^2)(1-kx^2)$.

The operation of points (x_1, y_1, z_1) and (x_2, y_2, z_2) is defined by

$$x=\frac{x_1y_2z_2+x_2y_1z_1}{1-kx_1^2x_2^2},$$

$$y=\frac{y_1y_2-x_1x_2z_1z_2}{1-kx_1^2x_2^2},$$

$$z=\frac{z_1z_2-kx_1x_2y_1y_2}{1-kx_1^2x_2^2}.$$

In the projective space, for (u_1, x_1, y_1, z_1), (u_2, x_2, y_2, z_2),

$$u=u_1{}^2u_2{}^2-kx_1{}^2x_2{}^2,$$

$$x=u_1x_1y_2z_2+u_2x_2y_1z_1,$$

$$y=u_1u_2y_1y_2-x_1x_2z_1z_2,$$

$$z=u_1u_2z_1z_2-kx_1x_2y_1y_2,$$

Total number of solutions is the sum of

$$(1+(1-x^2)^s)(1+(1-kx^2)^s),$$

where $s=(p-1)/2$ for $x=0, 1, \cdots, p-1$ and infinity. Namely

$$p-(-1/p)-(-k/p)-a(p,z)+(k/p),$$

where $z=4k/(k+1)^2$ and the last term (k/p) come from term of degree $2(p-1)$. And at infinity

$$(1+(-1/p))(1+(k/p))=1+(-1/p)+(k/p)+(-k/p).$$

Adding these terms, we have

$$n(p,z)=1+a(p,z)+p.$$

For $z \neq 0, 1$, discriminant $\neq 0$, so the number of solutions of

$$x^3+qx^2+rx=0$$

is 1 or 3. If p is odd prime, then total number of Legendre symbols $((x^3+qx^2+rx)/p) = \pm 1$ is even, so we have

$$a(p,z)=0 \pmod{2}.$$

This means that

$$n(p,z)=1+a(p,z)+p \neq p.$$

And the order of abelean groups associated to these curves are not p.

Conjecture 1. *Let $p(n,x)$ be the Legendre polynomial of degree n on Fp. Then for $p>23$, $n<p/2$,*

$$\forall x(p(n,x)^2<4p),$$

if and only if n is of the form

$$n=[p/k], \quad k=2, 3, 4, 6.$$

Some congruence relations such that for $p \neq 2, 5$, $x \neq \pm 1$,

$$p([p/2],x)=p+1 \pmod{4},$$

$$p([p/3],x)=-(p+1)/2 \pmod{3}$$

seems to hold. And the range of $p(n,x)$ is the set of elements such that $z^2<4p$ with the above condition for $n=[p/2], [p/3]$. For the renge of $p([p/4],x)$ is that, except for $p=3,7,11$, with the condition

$$p([p/4],x)=0 \pmod{2}$$

mentioned above. For the renge of $p([p/6],\mathrm{x})$, relatively small number of elements with $z^2<4p$ are exceptional.

The value x for which $p([p/k],x)=-1$ seems to be interesting with respect to factrization of integers.

Conjecture 2. *For every $p=1 \pmod{6}$, there exists*

$$x \neq \pm 1 \text{ such that}$$

$$p([p/3],x)=-1.$$

This is of course weaker than the above speculation about the range of $p([p/3], x)$.

2. Here we mention aout a series of prime numbers, such that if such prime number is a factor of some composite number, then the prime number can be found out by probabilitistic polynomial time computation. This means that each one trial needs polynomial time computation and for each trial with fixed positive possibility the factor can be computed.

As mentioned before, the abelian group associated to elliptic curve of the form

$$y^2 = x^3 + qx^2 + r$$

has the order

$$n(p, z) = 1 + a(p, z) + p$$

where $w(z) = a(p, z)$ and $z = -27r/4q^3$, and it satisfies Gaussian differential equation

$$z(z-1)\frac{d^2w}{dz^2} + (2z-1)\frac{dw}{dz} + \frac{5}{36}w = 0,$$

and for $n = [p/6]$, $n(n+1) = -5/36$ in Fp $(p \neq 2, 3)$. This equation is invariant under exchange of z and $1-z$. The fixed points are $z = 1/2, \infty$. These values are $p(n, 0)$ and $p(n, \infty)$ for $n = [p/6]$. If $p = -1$ (mod 4), then $p(n, 0) = 0$, and if $p = 1$ (mod 4), then $p = a^2 + b^2$ and $a = p(n, 0)/2$ is an integer. In this case $p(n, 0)$ can not be -1.

Now consider the value $p(n, \infty)$. Since $z = -27r/4q^3 = \infty$, this case corresponds to $r \neq 0$, $q = 0$, namely to

$$y^2 = x^3 + r.$$

Put $s = (p-1)/2$ and compute degree $p-1$ term of

$$(x^3 + r)^s.$$

If $p = -1$ (mod 6) then it is 0. If $p = 6n + 1$, then $s = 3n$,

$$a(p) = \frac{(3n)!}{n!(2n)!} r^n.$$

Notice that $r\hat{}n$ is a root of $x^6 - 1 = 0$. Each such root appears with probability 1/6 for random choice of r.

Now let

$$a = 1 \cdot 2 \cdots n, \; b = (n+1) \cdots (2n), c = (2n+1) \cdots (3n).$$

Then $(abc)^2 = \pm 1$, $2\hat{}n \cdot b^2 = c^2$ and

$$a(p) = hc/a, \quad h^6 = 1.$$

Theorem 3. *For $p=6n+1$, there is $b(p)$ such that*

$$a(p)^2+3b(p)^2=4p.$$

Let

$$\omega=\frac{-1\pm\sqrt{-3}}{2}$$

be cubic primitive root of 1. *Then in the ring* $Q[\omega]$,

$$\alpha=\frac{a(p)\pm b(p)\sqrt{-3}}{2}$$

have norm p and trace $a(p)$. *They are the solutions of*

$$x^2-a(p)x+p=0.$$

Proof. Let $p=6n+1$. Note that the number of solutions of $x^3=u$ is 0 or 3 for $u\neq 0$ and 1 for $u=0$. By the definition

$$a(p)=\sum_{x=0}^{p-1}\left(\frac{x^3-r}{p}\right),$$

we have the congruence

$$a(p)=\begin{cases}0, & r=0,\\ \pm 2 \pmod 6, & r \text{ is cubic residue},\\ \pm 1 \pmod 6, & r \text{ is cubic non-residue}.\end{cases}$$

Let e be primitive 6th root of 1. Then other $a(p)'$s are

$$a(p),\ ea(p),\ e^2a(p),\ -a(p),\ -ea(p),\ -e^2a(p)$$

and also

$$a(p)(1+e+e^2)=0 \pmod p.$$

Since the absolute value of square of each term $<2\sqrt{p}$, we have that

$$a(p)+ea(p)+e^2a(p)=O.$$

So take $a(p)$ of the form

$$a=a(p)=6t-2,$$

then we can put

$$b=ea(p)=6s+1,\ c=e^2a(p)=6u+1.$$

Now, we consider the term

$$((b+c)/2)^2+3((b-c)/6)^2.$$

This term is

$$(b^2+bc+c^2)/3=e^2a(1+e+e^2)=0 \pmod{p}.$$

So we have

$$((b+c)/2)^2+3((b-c)/6)^2=hp.$$

By b^2, $c^2<4p$, we have $h=0, 1, 2, 3$. But $h=0, 2, 3$ is impossible, so $h=1$. By putting

$$a(p)=a=-(b+c), \quad b(p)=(b-c)/3,$$

we get

$$a(p)^2+3b(p)^2=4p.$$

The order $n(p)$ of abelian group associated to the curve is expressed as

$$n(p)=(a(p)/2\pm 1)^2+3(b(p)/2)^2$$

for two even $a(p)'$s, and

$$n(p)=(a(p)\pm 1/2)^2+3(b(p)\pm 1/2)^2$$

for four odd $a(p)'$s. This means that they are norms of the sum of solutions of two equations

$$x^2-a(p)x+p=0,$$

$$x^6-1=0.$$

By this property if $n(p)=q$ is a prime number, then also $n(q)=p$ is possible by probability $1/6$. There are many such examples as

13—19, 31—43, 73—67—79—97—103, 109—127, 139—163—181,

199—211—241, 223—229, 277—283, 313—349—373, 337—367, ...

Corollary. *For given K, the prime p of the form $p=6m+1$ for which $a(p)=k$ is of the form*

$$p=(k/2)^2+3n^2$$

for even k, and

$$p=12n^2\pm 6kn\pm k^2$$

for odd k.

Proof. For even $k=a(p)$, it is already mentioned in the theorem. If $a(p)$ is odd, then multiplying primitive cubic root of 1, there are u, n such that

$$p=u^2+3n^2 \text{ and } k=a(p)=u\pm 3n.$$

Since $u=k\pm 3n$, we have

$$p=(k\pm 3n)^2+3n^2=12n^2\pm 6kn+k^2.$$

Notice that this is nessesary and sufficient condition. The case $k=-1$ is specially important with respect to factorization of integers, because in this case

$$n(p)=1+a(p)+p=p.$$

Theorem 4. *Let n_0 be a composit number with a prime factor of the form*

$$p=12n^2\pm 6n+1.$$

Then the factor p can be computed by probabilistic polynomial time operations.

Proof. We consider in the ring Z_{n_0}, namely the ring of residue classes modulo composite number n_0. Take (x_1, y_1) at random. Let $k=y_1{}^2-x_1{}^3$ and consider the curve

$$y^2=x^3+k.$$

Compute n_0+1 times of (x_1, y_1) in the associative operation determined rationally by the curve. Say this point (s, t). In order to compute inverse element mod n_0, we use Euclid's argorithm. If the number have common divisor with n_0, then we proceed the factorization one step further. So we may assume that all the numbers appear to compute the inverse is relatively prime to n_0. Let a_0 be the golden ratio, namely $(1+\sqrt{5})/2$. Then the number of needed divisions and multiplications are constant times of logarithm of n_0 with basis a_0. Any way they are no more than $\log(n_0)/\log(a_0)$. In order to compute n_0+1 times of (x_1, y_1), no more than $2\log_2(n_0)$ operations are necessary. Now we consider homomorphism

$$h: Z_{n_0}\rightarrow F_p.$$

Since the numbers appearing to compute (s, t) are relatively prime to n_0, the homomorphism h commute with rational operations needed here. The elliptic curve is now shifted to the curve

$$y^2=x^3+h(k)$$

over F_p. The order of the group is $p=n(p)$, and it is assumed that unknown prime p is a factor of n_0. This means that (s, t), namely n_0+1 times of (x_1, y_1), and the original point (x_1, y_1) are shifted to the same point by h, because n_0 is a multiple of p. So we compute greatest common divisor of x_1-s and n_0, namely (x_1-s, n_0) by Euclid's argorithm. This number is divisible by p with probability more than 1/6. Since this kind of operations are independent for other factors, we can find out p with some positive possibility for each randomely selected point (x_1, y_1). This is the meaning of probabilistic polynomial time (PP) computation.

Following are examples of primes $p=12n^2 \pm 6n+1$:

n	1	2	3	4	5	6	7	8	9	10	11	12	13
−	7	37	−	−	271	397	547	−	919	−	−	1657	1951
+	19	61	127	−	331	−	631	−	−	−	−	1801	−

Computation shows the density is roughly proportional to

$$(n^2 \log(n))^{-1}$$

with some fluctuations.

The case $p=7$, is the simplest case and the curve

$$y^2 = x^3 + 5$$

realizes order 7 group. Elements are

$$(3, 2), (3, 5), (5, 2), (5, 5), (6, 2), (6, 5), \infty$$

and (3, 2) is a generator of syclic group with

$$\infty, (3, 2), (5, 2), (6, 5), (6, 2), (5, 5), (3, 5)$$

as the transition table.

It seems to be very interesting to find out another series of prime numbers having algebraic density with the property of probabilitistic polynomial time complexty mentioned above. Another problem is the reseach of low degree polynomial with low absolute value residue mod p.

References

[1] R. P. Brent, An Inproved Monte Carlo Factorization Argorithm, *Nordisk Tidskrft for Informationsbehandling* (BIT) **20** (1980), 174—184.

[2] J. Brillhart, D. H. Lehmer, J. L. Selfridge, B. Tuckerman and S. S. Wagstaff, Jr., Factrization of b^n±1, b=2, 3, 4, 5, 6, 7, 10, 11, 12, up to High Powers, AMS, Providence R. I., 1983.

[3] H. Cohen, H. W. Lenstra, Jr., Primality Testing and Jacobi Sums, *Math. Comp.* **42** (1984), 297—330.

[4] H. Davenport, H. Hasse; Die Nullstellen der Kongruenz Zetafunction in gewissen zyklischen Fallen, *J. Reine und Angew. Math.*, **172**(1935), 151—182.

[5] C. F. Gauss; Disquisitiones Arithmeticae, Yale Univ. Press, New Heaven, 1966.

[6] K. Ireland, M. Rosen, A Classical Introduction to Modern Number Theory, GTM. 64, Springer-Verlag, 1982.

[7] D. E. Knuth; The Art of Computer Programming, vol. **2**, Addison-Wesley, 1981.

[8] H. W. Lenstra, Jr., Primality Testing Algorithms, *Seminaire Bourbaki* **33** (1980 — 81) No. 576, 243— 257.

[9] J. M. Pollard, A Monte Carlo Method for Factorization, Nordisk Tidskrift for Informationsbehandling (BIT), **15**(1975) 331—334.

[10] H. Riesel, Prime Numbers and Computer Methods for Factorization, *Progress in Mathematics* **57**, Birkhauser, 1985.

[11] A. Weil, Jacobi sums as "Grossencharactere", Trans. Am. Math. Soc., **73**(1952) 487— 495.

[12] A. Weil, Oeuvres Scientifiques, Collected Papers, Springer-Verlag, 1980.

Modular Forms of Weight 3/2 Related to Class Numbers*

Pei Dingyi

Department of Mathematics, GraduateSchool, Academia Sinica

Let the function $H(n)$, defined for positive integers n, denote the number of equivalence classes of quadratic forms of integral coefficients with discriminant $-n$, where the equivalence classes of $m(x^2+y^2)$ and $m(x^2+xy+y^2)$, $m\in\mathbb{Z}$, are counted with multiplicity 1/2 and 1/3 respectively. It is well known (for instance, see Hua Loo-Keng[1]) that $H(n)=0$ if $-n\equiv 2$ or 3 (mod 4) and

$$H(n)=\frac{\sqrt{|D|}}{\pi}L\left(1,\left(\frac{D}{\cdot}\right)\right)\sum_{e|f}e\prod_{p|e}\left(1-\left(\frac{D}{p}\right)p^{-1}\right)$$

$$=L\left(0,\left(\frac{D}{\cdot}\right)\right)\sum_{e|f}e\prod_{p|e}\left(1-\left(\frac{D}{p}\right)p^{-1}\right),$$

if $-n\equiv 0$ or 1 (mod 4), where $-n=Df^2$, D is the discriminant of $\mathbb{Q}(\sqrt{-n})$, f is a positive integer and p denots a prime. We define $H(0)=-1/12$. Then the series

$$\mathscr{H}(z)=\sum_{n=0}^{\infty}H(n)q^n,\quad z=x+iy,\ y>0,$$

where $q=e^{2\pi iz}$, has some relation with modular form of weight 3/2. $\mathscr{H}(z)$ is not a modular form itself. But the function

$$\mathscr{F}(z)=\mathscr{H}(z)+y^{-\frac{1}{2}}\sum_{n=-\infty}^{+\infty}\beta(4\pi n^2y)q^{-n^2},\tag{1}$$

where $\beta(x)$ is defined by

$$\beta(x)=\frac{1}{16\pi}\int_1^{\infty}u^{-3/2}e^{-xu}\,du\qquad(x\geqslant 0),$$

satisfies the transformation equation of modular form of weight 3/2 on $\Gamma_0(4)$, i.e.:

$$\mathscr{F}\left(\frac{az+b}{cz+d}\right)=\left(\frac{-1}{d}\right)^{-3/2}\left(\frac{c}{d}\right)(cz+d)^{3/2}\mathscr{F}(z),\ \text{for}\begin{pmatrix}a&b\\c&d\end{pmatrix}\in\Gamma_0(4)$$

* The project is supported by the Science Fund of the Chinese Academy of Sciences.

(see F. Hirzebruch and D. Zagier [2]). Since there is a non-analytic piece involving the powers q^{-n^2}, $\mathcal{F}(z)$ is a non-analytic form.

Let $N>1$ be a square-free positive odd integer. We use the notation.

$$L_N(s,\chi)=L(s,\chi)\prod_{p\mid N}(1-\chi(p)p^{-s}),\quad s\in\mathbb{C},$$

where χ is a character. For any non-negative integer n we define $H(n,N)=0$, if $-n\equiv 2$ or $3 \pmod 4$, and

$$H(n,N)=L_N\left(0,\left(\frac{D}{\cdot}\right)\right)\sum_{\substack{e\mid f\\(e,N)=1}} e\prod_{p\mid e}\left(1-\left(\frac{D}{p}\right)p^{-1}\right),$$

if $-n\equiv 0$ or $1 \pmod 4$ and $-n=Df^2\neq 0$ as mentioned above. Finally define

$$H(0,\mathrm{N})=\mathrm{L_N}(-1,\mathrm{id.})=\zeta(-1)\prod_{p\mid N}(1-p)=-\frac{1}{12}\prod_{p\mid N}(1-p).$$

In this note we shall prove that the function

$$\sum_{n=0}^{\infty}H(n,N)q^n$$

is a modular form of weight 3/2 on $\Gamma_0(4N)$. The non-analytic piece disappears in this case, when $N>1$. In fact we shall prove a little more general result.

For any natural number m with $m\mid N^\infty$, let χ_m be the character modulo $4N$ such that

$$\chi_m(k)=\left(\frac{m}{k}\right),\quad \text{for } (k,4N)=1.$$

For any positive divisor 1 of N we introduce the series

$$E(s,X_l,4N)(z)=y^{s/2}\sum_{r\in\Gamma_\infty\backslash\Gamma_0(4N)}X_l(d_r)j(r,z)^{-3}|j(r,z)|^{-2s}\qquad s\in\mathbb{C},\ \mathrm{Re}(s)>1/2,\ z=x+iy,\ y>0,$$

and

$$E'(s,\chi_l,4N)(z)=E(s,\chi_l,4N)(-1/4Nz)z^{-3/2},$$

where

$$\Gamma_\infty=\left\{\pm\begin{pmatrix}1&n\\0&1\end{pmatrix}\,\middle|\, n\in\mathbb{Z}\right\}$$

and

$$j(r,z)=\left(\frac{-1}{d}\right)^{-\frac{1}{2}}\left(\frac{c}{d}\right)(cz+d)^{\frac{1}{2}}\quad \text{for } r=\begin{pmatrix}a&b\\c&d_r\end{pmatrix}\in\Gamma_0(4).$$

We have

$$E(s,\chi_l,4N)(r(z))=\chi_l(d_r)j(r,z)^3E(s,\chi_l,4N)(z),$$

$$E'(s,\chi_{lN},4N)(r(z))=\chi_l(d_r)j(r,z)^3E'(s,\chi_{lN},4N)(z)$$

for any $r=\begin{pmatrix} a & b \\ c & d \end{pmatrix}\in\Gamma_0(4N)$, where $r(z)=(az+b)(cz+d)^{-1}$ (*see Shimura*[4]).
These functions are analytic in s when $\mathrm{Re}(s)>1/2$ and by analytic continuation we obtain functions $E(0, \chi_l, 4N)(z)$ and $E'(0, \chi_{lN}, 4N)(z)$ which are possibly not holomorphic in z but at least satisfy the transformation equation of modular form of weight 3/2 and "Nebentypus" χ_l on $\Gamma_0(4N)$.

We suppose $l=1$ at first. We shall see later that it is easy to obtain the result for general l from this special one. Put

$$E(N,z)=E(0,\mathrm{id},4N)(z) \text{ and } E'(N,z)=E'(0,\mathrm{X}_N,4N)(z).$$

We have already calculated the Fourier developments of $E(N,z)$ and $E'(N,z)$ in [3]. Let us introduce some notations used in [3] first. For a positive integer n and a prime p let $p^{h(p,n)}$ be the highest power which occurs in n. Now we define the notations $A(p,n)$ as follows: Put

$$A(2,n)=\begin{cases} 4^{-1}(1-i)(1-3\cdot 2^{-(1+h(2,n))/2}), & 2\nmid h(2,n), \\ 4^{-1}(1-i)(1-3\cdot 2^{-(1+h(2,n)/2)}), & 2\mid h(2,n),\ n/2^{h(2,n)}\equiv 1 \pmod 4, \\ 4^{-1}(1-i)(1-2^{-h(2,n)/2}), & 2\mid h(2,n),\ n/2^{h(2,n)}\equiv 3 \pmod 8, \\ 4^{-1}(1-i), & 2\mid h(2,n),\ n/2^{h(2,n)}\equiv 7 \pmod 8, \end{cases}$$

and, $p\neq 2$, (2)

$$A(p,n)=\begin{cases} p^{-1}-(1+p)p^{-(3+h(p,n))/2}, & 2\nmid h(p,n), \\ p^{-1}-2p^{-1-h(p,n)/2}, & 2\mid h(p,n),\ \left(\dfrac{-n/p^{h(p,n)}}{p}\right)=-1, \\ p^{-1}, & 2\mid h(p,n),\ \left(\dfrac{-n/p^{h(p,n)}}{p}\right)=1. \end{cases} \tag{3}$$

Define

$$\lambda(n,4N)=-4\pi(1+i)L_{4N}(1,\tilde{\chi}_{-n})L_{4N}^{-1}(2,\mathrm{id})\sum\mu(a)\tilde{X}_{-n}(a)(ab)^{-1},$$

where $\tilde{\chi}_{-n}$ is the primitive character determined by $\left(\dfrac{-n}{\ }\right)$, the last sum is extented over all positive integers a and b prime to $2N$ such that $(ab)^2$ divides n. Then $E(N,z)$ and $E'(N,z)$ have the following expansions:

$$E(N,z)=1+\sum_{n=1}^{\infty}\lambda(n,4N)\prod_{p|2N}A(p,n)n^{\frac{1}{2}}q^n-\prod_{p|N}(1+p)^{-1}g(z), \tag{4}$$

$$E'(N,z)=\sum_{n=1}^{\infty}\lambda(n,4N)n^{\frac{1}{2}}q^n-2(1+i)\prod_{p|N}p(1+p)^{-1}g(z), \tag{5}$$

where

$$g(z)=(\pi y^{\frac{1}{2}})^{-1}+4^{-1}\sum_{n=1}^{\infty}n\int_1^{\infty}v^{-1}(v-1)^{\frac{1}{2}}e^{-4\pi n^2 yv}dv\, q^{-n^2}.$$

Now for any positive divisor m of N, we define the functions $F(4m, 4N, z)$ and $F(m, 4N, z)$ by

$$F(4m,4N,z)=\sum_{d|N/m}\mu(d)E(md,z),$$

and

$$F(m,4N,z)=\sum_{d|m}\mu(d)E'(Nd/m,z),$$

respectively. Furthermore define the functions $\mathscr{H}(m, 4N, z)$ by

$$\mathscr{H}(m,4N,z)=\sum_{d|m}\frac{\mu(d)}{d}F(4m/d,4N,z)-\frac{i-1}{8}\sum_{d|m}\frac{\mu(d)}{d}F(m/d,4N,z).$$

By a straight forward calculation we obtain

$$\mathscr{H}(N,4N,z)=1+\sum_{n=1}^{\infty}\lambda(n,4N)\left(A(2,n)-\frac{i-1}{8}\right)\prod_{p|N}(A(p,n)-p^{-1})n^{\frac{1}{2}}q^n, \tag{6}$$

and, for $m|N$, $m\neq N$,

$$\mathscr{H}(m,4N,z)=\sum_{n=1}^{\infty}\lambda(n,4N)\left(A(2,n)-\frac{i-1}{8}\right)\prod_{p|m}(A(p,n)-p^{-1})n^{\frac{1}{2}}q^n. \tag{7}$$

When $-n\equiv 2$ or $3 \pmod 4$, we have $A(2, n)-(i-1)/8=0$ from (2). Write $-n=D_nf_n^2$ if $-n\equiv 0$ or $1 \pmod 4$, where f_n is a positive number and D_n is the discriminant of $\mathbb{Q}(\sqrt{-n})$, then we have in this case that

$$A(2,n)-\frac{i-1}{8}=\frac{3(1-i)}{8}\left(1-\frac{1-\left(\frac{D_n}{2}\right)}{2^{h(2,f_n)+1}\left(1-2^{-1}\left(\frac{D_n}{2}\right)\right)}\right), \tag{8}$$

and

$$\begin{aligned}\lambda(n,4N)&=-4\pi(1+i)\frac{L\left(1,\left(\frac{D_n}{\cdot}\right)\right)}{\zeta(2)}\prod_{p|2N}\frac{1-\left(\frac{D_n}{p}\right)p^{-1}}{1-p^{-2}}\\&\times\prod_{\substack{p|f\\p|2N}}\left(\sum_{l=0}^{h(p,f_n)}p^{-l}-\left(\frac{D_n}{p}\right)p^{-1}\sum_{l=0}^{h(p,f_n)-1}p^{-l}\right).\end{aligned} \tag{9}$$

When $p|N$, we also have, from (3), that

$$p^{h(p,f_n)}(A(p,n)-p^{-1})\left(1-\left(\frac{D_n}{p}\right)p^{-1}\right)(1-p^{-2})^{-1}=\left(1-\left(\frac{D_n}{p}\right)\right)(1-p)^{-1}. \quad (10)$$

Now for any positive divisor m of N introduce

$$H(n,m,N)=L_m\left(0,\left(\frac{D_n}{}\right)\right)\prod_{p|N/m}\left(1-\left(\frac{D_n}{p}\right)p^{-1}\right)p^{h(p,f_n)}$$

$$\times\sum_{\substack{e|f_n\\(e,N)=1,}} e\prod_{p|e}\left(1-\left(\frac{D_n}{p}\right)p^{-1}\right),$$

if $-n\equiv 0$ or $1 \pmod 4$, and

$$H(n,m,N)=0 \quad \text{if } -n\equiv 2 \text{ or } 3 \pmod 4,$$

$$H(0,m,N)=L_m(-1,\mathrm{id})\prod_{p|N/m}(1-p^{-2}).$$

Note that $H(n,N,N)$ is equal to $H(n,N)$ as defined above. Substituting (8), (9) and (10) into (6) and (7) we obtain

$$\mathscr{H}(N,4N,z)=H(0,N)^{-1}\sum_{n=0}^{\infty}H(n,N)q^n,$$

and

$$\mathscr{H}(m,4N,z)=H(0,m,N)^{-1}\sum_{n=1}^{\infty}H(n,m,N)q^n \quad (m|N,\ m\neq N).$$

They are (holomorphic) modular forms of weight 3/2 on $\Gamma_0(4N)$. This proves our conclusion.

Let l be a positive divisor of N. Using the Hecke operator $T(l)$ we know that the functions

$$H(0,N)^{-1}\sum_{n=0}^{\infty}H(nl,N)q^n$$

and

$$H(0,m,N)^{-1}\sum_{n=1}^{\infty}H(nl,m,N)q^n \quad (m|N,\ m\neq N)$$

are modular forms of weight 3/2 and character X_l on $\Gamma_0(4N)$.

References

[1] Hua Loo-Keng, Introduction to number theory, Springer-Verlag, 1982.

[2] F. Hirzebruch and D. Zagier, Intersection numbers of curves on Hilbert modular surfaces and modular forms of Nebentypus, *Inventions math.* **36** (1976), 57—113.

[3] Pei Dingyi, Eisenstein series of weight 3/2, I, *Trans. of Amer. Math. Soc.* **274** (1982), 573—606.

[4] G. Shimura, On modular forms of helf integral weight, *Ann. of Math.* **97** (1973), 440—481.

ON THE CONGRUENT NUMBER PROBLEM

Peng Tsu Ann
Department of Mathematics, National University of Singapore
Kent Ridge, Singapore 0511

Let A be a square–free positive integer. Suppose that A is a congruent number, i.e., suppose that there exist positive rational numbers X, Y, Z such that

$$X^2 + Y^2 = Z^2,$$
$$\frac{1}{2}XY = A.$$

It is easily verified that the rational points $(\frac{Z^2}{4}, \pm\frac{(X^2-Y^2)Z}{8})$ lie on the elliptic curve

$$y^2 = x^3 - A^2 x.$$

Let $x = \frac{Z^2}{4}$. Then we have

$$\left(\frac{X+Y}{2}\right)^2 = x + A,$$
$$\left(\frac{X-Y}{2}\right)^2 = x - A.$$

Assuming that $Y > X$ we get

$$(*) \qquad \begin{aligned} X &= \sqrt{x+A} - \sqrt{x-A}, \\ Y &= \sqrt{x+A} + \sqrt{x-A}, \\ Z &= 2\sqrt{x}. \end{aligned}$$

Now suppose that (x, y) is a rational point on the elliptic curve $y^2 = x^3 - A^2 x$. What conditions must x satisfy so that the numbers X, Y, Z given by the formulas $(*)$ are all rational? Koblitz stated in [1] (Proposition 2, p.7) that the following conditions are sufficient :

(i) x is the square of a rational number;
(ii) the denominator of x is even.

But this is false. Let $A = 5$. The integer 5 is congruent because we have $(\frac{3}{2})^2 + (\frac{20}{3})^2 = (\frac{41}{6})^2$ and $\frac{1}{2}XY = 5$. The rational point $(\frac{25}{4}, \frac{75}{8})$ lies on the elliptic curve $y^2 = x^3 - 25x$. Clearly $x = \frac{25}{4}$ satisfies the conditions (i) and (ii), but the formulas (*) give

$$X = \sqrt{\frac{25}{4} + 5} - \sqrt{\frac{25}{4} - 5} = \sqrt{5},$$
$$Y = \sqrt{\frac{25}{4} + 5} + \sqrt{\frac{25}{4} + 5} = 2\sqrt{5},$$
$$Z = 2\sqrt{\frac{25}{4}} = 5.$$

To make Koblitz's result true we need the following additional condition :

(iii) the numerator of x is prime to A.

In his proof Koblitz claimed that a certain Pythagorean triple is primitive. The claim only holds if the condition (iii) is satisfied. The rest of his proof goes through without change.

The conditions (i), (ii) and (iii) are also necessary, i.e., if A is a square-free positive integer and X, Y, Z are rational numbers such that

$$X^2 + Y^2 = Z^2,$$
$$\frac{1}{2}XY = A,$$

then the x-coordinate of $(\frac{Z^2}{4}, \pm\frac{(X^2 - Y^2)Z}{8})$ satisfies the conditions (i), (ii) and (iii). This follows immediately from

Proposition 1. Let $X^2 + Y^2 = Z^2$, $XY = 2A$, where X, Y, Z are positive rational numbers and A is a square-free positive integer. Let $X = m_1/n_1$, $Y = m_2/n_2$, $Z = m_3/n_3$, where m_i, n_i are positive integers with g.c.d. $(m_i, n_i) = 1$ for $i = 1, 2, 3$. Then the following hold :

(a) $(m_1, m_2) = 1$;
(b) $(m_3, A) = 1$;
(c) m_3 is odd.

Proof. We have

$$2A = \frac{m_1}{n_1} \cdot \frac{m_2}{n_2} = \frac{m_1}{n_2} \cdot \frac{m_2}{n_1}, \tag{1}$$

$$(n_2 m_1)^2 + (n_1 m_2)^2 = (m_3 \cdot \frac{n_1 n_2}{n_3})^2, \tag{2}$$

where m_1/n_2, m_2/n_1, n_1n_2/n_3 are all integers.

Let p be a prime such that $p \mid (m_1, m_2)$. Then $p \nmid n_1$ and $p \nmid n_2$. If p is odd, then it follows from (1) that $p^2 \mid A$, contrary to our assumption that A is square-free. Thus we must have $p = 2$. Then m_1 and m_2 are both even, so that n_1 and n_2 are both odd. Let $m_1 = 2m_1'$ and $m_2 = 2m_2'$. Then m_1' and m_2' must be both odd, otherwise A would not be square-free. It follows from (2) that

$$(n_2m_1')^2 + (n_1m_2')^2 = (\frac{m_3}{2} \cdot \frac{n_1n_2}{n_3})^2,$$

which yields

$$1 + 1 \equiv (\frac{m_3}{2} \cdot \frac{n_1n_2}{n_3})^2 \pmod 4.$$

This is impossible. Hence (a) holds.

Now suppose that p is a prime divisor of A. Then it follows from (1) that $p \mid m_1$ or $p \mid m_2$. If $p \mid m_3$, then, by (2), we have $p \mid m_1$ and $p \mid m_2$, contrary to (a). Hence $(m_3, A) = 1$, proving (b).

Finally, (1) and (a) show that exactly one of m_1 and m_2 is even. This implies that exactly one of n_2m_1 and n_1m_2 is odd. Thus

$$(n_2m_1)^2 + (n_1m_2)^2 \equiv 1 \pmod 4.$$

It then follows from (2) that m_3 is odd and (c) is proved.

Now denote by P the point

$$(x, y) = (\frac{Z^2}{4}, \frac{(X^2 - Y^2)Z}{8})$$

on the elliptic curve $y^2 = x^3 - A^2x$. Then it can easily be verified (using the usual addition formulas for points) that $P = 2Q$, where Q is the point

$$(-\frac{AY}{X+Z}, \frac{2A^2}{X+Z}).$$

This prompts one to ask the following question : if a point P on the elliptic curve $y^2 = x^3 - A^2x$ is the double of some rational point Q, is it always true that the numbers X, Y, Z given by the formulas (*) are rational? An affirmative answer is given in [1] (Proposition 19, p.46). We give below a direct and elementary proof of this.

Proposition 2. Let Q be a rational point of infinite order (i.e., Q is not the point 0, $(0,0)$, or $(\pm A, 0)$) on the elliptic curve $y^2 = x^3 - A^2x$ and let $P = 2Q = (x,y)$. Then the numbers X, Y, Z given by $(*)$ are rational.

Proof. Let $Q = (u,v)$. Then the addition formulas for points give

$$x = \left(\frac{u^2 + A^2}{2v}\right)^2.$$

It can be checked that

$$x + A = \left(\frac{u^2 - A^2 + 2Au}{2v}\right)^2,$$

$$x - A = \left(\frac{u^2 - A^2 - 2Au}{2v}\right)^2.$$

Thus the numbers X, Y, Z given by the formulas $(*)$ are rational.

Corollary. The x-coordinate of the double of any rational point of infinite order on $y^2 = x^3 - A^2x$ satisfies (i), (ii) and (iii).

Proof. That it satisfies (i) is trivial. That it satisfies (ii) and (iii) follows from Proposition 1.

It will be interesting to know whether the above corollary can be proved directly without using Propositions 1 and 2.

References

[1] N. Koblitz, *Introduction to Elliptic Curves and Modular Forms*, Graduate Texts in Mathematics 97, Springer–Verlag, 1984.

ASPECTS OF THE SMALL SIEVE

H.-E. Richert
Abt. für Mathematik, Universität Ulm
Oberer Eselsberg, D-7900 Ulm

1. Some one and a half decades ago, our book [29] on sieve methods was published. From there to modern sieve theory many a change has taken place: several important new ideas have been added, and the number of applications, also including many new ones, grew considerably.

Therefore I will take the opportunity to give a brief report on some selected, (personally biased) aspects of the small sieve development since. Before starting I would like to stress that it turned out that two old beliefs are true: Each problem deserves its own specific sieve treatment, and, even more important, the power of a sieve may be greatly enhanced by an injection of suitable analytical tools (Selberg [79]).

The following beautiful example for the latter was first given by Iwaniec and Jutila [50]. Due to the 'parity phenomenon' a proper sieve is not capable of getting results for primes only. Nevertheless, in [50] it was proved that

$$p_{n+1} - p_n \ll p_n^{\theta}$$

for $\theta \geqq \frac{13}{23}$, the best result so far. The starting point in this work is the linear sieve, enhanced by Iwaniec's bilinear form for the remainder term, but for a treatment of a certain portion the following analytical tools were employed: the mean value theorem for Dirichlet polynomials, the Halász-Montgomery inequality, and Jutila's weighted density theorem. Later on, in an improvement by Heath-Brown and Iwaniec [36] to $\theta > \frac{11}{20}$, even more ideas were added; presently the best result is [52]. We introduce some standard notations from sieve theory. Consider a finite sequence $\mathcal{A}$ of integers, a set $\mathcal{P}$ of primes and a real parameter $z \geqq 2$:

$$\mathcal{A} = \{a: \ldots\}, \quad \mathcal{P} \subset \mathbb{P}, \quad P(z) := \prod_{\substack{p<z \\ p\in\mathcal{P}}} p .$$

Then the basic object of small sieves is the sifting function, defined by

$$S(\mathcal{A},z) := S(\mathcal{A},\mathfrak{P},z) := |\{a\in\mathcal{A},\ (a,P(z)) = 1\}| = \sum_{a\in\mathcal{A}} \sum_{\substack{d|P(z)\\ d|a}} \mu(d)\ ,$$

i.e., on putting

$$\mathcal{A}_d := \{a\in\mathcal{A},\quad a\equiv 0 \bmod d\}\ ,$$

$$(1.1)\qquad S(\mathcal{A},z) = \sum_{d|P(z)} \mu(d)|\mathcal{A}_d| = \sum_{\substack{d\\ q(d)<z\\ (d,\overline{\mathfrak{P}})=1}} \mu(d)|\mathcal{A}_d|\ .^{+}$$

Arranging the right-hand side with respect to the greatest prime divisor of $d(>1)$ leads at once to the 'Buchstab formula' for the sifting function

$$(1.2)\qquad S(\mathcal{A},z) = S(\mathcal{A},z_1) - \sum_{\substack{z_1\le p<z\\ p\in\mathfrak{P}}} S(\mathcal{A}_p,p)\qquad \text{for}\quad 2\le z_1<z\ .$$

nature of remainder terms.

Assuming that

$$0 \leqq \omega(p) < p$$

and that there are constants $\kappa > 0$ and $A \geqq 1$ such that

$$\Big| \sum_{z_1 \leqq p < z_2} \frac{\omega(p)}{p} \log p - \kappa \log \frac{z_2}{z_1} \Big| \leqq A \qquad \text{for } 2 \leqq z_1 < z_2$$

(or even a one-sided condition only:

$$\prod_{z_1 \leqq p < z_2} \left(1 - \frac{\omega(p)}{p}\right)^{-1} \leqq \left(\frac{\log z_2}{\log z_1}\right)^{\kappa} \left(1 + \frac{A}{\log z_1}\right) \qquad \text{for } 2 \leqq z_1 < z_2 ,$$

cf. [47], [74], [15]), sieves can be classified with respect to κ (inf in the latter case), called the dimension of the sieve.

Under these assumptions one obtains a first general result:

$$S(\mathcal{A},z) \begin{matrix} \leqq \\ \geqq \end{matrix} XV(z) \left\{ \begin{matrix} F_\kappa \\ f_\kappa \end{matrix} \left(\frac{\log D}{\log z}\right) + o(1) \right\} + R(\mathcal{A},D,z) , \tag{1.4}$$

where $o(1)$ stands for a usually unimportant error term, the functions F_κ, f_κ in the main term are defined by certain differential-difference equations, and the remainder is of the form

$$R(\mathcal{A},D,z) = \sum_{\substack{d<D \\ d \mid P(z)}} c_d \, r(\mathcal{A},d) \tag{1.5}$$

with coefficients $|c_d| \leqq 1$. $D(\geqq z)$ is a parameter, called the level (of uniform distribution in arithmetic progressions, cf. (1.3)) will be chosen in such a way that the remainder becomes of less order than the main term:

$$R(\mathcal{A},D,z) = o(XV(z)) \tag{1.6}$$

(in applications the values of F_κ, f_κ are bounded and bounded away from zero). As to the quality of these results, and focussing on the main terms only, Selberg [80] has given examples in the cases $\kappa = 1$ and $\kappa = \frac{1}{2}$, which show that (1.4) cannot be improved upon, and for the other values of $\kappa(\leq 1)$, it follows from the definitions of F_κ and f_κ, that (1.4) cannot be improved by further applications of Buchstab's procedure.

The above results can be proved by the Buchstab-Rosser sieve, first fully developed and published by Iwaniec [47]. It rests on combinatorial means, which amounts to a careful selection of terms in an iterated Buchstab formula (1.2).

A second method of proving a result of type (1.4) is based on Selberg's Λ^2-sieve, i.e. one starts by an injection of Selberg's well-known expression

$$\left(\sum_d \lambda_d\right)^2 \tag{1.7}$$

and improves the resulting estimates step by step by Buchstab's formula. For general κ this becomes very complicated, and only step one and step two were carried out by Ankeny-Onishi [2] and Porter [73], respectively; in 1973 Diamond and Jurkat (unpublished) worked out numerically ten steps for some values of κ in $\frac{1}{2} < \kappa \leq 2$. However, in the most important case, the linear sieve, i.e. $\kappa = 1$, a sufficiently long iteration of (1.2) can be executed, cf. [53], [29], and this leads to essentially the same result as in (1.4). Moreover, in this case Motohashi [65], [66] was the first to notice that the latter approach could also be based on combinatorial arguments. In fact, we may now see the following 'sieve identity' at the basis of all results for $S(\mathcal{A}, z)$ of type (1.4):

Let $\chi(.)$ be an arithmetic function, $\chi(1) = 1$, and define [†]

$$\bar{\chi}(d) = \chi\left(\frac{d}{p(d)}\right) - \chi(d) \quad \text{for } d > 1, \qquad \bar{\chi}(1) = 0.$$

Then, for any arithmetic function $\phi(.)$,

$$\sum_{d|P(z)} \mu(d)\phi(d) = \sum_{d|P(z)} \mu(d)\chi(d)\phi(d) + \sum_{d|P(z)} \mu(d)\bar{\chi}(d) \sum_{t|P(p(d))} \mu(t)\phi(dt),$$

cf. [47], Lemma 1; [67], Theorem 6; and in the above general form [30], Lemma 2.

[†] p(d) denotes the least prime factor of d.

Iwaniec [45], [51] gave the (conditionally) ideal, i.e. Buchstab-invariant, functions and conditions to be satisfied by a Selberg-sieve result for $\kappa > 1$. Based on this, Diamond-Halberstam-Richert [15] succeeded via the combinatorial way in obtaining the Selberg analogue of (1.4) for all $\kappa > 1$. The resulting functions F_κ, f_κ, however, differ from those obtained by the Buchstab-Rosser sieve. The latter are the better ones for $\kappa < 1$, whereas Selberg's sieve gives superior results for $\kappa > 1$, and for $\kappa = 1$ both coincide (see Iwaniec [45]).

2. The combinatorial method for sieve estimates consists of iterating Buchstab's formula, and then, upon using the trivial estimate $S(\mathcal{A},z) \geq 0$, omitting judiciously certain terms from the resulting expression. One method of improvement is a slight deviation from the pattern which led to the results mentioned before. We do not fully understand yet, under what conditions this can be performed, and then what the optimal procedure would be. This was applied in the paper of Heath-Brown and Iwaniec [36] mentioned at the beginning. As another example where this idea (among many others) was vitally used, we mention the paper of Fouvry [18], where he succeeded in proving that with some $\delta > 0$ and $c_0(\delta) > 0$

$$|\{p \leq x, \quad p \equiv 2 \bmod 3, \quad q(p-1) \geq x^{2/3+\delta}\}| \geq c_0 \frac{x}{\log x} \qquad \text{for } x \geq x_0(\delta).$$

This in turn, by a criterion of Adleman and Heath-Brown [1] proved that the so-called first case of Fermat's Last Theorem holds true for infinitely many primes, i.e.

$$x^p + y^p = z^p, \qquad x,y,z \in \mathbb{Z}, \qquad p \mid xyz,$$

has no solution for infinitely many $p \in \mathbb{P}$. (Remember that the famous result of Faltings states that $x^n + y^n = z^n$ has for each $n \geq 3$ at most finitely many primitive solutions ($(x,y,z) = 1$, $xyz \neq 0$).)
An often used straight enhancement of the power of (1.4) has its origin in a paper of Kuhn [56], where he inserted constant weights into $S(\mathcal{A},z)$. Mostly used are logarithmic weights, by which $S(\mathcal{A},z)$ is replaced by an expression

$$S(\mathcal{A},z) - \lambda \sum_{\substack{z \leq p < z_2 \\ p \in \mathcal{P}}} \left(1 - \frac{\log p}{\log z_2}\right) S(\mathcal{A}_p, z) ,$$

For $\kappa = 1$ Greaves [25], [26], [27] developed a method by which the S's can be estimated simultaneously. His method was extended by Halberstam-Richert [30] to $\frac{1}{2} \leq \kappa \leq 1$; and for $\kappa = 1$ Iwaniec's bilinear form of the remainder term [48] was incorporated [31]. An application gave the existence of a P_2 † in

$$x < P_2 \leq x + x^{\theta}, \quad x \geq x_0 \quad \text{with} \quad \theta < 0.45 .$$

Other weights in the linear sieve are due to Buchstab [9] (analyzed and simplified by Rotermund [76] and Laborde [57]) and Chen, e.g. [13], [14]. I also mention another idea of Chen [12] where, after an application of (1.2), he does not estimate each term in the sum in (1.2) individually, but rather deals with short segments of that sum. When working with Buchstab-invariant estimates, this has no influence on the main terms, but results in remarkable cancellations of the remainder terms, see Iwaniec [46] .

Apart from the use of the classical sifting function $S(\mathcal{A},z)$ and the preceding remarks, there are various different sifting procedures of small sieve character, which cannot be discussed here. They reach from an ad hoc insertion of Selberg's factor (1.7) and more sophisticated expressions cf. [5], Théorème 18, [78] over special ideas of Hooley's e.g. [38], [39], [40], [41], Montgomery's sieve [58], [59], Bombieri's asymptotic sieve [4], [21], Selberg's $\tau(n)$-method [81], [62], the sieve of Erdös-Ruzsa [16], [77], Heath-Brown's square sieve [33], his sieve for arbitrary functions [35] up to Selberg's introduction of pseudocharacters [82], [64] and Barban-Vehov's method [3], [60], [55] with important applications in analysis, e.g. to zero-density estimates for Dirichlet series.

3. We now turn to the remainder term (1.5), where in view of (1.3) and (1.6) an estimate

$$\sum_{\substack{d<D \\ d \mid P(z)}} c_d\left(|\mathcal{A}_d| - \frac{\omega(d)}{d} X\right) = o(XV(z)) \tag{3.1}$$

for a level D, as large as possible, is required. For a long time, the term in (1.5) was just estimated by $|r(\mathcal{A},d)|$, often producing a unsatisfactory result. Great progress was made (after some predecessors) by the Bombieri-Vinogradov theorem

---

† P_r denotes an integer with a total number of prime divisors $\leq r$.

For any $A > 0$ there is a $B = B(A) > 0$ such that

$$(3.2)\qquad \sum_{q\leq Q} \max_{y\leq x} \max_{(\ell,q)=1} \left|\pi(y;q,\ell) - \frac{\mathrm{li}(y)}{\phi(q)}\right| \ll x \log^{-A} x$$

for $Q = x^{1/2}\log^{-B} x$.

For shifted sequences of primes this can be immediately applied, and leads via a weighted linear sieve at once to (3.3) below with a P_3 instead of P_2. Jutila [54] was the first to prove such a mean value theorem for short intervals, here the most recent, stronger result is due to Perelli-Pintz-Salerno [71].

In his famous proof for the existence of solutions of the equation

$$(3.3)\qquad N = p + P_2$$

for even $N \geq N_0$, Chen [10], [11] introduced a switching principle, which extends the range of applications for a sieve beyond the normal limit. This was also, independently, discovered by Iwaniec [43] (see [19]), and has been used many times since. A remark concerning limits of this method can be found in Fouvry [17], p. 632.

The 'switching' acts as if (for the switched part) (3.2) was true with some $Q > x^{1/2}$. In pursuing this observation Pan-Ding [68], [69], and about simultaneously Fujii [22], [23], [24] found several generalizations of (3.2), e.g. [68]:

For any $A > 0$ there is a $B = B(A) > 0$ such that

$$(3.4)\qquad \sum_{q\leq x^{1/2}\log^{-B}x} \max_{y\leq x} \max_{(\ell,q)=1} \left| \sum_{\substack{a\leq x^{1-\varepsilon}\\(a,q)=1}} f(a)\left(\pi(y;a,q,\ell) - \frac{\pi(y;a,1,1)}{\phi(q)}\right)\right| \ll x \log^{-A} x,$$

where

$$\pi(y;a,q,\ell) = \left|\{p \leq \frac{y}{a}\ ,\ ap \equiv \ell \bmod q\}\right| \ ,$$

and f is any bounded, real-valued function.

Pan [68] gave, apart from a simpler proof for (3.3) and similar results (see also [70], [24]), some further applications of this mean value theorem. Heath-Brown [34], in his well-known improvement of the work of Gupta and Ram Murty [28] has also, besides an extension of a theorem of Bombieri-Friedlander-Iwaniec [6], used (3.4) for estimating the remainders in his sieve. Artin conjectured in 1927 that any integer $a \neq -1$, which is not a perfect square, is a primitive root for infinitely many primes. Corollaries from the work of Heath-Brown are: Artin's conjecture fails for a) at most three squarefree integers $a > 1$, b) at most two $a \in \mathbb{P}$, c) $\ll \log^2 x$ numbers $|a| \leq x$.

In view of the many successful applications of (3.2), mean value theorems of this type are of great interest not only for the characteristic function of the primes or $\Lambda(n)$, but also for more general sequences $f(n)$, f.e. of the type

$$\sum_{q \leq Q} \max_{(\ell,q)=1} \Big| \sum_{\substack{n \leq x \\ n \equiv \ell \bmod q}} f(n) - \frac{1}{\phi(q)} \sum_{\substack{n \leq x \\ (n,q)=1}} f(n) \Big| \tag{3.5}$$

in order to obtain, for problems which correspond to $f(n)$, a higher level D in (3.1). Once the large sieve for Dirichlet's characters was available, several such results were obtained, especially after Motohashi [63] had found his induction principle, which roughly states that if for two sequences f and g, satisfying a 'Siegel-Walfisz condition', a mean value theorem of the type (3.5) holds, then the same is true for their Dirichlet convolution $f * g$.

4. Great progress was made by Iwaniec [48]. We may say, that Iwaniec with the discovery, that the remainder term in the linear sieve can be given a bilinear form, has started a new chapter in sieve theory. As he stated, Motohashi's penetrating paper on the Brun-Titchmarsh theorem [61] inspired him to this work. To be precise: For $\kappa = 1$, (1.4) can be replaced by

Let $0 < \varepsilon < \frac{1}{8}$, $M > 1$, $N > 1$. Then for any $z \leq (MN)^{1/2}$

$$S(\mathcal{A}, z) \lesseqgtr XV(z) \left\{ \begin{matrix} F \\ f \end{matrix} \left(\frac{\log (MN)}{\log z} \right) + E \right\} \pm R(\mathcal{A}, M, N, z) ,$$

where F, f are the well-known functions of the linear sieve [53]:

$$F(u) = \frac{2e^\gamma}{u}, \qquad f(u) = 0 \qquad \text{for } 0 < u \leq 2 ,$$

$$(uF(u))' = f(u-1), \qquad (uf(u))' = F(u-1) \qquad \text{for } u \geq 2 ,$$

(4.1) $$R(\mathcal{A},M,N,z) = \sum_{\ell<\exp(8\varepsilon^{-3})} \sum_{m<M} \sum_{\substack{n<N \\ mn|P(z)}} a_{m,\ell}\, b_{n,\ell}\, r(\mathcal{A},mn) ,$$

$$E \ll \varepsilon + \varepsilon^{-8} e^{A} \log^{-1/3}(MN), \qquad |a_{m,\ell}| \le 1, \quad |b_{n,\ell}| \le 1 ,$$

and $a_{m,\ell}$, $b_{n,\ell}$ depend at most on ε, M and N, [45].

Recently, Pomykała [72] has obtained a similar result for $\frac{1}{2} < \kappa < 1$; Motohashi [67], Lemma 16, proved a refinement of Iwaniec's basic lemma. This general form of the remainder permits us to inject various efficient analytical tools for a treatment of (4.1) which leads to $R(\mathcal{A},M,N,z) = o(XV(z))$:

a) Expansion of $r(\mathcal{A},mn)$ into a Fourier series. Then, after application of Cauchy's inequality, one is left with exponential sums, which can be dealt with in various ways. Here the recent new developments, starting with the papers Bombieri-Iwaniec [7], [8] and Huxley-Watt [42], regarding estimations of exponential sums should be mentioned.

b) Using ideas of Linnik's dispersion method, which is in some instances more powerful than the large sieve, and then employ estimates for Kloosterman sums.

c) Expressing the remainders via Perron's formula by Dirichlet series and Dirichlet polynomials, and applying their mean value theorems and the Halász-Montgomery-Huxley inequality for Dirichlet polynomials.

All these methods for estimating remainder terms had been, at least in parts, used before (cf. Chen [12], [14], Hooley [37], Motohashi [61]). However, now the greater flexibility by the variables M,N in (4.1), instead of the single D in (3.1), render more powerful results, therefore yielding an enlarged level D, sometimes even $>|\mathcal{A}|$.

Nowadays, following an idea of Selberg [82], a typical general procedure is, roughly speaking, the following: insert a 'test function' f for smoothing, and apply Poisson's summation formula. The resulting expression, containing the Fourier transform $\hat{f}$, can either be treated by bounds for $\hat{f}$ or, even better, by employing estimates for sums over Kloosterman sums. For the use of the latter, which have become an essential tool in sieve applications, the history, the backgrond in the theory of automorphic functions, and the literature we refer to the beautiful article of Iwaniec [49].

These techniques, the use of bilinear forms and the large sieve in an extended version of Motohashi's principle have also led to mean value theorems of type (3.5) for a wide class of arithmetic functions f(n). The crucial point being that f can be represented by a linear combination of convolutions $\alpha * \beta$ of sequences which satisfy the conditions of the induction principle. In the case $f(n) = \Lambda(n)$, this can be achieved by an application of Heath-Brown's identity [32]. For details, conjectures and further references we refer to the important papers of Bombieri-Friedlander-Iwaniec [6], the survey article of Bombieri [5], Appendix, and to Friedlander-Granville [20] for a first, surprising result in the opposite direction.

References

1. Adleman, L. M. and Heath-Brown, D. R.: The first case of Fermat's last theorem. Invent. math. 79 (1985), 409-416.

2. Ankeny, N. C. and Onishi, H.: The general sieve. Acta Arith. 10 (1964), 31-62.

3. Barban, M. B. and Vehov, P. P.: An extremal problem. Trans. Moscow Math. Soc. 18 (1968), 91-99.

4. Bombieri, E.: The asymptotic sieve. Rend. Accad. Naz. XL (5) 5 (1976), 243-269.

5. Bombieri, E.: Le grand crible dans la théorie analytique des nombres. Soc. Math. France, Astérisque no. 18, 2nd ed. 1987.

6. Bombieri, E., Friedlander, J. B. and Iwaniec, H.: Primes in arithmetic progressions to large moduli. I-III. I: Acta Math. 156 (1986), 203-251. II: Math. Ann. 277 (1987), 361-393. III: to appear.

7. Bombieri, E. and Iwaniec, H.: On the order of $\zeta(\frac{1}{2} + it)$. Ann. Scuola Norm. Sup. Pisa Cl. Sci. (4) 13 (1986), 449-472.

8. Bombieri, E. and Iwaniec, H.: Some mean-value theorems for exponential sums. Ann. Scuola Norm. Sup. Pisa Cl. Sci. (4) 13 (1986), 473-486.

9. Buchstab, A.A.: Combinatorial intensification of the sieve method of Eratosthenes. (Russian) Uspehi Mat. Nauk 22 (1967), no. 3 (135), 199-226 = Russian Math. Surveys 22 (1967), no. 3, 205-233.

10. Chen, J.: On the representation of a large even integer as the sum of a prime and the product of at most two primes. (Chinese) Kexue Tongbao 17 (1966), 385-386.

11. Chen, J.: On the representation of a larger even integer as the sum of a prime and the product of at most two primes. Sci. Sinica 16 (1973), 157-176.

12. Chen, J.: On the distribution of almost primes in an interval.
Sci. Sinica 18 (1975), 611-627.

13. Chen, J.: On the Goldbach's problem and the sieve methods.
Sci. Sinica 21 (1978), 701-739.

14. Chen, J.: On the distribution of almost primes in an interval. II.
Sci. Sinica 22 (1979), 253-275.

15. Diamond, H., Halberstam, H. and Richert, H.-E.: Combinatorial sieves of dimension exceeding one.
J. Number Theory 28 (1988), 306-346.

16. Erdös, P. and Ruzsa, I. Z.: On the small sieve. I. Sifting by primes.
J. Number Theory 12 (1980), 385-394.

17. Fouvry, E.: Répartition des suites dans les progressions arithmétiques.
Acta Arith. 41 (1982), 359-382.

18. Fouvry, E.: Théorème de Brun-Titchmarsh; application au théorème de Fermat.
Invent. math. 79 (1985), 383-407.

19. Fouvry, E. and Grupp, F.: On the switching principle in sieve theory.
J. r. ang. Math. 370 (1986), 101-126.

20. Friedlander, J. B. and Granville, A.: Limitations to the equi-distribution of primes. I.
to appear

21. Friedlander, J. B. and Iwaniec, H.: On Bombieri's asymptotic sieve.
Ann. Scuola Norm. Sup. Pisa (4) 5 (1978), 719-756.

22. Fujii, A.: A local study of some additive problems in the theory of numbers.
Proc. Japan Acad. 52 (1976), 113-115.

23. Fujii, A.: Some remarks on Goldbach's problem.
Acta Arith. 32 (1977), 27-35.

24. Fujii, A.: Some additive problems of numbers.
Banach Center Publications (Warsaw 1983), 17 (1985), 121-141.

25. Greaves, G.: A weighted sieve of Brun's type.
Acta Arith. 40 (1982), 297-332.

26. Greaves, G.: A comparison of some weighted sieves.
Banach Center Publications (Warsaw 1983), 17 (1985), 143-153.

27. Greaves. G.: A weighted linear sieve and Selberg's λ^2-method.
Acta Arith. 47 (1986), 71-96.

28. Gupta, R. and Ram Murty, M.: A remark on Artin's conjecture.
Invent. math. 78 (1984), 127-130.

29. Halberstam, H. and Richert, H.-E.: Sieve methods.
Academic Press, London-New York-San Francisco. 1974.

30. Halberstam, H. and Richert, H.-E.: A weighted sieve of Greaves' type. I.
Banach Center Publications (Warsaw 1983) 17 (1985), 155-182.

31. Halberstam, H. and Richert, H.-E.: A weighted sieve of Greaves' type. II. Banach Center Publications (Warsaw 1983) 17 (1985), 183-215.

32. Heath-Brown, D. R.: Prime numbers in short intervals and a generalized Vaughan identity. Can. J. Math. 34 (1982), 1365-1377.

33. Heath-Brown, D. R.: The square sieve and consecutive square-free numbers. Math. Ann. 266 (1984), 251-259.

34. Heath-Brown, D. R.: Artin's conjecture for primitive roots. Quart. J. Math. Oxford (2), 37 (1986), 27-38.

35. Heath-Brown, D. R.: The number of primes in a short interval. J. r. ang. Math. 389 (1988), 22-63.

36. Heath-Brown, D. R. and Iwaniec, H.: On the difference between consecutive primes. Invent. math. 55 (1979), 49-69.

37. Hooley, C.: On the greatest prime factor of a quadratic polynomial. Acta Math. 117 (1967), 281-299.

38. Hooley, C.: On the Brun-Titchmarsh theorem. J. r. ang. Math. 255 (1972), 60-79.

39. Hooley, C.: On the Brun-Titchmarsh theorem. II. Proc. London Math. Soc. (3) 30 (1975), 114-128.

40. Hooley, C.: Applications of sieve methods to the theory of numbers. Cambridge University Press. 1976.

41. Hooley, C.: On the representations of a number as the sum of four cubes. I. Proc. London Math. Soc. (3) 36 (1978), 117-140.

42. Huxley, M. N. and Watt, N.: Exponential sums and the Riemann zeta function. Proc. London Math. Soc. (3) 57 (1988), 1-24.

43. Iwaniec, H.: Primes of the type $\phi(x,y)+A$ where ϕ is a quadratic form. Acta Arith. 21 (1972), 203-234.

44. Iwaniec, H.: The sieve of Eratosthenes-Legendre. Ann. Scuola Norm. Sup. Pisa (4) 4 (1977), 257-268.

45. Iwaniec, H.: Rosser's sieve-bilinear forms of the remainder terms - some applications. Recent progress in analytic number theory, Vol. 1 (Durham 1979), 203-230.

46. Iwaniec, H.: Sieve methods. Proc. International Congr. Math. (Helsinki, 1978), 357-364.

47. Iwaniec, H. Rosser's sieve. Acta Arith. 36 (1980), 171-202.

48. Iwaniec, H.: A new form of the error term in the linear sieve. Acta Arith. 37 (1980), 307-320.

49. Iwaniec, H.: Promenade along modular forms and analytic number theory. University of Texas Press, Austin 1985, 221-303.

50. Iwaniec, H. and Jutila, M.: Primes in short intervals.
Arkiv mat. 17 (1979), 167-176.

51. Iwaniec, H., van de Lune, J. and te Riele, H. J. J.: The limits of Buchstab's iteration sieve.
Indag. Math. (4) 42 (1980), 409-417.

52. Iwaniec, H. and Pintz, J.: Primes in short intervals.
Mh. Math. 98 (1984), 115-143.

53. Jurkat, W. B. and Richert, H.-E.: An improvement of Selberg's sieve method. I.
Acta Arith. 11 (1965), 217-240.

54. Jutila, M.: A statistical density theorem for L-functions with applications.
Acta Arith. 16 (1969), 207-216.

55. Jutila, M.: On a problem of Barban und Vehov.
Mathematika 26 (1979), 62-71, 306.

56. Kuhn, P.: Zur Viggo Brun'schen Siebmethode. I.
Norske Vid. Selsk. Forh. Trondhjem 14 (1941), no. 39, 145-148.

57. Laborde, H.: Buchstab's sifting weights.
Mathematika 26 (1979), 250-257.

58. Montgomery, H. L.: A note on the large sieve.
J. London Math. Soc. 43 (1968), 93-98.

59. Montgomery, H. L. and Vaughan, R. C.: The large sieve.
Mathematika 20 (1973), 119-134.

60. Motohashi, Y.: On a problem in the theory of sieve methods. (Japanese)
Res. Inst. Math. Sci. Kyoto Univ. Kōkyūroku 222 (1974), 9-50.

61. Motohashi, Y.: On some improvements of the Brun-Titchmarsh theorem.
J. Math. Soc. Japan 26 (1974), 306-323.

62. Motohashi, Y.: On almost-primes in arithmetic progressions. III.
Proc. Japan Acad. 52 (1976), 116-118.

63. Motohashi, Y.: An induction principle for the generalization of Bombieri's prime number theorem.
Proc. Japan Acad. 52 (1976), 273-275.

64. Motohashi, Y.: Primes in arithmetic progressions.
Invent. math. 44 (1978), 163-178.

65. Motohashi, Y.: On the linear sieve. I.
Proc. Japan Acad. 56 (1980), 285-287.

66. Motohashi, Y.: On the linear sieve. II.
Proc. Japan Acad. 56 (1980), 386-388.

67. Motohashi, Y.: Lectures on the sieve method and prime number theory.
Tata Institute for Fundamental Research, Bombay. 1983.

68. Pan, C.-D.: A new mean value theorem and its applications.
Recent progress in analytic number theory, Vol. 1 (Durham 1979), 275-287.

69. Pan, C.-D. And Ding, X.: A new mean value theorem.
Sci. Sinica 1979, Special Issue II on Math., 149-161.

70. Pan, C.-D., Ding X. and Wang, Y.: On the representation of every large even integer as a sum of a prime and an almost prime.
Sci. Sinica 18 (1975), 599-610.

71. Perelli, A., Pintz, J. and Salerno, S.: Bombieri's theorem in short intervals. II. Invent. math. 79 (1985), 1-9.

72. Pomykała, H.: Bilinear form of the remainder term in Rosser-Iwaniec sieve of dimension $\kappa \in (\frac{1}{2},1)$.
Acta Arith. to appear

73. Porter, J. W.: On the non-linear sieve.
Acta Arith. 29 (1976), 377-400.

74. Rawsthorne, D. A.: Selberg's sieve with a one sided hypothesis.
Acta Arith. 41 (1982), 281-289.

75. Richert, H.-E.: Selberg's sieve with weights.
Mathematika 16 (1969), 1-22.

76. Rotermund, R.: Fast-Primzahlen in Intervallen und arithmetischen Progressionen.
Dissertation Ulm, 1974.

77. Ruzsa, I. Z.: On the small sieve. II. Sifting by composite numbers.
J. Number Theory 14 (1982), 260-268.

78. Salerno, S.: A note on Selberg sieve.
Acta Arith. 45 (1986), 279-288.

79. Selberg, A.: On elementary methods in prime number-theory and their limitations.
11. Shand. Mat. Kongr., Trondheim 1949, 13-22.

80. Selberg, A.: The general sieve method and its place in prime number theory.
Proc. Internat. Congress Math., Cambridge, Mass. 1 (1950), 286-292.

81. Selberg, A.: Twin prime problem.
Manuscript 16 pp. and Appendix 7 pp. (unpublished)

82. Selberg, A.: Remarks on sieves.
Proc. Number Theory Conference (Univ. Colorado, Boulder, Co. 1972), 205-216.

83. Wolke, D.: Eine Bemerkung zum Sieb des Eratosthenes.
Mh. Math. 78 (1974), 264-266.

On the Number of Good Simultaneous Approximations to Algebraic Numbers*

Wolfgang M. Schmidt
Department of Mathematics, University of Colorado Boulder, U.S.A

1. Introduction.

The present work is a continuation of [2] by J. Mueller and the author, where approximations to a single number had been considered. Again it will be convenient to begin with approximations to real numbers in general. Given $\underline{\underline{\xi}}=(\xi_1,\cdots,\xi_n)$ in $\mathbb{R}^n$, we are interested in rational approximations $\underline{\underline{x}}=(x_1/y,\cdots,x_n/y)$ with $|\xi_i-(x_i/y)|<y^{-1-(1/n)-\delta}(i=1,\cdots,n)$ where $\delta>0$. More generally, given $\delta>0$, $M>0$, a (δ, M)-*approximation* to $\underline{\underline{\xi}}$ will be a rational point $\underline{\underline{x}}=(x_1/y,\cdots,x_n/y)$ with $y>0$ and

$$|\xi_i-\frac{x_i}{y}|<My^{-1-(1/n)-\delta}\quad(i=1,\cdots,n). \tag{1}$$

Simply to count solutions of (1) with $y\leqslant B$ is not a good analogue of what was done in [2] in the case $n=1$. For suppose $\underline{\underline{\xi}}=(\xi_1,0,\cdots,0)$, and take $\underline{\underline{x}}=(x_1/y,0,\cdots,0)$; then (1) reduces to

$$|\xi_1-(x_1/y)|<My^{-1-(1/n)-\delta}, \tag{2}$$

i.e., a one-dimensional problem. In fact when $\delta<1-(1/n)$, then for almost all $\xi_1\in\mathbb{R}$ (in the sense of Lebesgue measure), the number of *primitive* solutions, i.e. solutions of (2) with g.c.d. $(x_1,y)=1$, and with $y\leqslant B$, is asymptotically équal to

$$\sim\frac{2M}{\zeta(2)}\sum_{y=1}^{B}y^{-(1/n)-\delta}\sim\frac{2M}{\zeta(2)(1-(1/n)-\delta)}B^{1-(1/n)-\delta},$$

e.g., by work of Szusz [6]. This is very different from the loglog type bound of [2]. Another bad feature is that we are apparentyl reduced to a 1-dimensional problem such as in (2).

The following appears to be a better generalization of the one-dimensional case as studied in [2]. A set of points in $\mathbb{R}^n$ will be called *independent* if every affine subspace of dimension d where $0\leqslant d<n$ contains at most $d+1$ points of the set. In particular, any $n+1$ points of the set will span $\mathbb{R}^n$ as an affine space. **We will**

* Written with partial support from NSF grant DMS-8603093.

estimate the cardinality of independent sets of approximations. Note that for $n=1$, a set of points will be independent if the points are distinct, so that one is led to counting rational approximations x/y in reduced form.

Before formulating our results we have to introduce some notation.

$$f(x):=\frac{1}{x}+\frac{1}{x^2}+\cdots+\frac{1}{x^n}$$

for $x>0$ is decreasing and it assumes every positive real value. Thus given $\delta>0$, there is a unique $\rho=\rho(\delta)>0$ with $f(\rho)=(\frac{1}{n}+\delta)^{-1}$. This number ρ satisfies

$$(\frac{1}{n}+\delta)(\rho^{-1}+\rho^{-2}+\cdots+\rho^{-n})=1. \tag{3}$$

Since

$$f(1+\delta)=\delta^{-1}(1-(1+\delta)^{-n})\geqslant\delta^{-1}(1-(1+n\delta)^{-1})=(\frac{1}{n}+\delta)^{-1},$$

we have $\rho(\delta)\geqslant 1+\delta$. Equality holds when $n=1$; in general there is an expansion $\rho(\delta)=1+(2n/(n+1))\delta+\cdots$. Write

$$L=\log\rho. \tag{4}$$

THEOREM 1. *Given* $\xi\in\mathbb{R}^n$ *and* $\delta>0$, $M\geqslant 1$, *any independent set of* (δ,M)-*approximations with denominators* $\leqslant B$ *has cardinality at most*

$$t=L^{-1}\log^+\log B+5n!(2eM)^n(1+\delta^{-1}). \tag{5}$$

Here $\log^+ x$ is $\log x$ when $x\geqslant 1$, and $\log^+ x=0$ when $0<x<1$. A particular consequence of the Theorem is as follows. For rational $x=(x_1/y,\cdots,x_n/y)$ with $y>0$ and g.c.d. $(x_1,\cdots,x_n,y)=1$, write $y=y(x)$. Now (if there exist such at all) let x_1 be the (δ,M)-approximation to ξ with the least value of $y=y(x_1)$. Next, let x_2 be the (δ,M)-approximation such that x_1, x_2 are independent, and with $y=y(x_2)$ minimal for all x_2 with this property. And so on. In this way we obtain sequence $x_1,x_2,\cdots$ of independent (δ,M)-approximations which may either terminate or continue at infinitum. We will call this the *minimal sequence* of (δ, M)-approximations. Our theorem says that the number of members of this sequence with $y(x)\leqslant B$ does not exceed t.

It follows that the set of all (δ,M)-approximations with denominators $\leqslant B$ lies in the union of certain t affine hyperplanes.

Theorem 1 is in some sense best possible:

THEOREM 2. *Given* n, $\delta>0$, $M>0$, *there is a point* $\underline{\xi}$ *in* $\mathbb{R}^n$ *with* $1,\xi_1,\cdots,\xi_n$ *linearly independent over* $\mathscr{R}$, *and with an infinite minimal sequence of* (δ,M)-*approximations, such that for any* $B>1$, *the number of elements of this sequence with* *denominators* $\leqslant B$ *is*

$$\geqslant L^{-1}\log^{+}\log B - c_1(n, \delta, M). \tag{6}$$

The construction of ξ will be the most difficult part of the paper. In contrast to the one-dimensional case, an appeal to continued fractions is not possible. The constant $c_1 = c_1(n, \delta, M)$, as well as constants $c_2, c_3, \cdots$ below, could be explicitly computed.

We now turn to approximations to algebraic numbers. Let $\alpha = (\alpha_1, \cdots, \alpha_n)$ have components in an algebraic number field K of degree r.

We define the height $H = H(\alpha)$ as the maximum of the heights $H(\alpha_1) \cdots, H(\alpha_n)$, where $H(\alpha)$ is the absolute Mahler height as in [1], [2]. (We could have chosen some other height, such as the affine height of α, given as the product of $\max(1, |\alpha_1|_v, \cdots, |\alpha_n|_v)$ over absolute values $|\cdot|_v$, but in the results to be stated the term $\log^{+}\log H$ is not very sensitive to such differences of definition, so that the results would remain true with such other heights.)

THEOREM 3. *Let* α *be as above. Suppose that* $\delta > 0, M > 0$. *Then any independent set of* (δ, M)*-approximations to* α *has cardinality at most*

$$s = L^{-1}\log^{+}\mathrm{lgo}H + c_2(n, r, \delta, M). \tag{7}$$

Thus these rational approximations lie in the union of s affine hyperplanes.

Some time ago [3] I proved that under the additional hypothesis that $1, \alpha_1, \cdots, \alpha_n$ are linearly independent over $\mathscr{R}$, the set of (δ, M)-approximations is finite. But while it is possible to estimate the minimal number of affine hyperplanes containing all the (δ, M)-approximations, we are for $n > 1$ not able at present to give any estimate whatsoever for the number of these approximations. Perhaps the title of the paper is promising too much; we are asking the reader to forgive us. As was pointed out elsewhere [5], for $n > 1$ it is extremely unlikely that we will be able to estimate the number of all the (δ, M)-approximations, unless we have an effective version of Roth's Theorem, and of its generalization to simultaneous approximation.

Theorem 3 is in some sense best possible:

THEOREM 4. *Let* K *be an algebraic number field of degree* $r > n$, *and let* $\delta > 0, M > 0$. *Then there are infinitely many* $\alpha \in K^n$ *with* $1, \alpha_1, \cdots, \alpha_n$ *linearly independent over* $\mathscr{R}$, *and having a set of independent* (δ, M)*-approximations of cardinality*

$$\geqslant L^{-1}\log^{+}\log H(\alpha) - c_3(n, K, \delta, M).$$

As we said above, the hardest part will be Theorem 2. Theorem 3 is independent of that Theorem, but it will be (easily) deduced from my recent effective version of the Subspace Theorem [4]. Theorem 4 is an easy consequence of Theorem 2. The proof of Theorem 2 will be postponed to the end of the paper.

2. Proof of Theorem 1.

LEMMA 1. *Let* $y_0, y_1, \cdots$ *be numbers with* $y_\nu \geqslant e(nM)^{1/\delta}$ $(\nu \geqslant 0)$ *and*

$$y_\nu \geqslant \frac{1}{n!M}(y_{\nu-1}y_{\nu-1}\cdots y_{\nu-n})^{(1/n)+\delta} \quad (\nu \geqslant n).$$

Then

$$y_\nu \geqslant e^{\rho^{\nu+1-n}} \quad (\nu \geqslant 0). \tag{8}$$

Proof. Put $\eta_\nu = \log y_\nu$; then

$$\eta_\nu \geqslant \left(\frac{1}{n}+\delta\right)(\eta_{\nu-1}+\cdots+\eta_{\nu-n}) - \log(n!M)\ (\nu \geqslant n).$$

Thus $\theta_\nu := \eta_\nu - (n\delta)^{-1}\log(n!M)$ has

$$\theta_\nu \geqslant \left(\frac{1}{n}+\delta\right)(\theta_{\nu-1}+\cdots+\theta_{\nu-n}),$$

and $\lambda_\nu := \rho^{-\nu\theta_\nu}$ has

$$\lambda_\nu \geqslant \left(\frac{1}{n}+\delta\right)(\lambda_{\nu-1}\rho^{-1}+\cdots+\lambda_{\nu-n}\rho^{-n})\ (\nu \geqslant n). \tag{9}$$

By hypothesis on y_ν we have $\eta_\nu \geqslant 1+\delta^{-1}\log(nM) \geqslant 1+(n\delta)^{-1}\log(n!M)$, so that $\theta_\nu \geqslant 1$ $(\nu = 0,1,\cdots)$, and each of $\lambda_0, \lambda_1, \cdots, \lambda_{n-1}$ is $\geqslant \rho^{1-n}$. Applying (9) successively for $\nu = n, n+1, \cdots$, and recalling (3), we see that $\lambda_\nu \geqslant \rho^{1-n}$ for each $\nu \geqslant 0$. Therefore $\theta_\nu \geqslant \rho^{\nu+1-n}$, and (8) holds.

LEMMA 2. *Let* $x_0, x_1, \cdots, x_n$ *be* (δ, M)*-approximations to* ξ *with* $y(x_0) \leqslant \cdots \leqslant y(x_n)$. *Suppose they are not all contained in an affine hyperplane. Then*

$$y(x_n) \geqslant \frac{1}{n!M}(y(x_0)y(x_1)\cdots y(x_n))^{(1/n)+\delta}.$$

Proof. Write $x_i = (x_{i1}/y_i, \cdots, x_{in}/y_i)$ where $y_i = y(x_i)(i=0,\cdots,n)$. Since $x_0, \cdots, x_n$ are not contained in a hyperplane, the following determinant is not zero:

$$\begin{vmatrix} 1 & \frac{x_{01}}{y_0} \cdots \frac{x_{0n}}{y_0} \\ \cdots & \\ 1 & \frac{x_{n1}}{y_n} \cdots \frac{x_{nn}}{y_n} \end{vmatrix} = \begin{vmatrix} 1 & \frac{x_{01}}{y_0} - \alpha_1 & \cdots & \frac{x_{0n}}{y_0} - \alpha_n \\ \cdots & & & \\ 1 & \frac{x_{n1}}{y_n} - \alpha_1 & \cdots & \frac{x_{0n}}{y_0} - \alpha_n \end{vmatrix} \tag{10}$$

On the one hand, this determinant has absolute value $\geqslant (y_0, y_1 \cdots y_n)^{-1}$. On the other hand, in view of the definition (1) of (δ, M)-approximations, it has absolute value

$$\leqslant n!M(y_0 y_1 \cdots y_n)^{-1-(1/n)-\delta}.$$

We obtain

$$y_n \geqslant \frac{1}{n!M}(y_0 y_1 \cdots y_{n-1})^{(1/n)+\delta}. \tag{11}$$

Now let S be an independent set of (δ, M)-approximations with denominators $\leqslant B$. Write $S = S_1 \bigcup S_2$, where S_1 consists of approximations with

$$y(x) \geqslant e(nM)^{1/\delta} = C,$$

say, the so-called "large approximations", and S_2 consists of approximations with $y(x) < C$, the "small approximations".

We will show that S_1 has cardinality

$$|S_1| \leqslant L^{-1} \log^+ \log B + n.$$

We may suppose that $|S_1| > n$. Order the elements of S_1 as $x_0, x_1, \cdots, x_\nu$ with $y(x_0) \leqslant \cdots \leqslant y(x_\nu)$. By the independence of our set, Lemma 2 says that for $\nu \geqslant n$ we have

$$y_\nu := y(x_\nu) \geqslant \frac{1}{n!M}(y_{\nu-1} \cdots y_{\nu-n})^{(1/n)+\delta}.$$

Thus by Lemma 1 we have (8). Therefore $e^{\rho^{\nu+1-n}} \leqslant B$ and

$$\nu + 1 \leqslant L^{-1} \log^+ \log B + n.$$

It remains for us to estimate the cardinality of the set S_2 of small approximations with denominators $< C$. For non-negative u, let $S(u)$ consist of the elements of S with $e^u \leqslant y < e^{u+1}$. We claim that

$$|S(u)| \leqslant n + (2M)^n n! e^{n+1-un\delta}. \tag{12}$$

This is true for $n=1$ by [2, § 2], so that we may suppose that $n \geqslant 2$. The claim (12) is true when $|S(u)| \leqslant n$. Otherwise, pick $z, \cdots, z_{n-1}$ in $S(u)$; by independence, they will span an $(n-2)$-dimensional affine linear subspace A. Every $w \in \mathbb{R}^n$ may uniquely be written as $w = w^A + w^B$, where $w^A \in A$, and w^B lies in the 2-dimensional subspace B of $\mathbb{R}^n$ orthogonal to A. (When $n=2$, this subspace is $\mathbb{R}^2$ itself.) Introduce polar coordinates r, φ in B. Say $S(u)$ consists of $z_1, \cdots, z_{n-1}, w_1, \cdots, w_l$. We may suppose that $w_1, \cdots, w_l$ are ordered such that $0 < \varphi(w_1^B) < \cdots < \varphi(w_l^B) < 2\pi$. (Note that $\varphi(w_i^B)$ is well defined since $w_i \notin A$ by the independence of S. We have strict inequalities, since $z_1, \cdots, z_{n-1}, \underline{\underline{w}}_i, \underline{\underline{w}}_{i+1}$ span $\mathbb{R}^n$ as an affine space.)

Put $\psi_i = \varphi(w^B_{i+1}) = \varphi(w^B_i)$ when $a \leqslant i < l$, and $\psi_l = \varphi(w^B_1) - \varphi(w^B_l) + 2\pi$. Then $\psi_1 + \cdots + \psi_l = 2\pi$, and there is at most one i_0 with $\psi_{i0} > \pi$. The simplices Δ_i with vertices $z_1, \cdots, z_{n-2}, \underline{\underline{w}}_i, \underline{\underline{w}}_{i+1}$ where $1 \leqslant i \leqslant l$ and $i \neq i_0$ (and where we set $\underline{\underline{w}}_{l+1} = \underline{\underline{w}}_1$) have mutually disjoint interiors. The determinant analogous to (10) formed from $z_1, \cdots, z_{n-2}, \underline{\underline{w}}_i, \underline{\underline{w}}_{i+1}$ has absolute value

$$\geqslant e^{-(u+1)(n+1)}$$

so that Δ_i has a volume at least $(n!)^{-1}$ times this amount, and the simplices Δ_i $(1 \leqslant i \leqslant l;\ i \neq i_0)$ together have volume

$$\geqslant \frac{l-1}{n!} e^{-(u+1)(n+1)}. \tag{13}$$

Now for x in S(u),

$$|\xi_i - \frac{x_i}{y}| < My^{-1-(1/n)-\delta} \leqslant Me^{-u(1+(1/n)+\delta)} \quad (i = 1, \cdots, n),$$

so that x lies in a cube of volume

$$(2M)^n e^{-u(n+1+n\delta)}.$$

Comparison with (13) yields

$$l \leqslant 1 + (2M)^n n! e^{n+1-un\delta},$$

establishing our claim about $|S(u)|$.

The small approximations are contained in the union of $S(0), \cdots, S(q)$ with $q = 1 + [\delta^{-1}\log(nM)]$. Their number is

$$\begin{aligned}
&\leqslant \sum_{u=0}^{q} |S(u)| \leqslant \sum_{u=0}^{q} (n + (2M)^n n! e^{n+1-un\delta}) \\
&\leqslant n(q+1) + en!(2eM)^n \sum_{u=0}^{\infty} e^{-un\delta} \\
&= nq + n + en!(2eM)^n (1 - e^{-n\delta})^{-1} \\
&< \delta^{-1} n \log(nM) + 2n + en!(2eM)^n (1 + (n\delta)^{-1}) \\
&< 4n!(2eM)^n (1 + \delta^{-1}),
\end{aligned}$$

since $M \geqslant 1$. Theorem 1 follows.

3. Proof of Theorem 3.

When $x = (x_1/y, \cdots, x_n/y)$ is a (δ, M)-approximaton to $\alpha = (\alpha_1, \cdots, \alpha_n)$, then

$|x_i| \leqslant |\alpha_i| y + M \leqslant H^r y + My$, so that $\underline{\underline{X}} := (x_1, \cdots, x_n, y)$ has $|\underline{\underline{X}}| \leqslant (1 + n(H^r + M)^2)^{1/2} y < y^2$, when $y > 1 + n(H^r + M)$. Now (1) with $\underline{\underline{\xi}} = \underline{\underline{\alpha}}$ yields

$$\alpha_1 y - x_1 | \cdots | \alpha_n y - x_n | y < y^{-n\delta} < |\underline{\underline{X}}|^{-\delta}$$

if we suppose that $n \geqslant 2$, which is allowed in view of [2]. We have here an inequality involving the $n+1$ linear forms $\alpha_1 y - x_1, \cdots, \alpha_n y - x_n, y$. By the Theorem in [4], the solutions of this inequality with $|\underline{\underline{X}}| \geqslant \max((n!)^{8/\delta}, H)$ lie in the union of not more than $c_4(n, r, \delta)$ proper subspaces of $\mathbb{R}^{n+1}$ (where c_4 is explicitly given). The corresponding rational points $\underline{\underline{x}}$ lie in the union of not more than $c_4(n, r, \delta)$ proper affine subspaces (i.e., affine hyperplanes) of $\mathbb{R}^n$. Any independent set of such points has cardinality not exceeding $nc_4(n, r, \delta)$. There remain the approximations with $y \leqslant \max(1 + n(H^r + M), (n!)^{8/\delta}) = B$, say. By Theorem 1, any independent set of such approximations has cardinality at most $c_5(n, r, \delta, M) + L^{-1}\log^+\log H$. (Note that it suffices to prove Theorem 3 for $M \geqslant 1$, so that we need to invoke Theorem 1 for $M \geqslant 1$ only.)

4. Deduction of Theorem 4 from Theorem 2.

Let $\underline{\underline{\beta}} \in K^n$ be given with $1, \beta_1, \cdots, \beta_n$ linearly independent over $\mathscr{R}$. We will construct $\underline{\underline{\alpha}}$ of the type

$$\alpha_i = \beta_i + \frac{a_i}{b} \quad (i = 1, \cdots, n).$$

Let $\underline{\underline{\xi}} \in \mathbb{R}^n$ be as in Theorem 2, but with $M/2$ in place of M. Pick a large integer b and choose integers $a_1, \cdots, a_n$ with

$$|\alpha_i - \xi_i| \leqslant \frac{1}{2b} \quad (i = 1, \cdots, n).$$

Put $B = (bM)^\lambda$ with $\lambda = (1 + (1/n) + \delta)^{-1}$. By Theorem 2, there is a set of at least

$$L^{-1}\log^+\log B - c_1(n, \delta, M/2)$$

independent $(\delta, M/2)$-approximations to ξ with denominators $\leqslant B$. For these approximations $(x_1/y, \cdots, x_n/y)$ we have

$$|\alpha_i - \frac{x_i}{y}| \leqslant |\alpha_i - \xi_i| + |\xi_i - \frac{x_i}{y}| < \frac{1}{2b} + \frac{M}{2y^{1+(1/n)+\delta}} \leqslant \frac{M}{y^{1+(1/n)+\delta}},$$

since $y \leqslant B = (bM)^\lambda$. On the other hand, it is easily seen that $H(\alpha_i) \leqslant c_6(\beta) b$ $(i = 1, \cdots, n)$ (see, e.g., (16) in [2]). Therefore $H(\alpha) \leqslant c_6(\beta) b$ and

$$\log^+\log B \geqslant \log^+\log H(\alpha) - c_7(\beta, \delta, M).$$

4.1. Proof of Theorem 2, beging. We will construct an independent sequence

$$\underline{\underline{x}}_1, \underline{\underline{x}}_2, \cdots$$

of rational points with a limit $\underline{\underline{\xi}}$ such that each $\underline{\underline{x}}_\nu$ is a (δ, M)-approximation to $\underline{\underline{\xi}}$. The sequence has to be constructed such that each $\underline{\underline{x}}_{\nu+1}$ is very close to the preceding element x_ν.

It is not easy to deal directly with rational points $\underline{\underline{x}} = (x_1/y_1, \cdots, x_n/y)$. We therefore will deal with points $\underline{\underline{X}} = (x_1, \cdots, x_n, y)$ with integer components. We will construct a sequence

$$\underline{\underline{X}}_1, \underline{\underline{X}}_2, \cdots$$

of integer points in $\mathbb{R}^k$ where

$$\boxed{k = n + 1} \tag{14}$$

The condition that $\underline{\underline{x}}_{\nu+1}$ be close to $\underline{\underline{x}}_\nu$ translates into a condition that $\underline{\underline{X}}_\nu \Lambda \underline{\underline{X}}_{\nu+1}$ should be small, where Λ denotes the exterior product. Now $\underline{\underline{X}}_\nu \Lambda \underline{\underline{X}}_{\nu+1}$ lies in $\mathbb{R}^l$ with $l = \left[\begin{matrix} k \\ 2 \end{matrix}\right]$, and the condition will be that $|X_\nu \Lambda X_{\nu+1}|$ should be small as compared to $|\underline{\underline{X}}_\nu||\underline{\underline{X}}_{\nu+1}|$, where all our norms are Euclidean. Now it is not feasible to keep track only of $|X_{\nu+1} \Lambda X_\nu|$, but it turns out that we will have to deal with

$$|\underline{\underline{X}}_{\nu+1} \Lambda \underline{\underline{X}}_\nu|, |\underline{\underline{X}}_{\nu+1} \Lambda \underline{\underline{X}}_\nu \Lambda \underline{\underline{X}}_{\nu-1}|, \cdots, |\underline{\underline{X}}_{\nu+1} \Lambda \underline{\underline{X}}_\nu \Lambda \cdots \Lambda \underline{\underline{X}}_{\nu-n+1}|.$$

We will begin with two simple lemmas, the first of which is a variation on Lemma 1.

LEMMA 3. *Let* $\delta > 0, P > 0$ *be given, and let* $C_1, C_2, \cdots$ *be a sequence of positive numbers with the recursive relation*

$$C_\nu = P^\nu (C_{\nu-1} \cdots C_{\nu-n})^{(1/n)+\delta} \quad (v > n). \tag{15}$$

Then

(a) $C_\nu < e^{E\rho^\nu}$ $(\nu > 0)$

where $E = E(C_1, \cdots, C_n, \delta, P)$.

(b) *Given* $F > 0$, *we have*

$$C_\nu > e^{F\rho^\nu} \; \nu > 0),$$

provided $\min(C_1, \cdots, C_n) > c_8(F, \delta, n, P)$.

(c) *Given* $G > 0$, *the quotients* $Q_v = C_{v+1}/C_v$ *have*

$$Q_v > e^{G\rho^v} \quad (v > 0),$$

provided $\min(Q_1,\cdots,Q_v)>c_9(G,\delta,n,P)$.

Proof. Write $c_v=\log C_v$ and $\lambda_v=c_v\rho^{-v}$. Then

$$c_v=\left(\frac{1}{n}+\delta\right)(c_{v-1}+\cdots+c_{v-n})+v\log P$$

and

$$\lambda_v=\left(\frac{1}{n}+\delta\right)(\lambda_{v-1}\rho^{-1}+\cdots+\lambda_{v-n}\rho^{-n})+v\rho^{-v}\log P \quad (v>n). \tag{16}$$

Therefore, writing $\mu_v=\max(\lambda_v,\lambda_{v-1},\cdots,\lambda_{v-n+1})$, we have from (3) that $\lambda_v\leqslant\mu_{v-1}+v\rho^{-v}\log P$ and $\mu_v\leqslant\mu_{v-1}+v\rho^{-v}|\log P|$ $(v>n)$. Since $\sum_v v\rho^{-v}$ is convergent, we have $\mu_v<E$ where $E=E(\mu_n,\rho,P)=E(C_1,\cdots,C_n,\delta,P)$. Now (a) follows.

Writing $\kappa_v=\min(\lambda_v,\lambda_{v-1},\cdots,\lambda_{v-n+1})$, we have from (16) that $\lambda_v\geqslant\kappa_{v-1}-v\rho^{-v}|\log P|$. Thus if $\kappa_n>F+\sum_{v=n}^{\infty}v\rho^{-v}|\log P|$, then each $\mu_v>F$. The condition on κ_n holds if $C_1,\cdots,C_n$ are large, say if they exceed $c_8(F,\delta,n,P)$. This establishes (b).

Finally by (15),

$$Q_v=P\left[\frac{C_v\cdots C_{v-n+1}}{C_{v-1}\cdots C_{v-n}}\right]^{(1/n)+\delta}=P(C_v/C_{v-n})^{(1/n)+\delta}=P(Q_{v-1}\cdots Q_{v-n})^{(1/n)+\delta}\ (v>n).$$

Thus the Q_v satisfy a recursive relation similar to the C_v themselves. Therefore (c) follows by an adaptation of (b).

LEMMA 4. *Let* Λ *be a lattice in* $\mathbb{R}^n$ *with basis* $a_1,\cdots,a_n$. *Put*

$$\Delta_0=1,\ \Delta_u=|a_1\wedge\cdots\wedge a_u|\quad(u=1,\cdots,n).$$

so that in particular $\Delta_n=\det\Lambda$. *Suppose that*

$$\Delta_u/\Delta_{u-1}\leqslant4^{u-n}(\Delta_n/\Delta_{n-1})N^{-1}\quad(u=1,\cdots,n-1), \tag{17}$$

where N *is a natural number. Then given any* N *hyperplanes in* $\mathbb{R}^n$, *there is a basis* $a_1,\cdots,a_{n-1},b$ *of* Λ *where* b *does not lie in any of these hyperplanes and where*

$$\Delta_n/\Delta_{n-1}\leqslant|b|\leqslant2\Delta_n/\Delta_{n-1} \tag{18}$$

and

$$\frac{1}{2}(\Delta_n/\Delta_{n-1})\Delta_u\leqslant|b\wedge a_1\wedge\cdots\wedge a_u|\leqslant2(\Delta_n/\Delta_{n-1})\Delta_u\quad(u=1,\cdots,n-1). \tag{19}$$

Proof. We first construct a certain basis $b_1,\cdots,b_n$ where b_u is of the type $b_u=c_{u1}a_1+\cdots+c_{u,u-1}a_{u-1}+a_u$ $(u=1,\cdots,n)$. Set $b_1=a_1$. Then $|b_1|=\Delta_1=\Delta_1/\Delta_0$. Suppose $u<n$ and we already have constructed $b_1,\cdots,b_u$ with $|b_1|=\Delta_1/\Delta_0$,

$$\Delta_j/\Delta_{j-1} \leqslant |b_j| < \Delta_j/\Delta_{j-1} + (\Delta_{j-1}/\Delta_{j-2} + 2\Delta_{j-2}/\Delta_{j-3} + \cdots + 2^{j-2}\Delta_1/\Delta_0)(2 \leqslant j \leqslant u). \quad (20)$$

Write $a_{u+1} = \lambda_1 b_1 + \cdots + \lambda_u b_u + c$ where $\lambda_1, \cdots, \lambda_u$ are real and c is orthogonal to $b_1, \cdots, b_u$. We now choose

$$b_{u+1} = q_1 b_1 + \cdots + q_u b_u + \cdots + q_u b_u + a_{u+1} = (\lambda_1 + q_1) b_1 + \cdots + (\lambda_u + q_u) b_u + c, \quad (21)$$

with integers $q_1, \cdots, q_u$ yet to be determined. When $u+1 < n$, we choose these integers with $|\lambda_j + q_j| \leqslant \frac{1}{2}$ $(j = 1, \cdots, u)$. Then $b_1, \cdots, b_{u+1}$ generate the same group as $a_1, \cdots, a_{u+1}$, and

$$|b_{u+1}| \geqslant |c| = |b_1 \wedge \cdots \wedge b_u \wedge a_{u+1}| / |b_1 \wedge \cdots \wedge b_u| = \Delta_{u+1}/\Delta_u. \quad (22)$$

On the other hand, $|b_{u+1}| \leqslant |c| + |b_1| + \cdots + |b_u|$, so that by the truth of (20) for $j = 1, \cdots, u$ we see its truth for $u+1$.

When $u+1 = n$, i.e., when $u = n-1$, as $(q_1, \cdots, q_u)$ runs through $\mathbb{R}^u$, the points $q_1 b_1 + \cdots + q_u b_u + a_{u+1}$ will run through an affine hyperplane $\mathscr{H}$ of $\mathbb{R}^n$ not containing the origin. The intersection of $\mathscr{H}$ with a hyperplane (i.e., a subspace of $\mathbb{R}^n$ of codimension 1) will have dimension $\leqslant n-2$. Therefore we can choose integers $q_1, \cdots, q_u$ with $|\lambda_j + q_j| \leqslant N$ $(j = 1, \cdots, u)$, such that $b_n = b_{u+1}$ as given by (21) avoids any N given hyperplanes. Now (22) holds also for $u+1 = n$, so that $|b_n| \geqslant \Delta_n/\Delta_{n-1}$. Also, $|b_n| \leqslant |c| + N|b_1| + \cdots + N|b_{n-1}|$ with $|c| = \Delta_n/\Delta_{n-1}$ so that from (20) with $u = n-1$ we get

$$|b_n| \leqslant \Delta_n/\Delta_{n-1} + N(\Delta_{n-1}/\Delta_{n-2} + 2\Delta_{n-2}/\Delta_{n-3} + \cdots + 2^{n-2}\Delta_1/\Delta_0).$$

We now set $b = b_n$. Then (18) follows in view of (17). The assertions of the lemma hold for b, which is clear except perhaps for the left inequality in (19). But for $1 \leqslant u \leqslant n-1$,

$$|b \wedge a_1 \wedge \cdots \wedge a_u| \geqslant |c \wedge a_1 \wedge \cdots \wedge a_u| - N \sum_{j=1}^{n-1} |b_j| |a_1 \wedge \cdots \wedge a_u|.$$

Here

$$|c \wedge a_1 \wedge \cdots \wedge a_u| = |c \wedge b_1 \wedge \cdots \wedge b_u| = |c| |b_1 \wedge \cdots \wedge b_u| = |c| \Delta_u,$$

so that

$$|b \wedge a_1 \wedge \cdots \wedge a_u| \geqslant (|c| - \sum_{j=1}^{n-1} |b_j|) \Delta_u.$$

Now c (perhaps we should have written c_n!) has $|c| = \Delta_n/\Delta_{n-1}$. Thus from (17) and from (20) with $j = 1, \cdots, n-1$ we obtain $|b \wedge a_1 \wedge \cdots \wedge a_u| \geqslant \frac{1}{2}(\Delta_n/\Delta_{n-1})\Delta_u$.

5. The Main Lemma.

The angle $\varphi=\varphi(X,Y)$ between points $X\neq 0$ in $\mathbb{R}^k$ is defined by $|X\wedge Y|=|X||Y|\sin\varphi$, with $0\leqslant\varphi\leqslant\pi/2$ if the inner product $XY\geqslant 0$, and $\pi/2<\varphi\leqslant\pi$ if $XY<0$. We will single out the particular point $E=(0,\cdots,0,1)$.

Let $P>1$ be given. We define $C_1, C_2, \cdots$ by

$$C_i=Q^i \quad (i=1,\cdots,n) \tag{23}$$

and by (15), where $Q>1$ is a new parameter.

LEMMA 5. *With* $Q>Q_0(P,n,\delta)$ *and with* $C_1, C_2, \cdots$ *as above, there is a sequence of integer points*

$$X_1, X_2, \cdots$$

in $\mathbb{R}^k$ *where* $k=n+1$, *such that any* k *consecutive points are a basis of* $\mathbb{Z}^k$ *and any* k *points span* $\mathbb{R}^k$, *and such that*

$$|X_v|\leqslant 8^v C_{v-1}C_{v-2}\cdots C_{v-n} \quad (v\geqslant k), \tag{24}$$

$$\varphi(X_v,E)<\frac{1}{20}(1-2^{n-v}) \quad (v\geqslant k), \tag{25}$$

and the quantities

$$\Delta_u^v:=|X_v\wedge X_{v-1}\wedge\cdots\wedge X_{v-n}|/|X_v| \quad (v\geqslant k;\ 1\leqslant u\leqslant n) \tag{26}$$

have

$$(8^v C_{v-1}C_{v-2}\cdots C_{v-u})^{-1}\leqslant\Delta_u^v\leqslant(C_{v-1}C_{v-2}\cdots C_{v-u})^{-1} \quad (v\geqslant k;\ 1\leqslant u\leqslant n). \tag{27}$$

Proof. In our construction we will denote the hyperplane orthogonal to X_v by A_v, and vectors in the n-dimensional space A_v by $a^v, b^v, \cdots$. But vectors in $\mathbb{R}^k$ in general will be denoted by $X, Y, \cdots$.

Pick primes $p_1,\cdots,p_n$ with

$$2C_i\leqslant p_i\leqslant 4C_i \quad (i=1,\cdots,n);$$

when $Q>2$, these primes will be distinct, Put

$$Y_1=(p_1,0,\cdots,0,1),\cdots,Y_n=(0,0,\cdots,p_n,1).$$

Let $X_k=Y_1\wedge\cdots\wedge Y_n$, and let A_k be the hyperplane spanned by $Y_1,\cdots,Y_n$; then X_k, A_k are orthogonal. By changing X_k into minus itself if necessary we get

$$X_k=(\pm p_2p_3\cdots p_n,\cdots,\pm p_1p_2\cdots p_{n-1},p_1p_2\cdots p_n).$$

Since $p_i\geqslant 2Q>60n$ when Q is large, we get $|E\wedge X_k|<\frac{1}{60}|X_k|$, and $\varphi(E,X_k)<1/40$, so that (25) holds with $v=k$.

A point $x_1Y_1 + \cdots + x_nY_n$ lies in $\mathbb{Z}^k$ precisely when $p_1x_1, \cdots, p_nx_n$ and $x_1 + \cdots + x_n$ lie in $\mathbb{Z}$, and since $p_1, \cdots, p_n$ are distinct primes, this happens only when $x_1, \cdots, x_n$ are in $\mathbb{Z}$. Therefore $Y_1, \cdots, Y_n$ is a basis of the lattice Γ_k in A_k consisting of integer points in A_k. Let Λ_k be the lattice dual to Γ_k in A_k. Thus Λ_k consists of points $a^k \in A_k$ such that the inner product $v^k a^k \in \mathbb{Z}$ for every $v^k \in \Gamma_k$. It is well known that Λ_k consists of the orthogonal projections of the points $X \in \mathbb{Z}^k$ on A_k.

Pick $a_1^k, \cdots, a_n^k$ in A_k with $a_i^k Y_j = \delta_{ij}$ $(1 \leqslant i, j \leqslant n)$. Then $a_1^k, \cdots, a_n^k$ is a basis of Λ_k. Now if $Y_1, \cdots, Y_n$ were mutually orthogonal, we would have $a_u^k = |Y_u|^{-2} Y_u$, and

$$|a_{k-1}^k \wedge \cdots \wedge a_{k-u}^k| = |a_n^k \wedge a_{n-1}^k \wedge \cdots \wedge a_{n+1-u}^k| = (|Y_n| \cdots |Y_{n+1-u}|)^{-1}$$

$(u = 1, \cdots, n)$. As it is, when Q and hence $C_1, \cdots, C_n$ and $p_1, \cdots, p_n$ are large, then $Y_1, \cdots, Y_n$ are almost orthogonal, and

$$\frac{1}{2} (|Y_{k-1}| \cdots |Y_{k-u}|)^{-1} \leqslant |a_{k-1}^k \wedge \cdots \wedge a_{k-u}^k| \leqslant 2(|Y_{k-1}| \cdots |Y_{k-u}|)^{-1} \ (u = 1, \cdots, n).$$

Moreover,

$$2^u C_{k-1} \cdots C_{k-u} \leqslant p_{k-1} \cdots p_{k-u} \leqslant |Y_{k-1}| \cdots |Y_{k-u}| \leqslant 2 p_{k-1} \cdots p_{k-u} \leqslant 2^{1+2u} C_{k-1} \cdots C_{k-u}$$
$$(u = 1, \cdots, n),$$

so that if we set

$$\Delta_u^k = |a_{k-1}^k \wedge \cdots \wedge a_{k-u}^k|,$$

then

$$(4^k C_{k-1} \cdots C_{k-u})^{-1} \leqslant \Delta_u^k \leqslant (C_{k-1} \cdots C_{k-u})^{-1} \quad (u = 1, \cdots, n).$$

Let $X_1, \cdots, X_n$ be integer points whose orthogonal projections on A_k are $a_1^k, \cdots, a_n^k$, respectively. Since X_k is primitive (i. e., its coordinates have no common factor > 1), the points $X_1, \cdots, X_k$ are a basis of $\mathbb{Z}^k$. When Q is sufficiently large we have

$$|X_k| \leqslant 2 p_1 \cdots p_n \leqslant 2^k C_1 \cdots C_n = 2^k C_{k-1} \cdots C_{k-n}.$$

Further for $1 \leqslant u \leqslant k$,

$$\Delta_u^k = |a_{k-1}^k \wedge \cdots \wedge a_{k-u}^k| = |X_k||a_{k-1}^k \wedge \cdots \wedge a_{k-u}^k|/|X_k| = |X_k \wedge X_{k-1} \wedge \cdots \wedge X_{k-u}|/|X_k|.$$

It follows that $X_1, \cdots, X_k$ satisfy the conditions of the lemma, as far as they apply only to them.

Suppose now that $X_1, \cdots, X_v$ where $v \geqslant k$ have been constructed with the desired properties; we proceed to construct X_{v+1}. Let A_v be the hyperplane orthogonal to X_v, and let Λ_v be the orthogonal projection of $\mathbb{Z}^k$ on A_v.Then Λ_v is a lattice in A_v of determinant $\det \Lambda_v = |X_v|^{-1}$. Let $a^v_{k-1}, \cdots, a^v_{k-n}$(that is, $a^v_n, \cdots, a^v_1$) be the respective projections of $x_{v-1}, \cdots, X_{v-n}$ on A_v. Then $a^v_n, \cdots, a^v_1$ is a basis of Λ_v, and the quantities Δ^v_u of (26) have

$$\Delta^v_u = |a^v_{k-1} \wedge \cdots \wedge a^v_{k-u}| \quad (u=1, \cdots, n).$$

Also set $\Delta^v_0 = 1$. From the validity of (27) for v we get

$$\Delta^v_n/\Delta^v_{n-1} \geqslant 1/(8^v C_{v-n}),$$

$$\Delta^v_u/\Delta^v_{u-1} \leqslant 8^v/C_{v-u} \quad (1 \leqslant u < n).$$

Thus for $1 \leqslant u < n$, we have from (c) of Lemma 3 that

$$(\Delta^v_u/\Delta^v_{u-1})/(\Delta^v_n/\Delta^v_{n-1}) \leqslant 2^{6v} C_{v-n}/C_{v-u} \tag{28}$$

$$= 2^{6v}(Q_{v-n} \cdots Q_{v-u-1})^{-1} \leqslant 2^{6v} Q^{-1}_{v-n} \leqslant 2^{6v} e^{-G\rho^{v-n}} \leqslant (4v)^{-n}$$

if G is sufficiently large and if $Q > c_9(G, \delta, n, P)$.

By (25), $\varphi(X_i, X_j) < \frac{1}{2}$ for $1 \leqslant i < j \leqslant v$. Hence X_v is not orthogonal to any of $X_1, \cdots, X_v$, and therefore none of $X_1, \cdots, X_v$ lines in A_v. Any n among $X_1, \cdots, X_v$ span a hyperplane H in $\mathbb{R}^k$. This gives $\begin{bmatrix} v \\ n \end{bmatrix}$ hyperplanes H. None of them can be A_v, and hence each of them intersects A_v in a space of dimension $k-2=n-1$, i.e. a hyperplane H' in A_v. By (28) and since $\begin{bmatrix} v \\ n \end{bmatrix} \leqslant v^n$, we may apply Lemma 4. More Preciesely, we apply Lemma 4 with $a^v_{k-1}, \cdots, a^v_{k-n}$ in place of $a_1, \cdots, a_n$, to find a basis $a^v_{k-1}, \cdots, a^v_{k-n+1}, b^v$ of Λ_v, such that (among other conditions) b^v lies outside the $N = [\,^v_n\,]$ hyperplanes H' of A_v above.

By definition of Λ_v, the point b^v is the projection of some $B_v \in \mathbb{Z}^k$ on A_v. In fact, there is such a B_v of the type $B_v = b^v + \lambda X_v$ with $|\lambda| \leqslant \frac{1}{2}$. We now pick

$$X_{v+1} = B_v + qX_v = b^v + (\lambda + q)X_v, \tag{29}$$

with the integer q yet to be determined. We have to pick X_{v+1} to lie outside the $\begin{bmatrix} v \\ n \end{bmatrix}$ hyperplanes H above. Since b^v does not lie in any of these hyperplanes, this means that q has to avoid at most $\begin{bmatrix} v \\ n \end{bmatrix} \leqslant v^n$ values.

Note that by Lemma 4,

$$|b^v| \leqslant 2\Delta^v_n/\Delta^v_{n-1} \leqslant 2 \cdot 8^v C^{-1}_{v-n} \leqslant 2 \cdot 8^v \cdot e^{-F\rho^{n-v}} < 1$$

in view of Lemma 3(b), if F is sufficiently large and $Q > c_8(F, \delta, n, P)$. Also,

$$|X_v\| \leqslant 8^v C_{v-1} \cdots C_{v-n} = 8^v Q_{v-1}^{-1} \cdots Q_{v-n}^{-1} C_v \cdots C_{v-n+1} \leqslant 8^v e^{-G\rho^{v-n}} C_v / \Delta_{n-1}^v$$

by (24),(27) and Lemma 3(c), provided $Q > c_9(G, \delta, n, P)$. Therefore

$$(2C_v/\Delta_{n-1}^v)(\max(|b^v|, |X_v|)^{-1} > 2 \cdot 8^v e^{G\rho^{v-n}} \geqslant v^n + 1,$$

if G was chosen large enough. We therefore may choose q in (29) such that

$$2C_v/\Delta_{n-1}^v \leqslant |X_{v+1}| \leqslant 4C_v/\Delta_{n-1}^v. \tag{30}$$

In fact, since we have more than v^n possibilities for q which give (30), we may choose q such that X_{v+1} lies outside the $\begin{bmatrix} v \\ n \end{bmatrix}$ hyperplanes H mentioned above. Thus any $k = n+1$ among $X_1, \cdots, X_{v+1}$ span $\mathbb{R}^k$. The projections of $X_{v+1}, X_{v-1}, \cdots, X_{v-n+1}$ on A_v are $b^v, a_{k-1}^v, \cdots, a_{k-n+1}^v$, which make up a basis of Λ_v, and therefore $X_{v+1}, X_v, X_{v-1}, \cdots, X_{v-n+1}$ are a basis of $\mathbb{Z}^k$.

We have from (30), and (27) for v, that

$$|X_{v+1}| \leqslant 4C_v \cdot 8^v C_{v-1} \cdots C_{v-n+1},$$

so that (24) holds for $v+1$. For $1 \leqslant u \leqslant n$,

$$\begin{aligned} \Delta_u^{v+1} &= |X_{v+1} \wedge X_v \wedge \cdots \wedge X_{v+1-u}| / |X_{v+1}| \\ &= |X_v \wedge b^v \wedge a_{k-1}^v \wedge \cdots \wedge a_{k-u+1}^v| / |X_{v+1}| \\ &= (|X| / |X_{v+1}|) |b^v \wedge a_{k-1}^v \wedge \cdots \wedge a_{k-u+1}^v|, \end{aligned}$$

where for $u = 1$ we interpret $b^v \wedge a_{k-1}^v \wedge \cdots \wedge a_{k-u+1}^v$ to be b^v. Since by Lemma 4,

$$\frac{1}{2} \Delta_n^v (\Delta_{n-1}^v)^{-1} \Delta_{u-1}^v \leqslant |b^v \wedge a_{k-1}^v \wedge \cdots \wedge a_{k-u+1}^v| \leqslant 2\Delta_n^v (\Delta_{n-1}^v)^{-1} \Delta_{u-1}^v,$$

and since $\Delta_n^v = \det \Lambda_v = |X_v|^{-1}$, we have

$$\frac{1}{2} \frac{\Delta_{u-1}^v}{\Delta_{n-1}^v |X_{v+1}|} \leqslant \Delta_u^{v+1} \leqslant 2 \frac{\Delta_{u-1}^v}{\Delta_{n-1}^v |X_{v+1}|} \qquad (1 \leqslant u \leqslant n).$$

Thus from (30) and from (27) for v we obtain

$$\frac{1}{8 \cdot 8^v C_v C_{v-1} \cdots C_{v-u+1}} \leqslant \frac{\Delta_{u-1}^v}{8C_v} \leqslant \Delta_u^{v+1} \leqslant \frac{\Delta_{u-1}^v}{C_v} \leqslant \frac{1}{C_v C_{v-1} \cdots C_{v-u+1}}.$$

Therefore (27) is true for $v+1$ also.

All that remains for us to do is to establish (25) for $v+1$. By (27) with $v+1$

and $n=1$, we have $|\underline{\underline{X}}_{\nu+1}\wedge\underline{\underline{X}}_\nu| = |\underline{\underline{X}}_{\nu+1}|\Delta_1^{\nu+1}\leqslant|\underline{\underline{X}}_{\nu+1}|C_\nu^{-1}$, so that

$$\varphi(\underline{\underline{X}}_{\nu+1},\underline{\underline{X}}_\nu)\leqslant\frac{\pi}{2}C_\nu^{-1}|\underline{\underline{X}}_\nu|^{-1}<\frac{1}{20}\cdot 2^{n-\nu-1}$$

by Lemma 3(b), if Q was chosen large enough. We get

$$\varphi(\underline{\underline{X}}_{\nu+1},\underline{\underline{E}})\leqslant\varphi(\underline{\underline{X}}_{\nu+1},\underline{\underline{X}}_\nu)+\varphi(\underline{\underline{X}}_\nu,\underline{\underline{E}})<\frac{1}{20}(1-2^{n-\nu}+2^{n-\nu-1})=\frac{1}{20}(1-2^{n-\nu-1}).$$

6. Proof of Theorem 2, Completion.

Write $\underline{\underline{X}}_\nu=(x_{\nu 1},\cdots,x_{\nu n},\ y_\nu)$. For $\nu\geqslant k$ we have $\varphi(\underline{\underline{X}}_\nu,E)<1/20$, so that

$$y_\nu\leqslant|\underline{\underline{X}}_\nu|<2y_\nu\quad(\nu\geqslant k). \tag{31}$$

Write $x_\nu=(x_{\nu 1}/y_\nu,\cdots,x_{\nu n}/y_\nu)$. The sequence $\underline{\underline{x}}_k,\underline{\underline{x}}_{k+1},\cdots$ is independent, i.e., any $n+1$ of its elements span $\mathbb{R}^n$ as an affine space — this is because any $n+1$ elements of $X_k, X_{k+1},\cdots$ span $\mathbb{R}^k$. We have

$$|\underline{\underline{x}}_\nu-\underline{\underline{x}}_{\nu+1}|^2=y_\nu^{-2}y_{\nu+1}^{-2}\sum_{i=1}^{n}(x_{\nu i}y_{\nu+1}-x_{\nu+1,i}\,y_\nu)^2\leqslant y_\nu^{-2}y_{\nu+1}^{-2}|\underline{\underline{X}}_\nu\wedge\underline{\underline{X}}_{\nu+1}|^2$$

and therefore

$$|x_\nu-x_{\nu+1}|\leqslant y_\nu^{-1}y_{\nu+1}^{-1}|\underline{\underline{X}}_{\nu+1}|\Delta_1^{\nu+1}<2y_\nu^{-1}C_\nu^{-1}\quad(\nu\geqslant k)$$

by (31), (26), (27). From Lemma 3(b) we see that C_ν grows rapidly, so that $\{\underline{\underline{x}}_\nu\}$tends to some limit $\underline{\underline{\xi}}\in\mathbb{R}^n$, and in fact

$$|\underline{\underline{\xi}}-\underline{\underline{x}}_\nu|<\sum_{\nu=\nu}^{\infty}2y_\mu^{-1}C_\mu^{-1}<3y_\nu^{-1}C_\nu^{-1}\quad(\nu\geqslant k)$$

if Q is large. Here $C_\nu=P^\nu(C_{\nu-1}\cdots C_{\nu-n})^{(1/n)+\delta}$ and $y_\nu\leqslant|X_\nu|\leqslant 8^\nu C_{\nu-1}\cdots C_{\nu-n}$, so that

$$C_\nu\geqslant P^\nu(8^{-\nu}y_\nu)^{(1/n)+\delta}\geqslant(P/8^{1+\delta})^\nu y_\nu^{(1/n)+\delta}$$

and

$$|\underline{\underline{\xi}}-\underline{\underline{x}}_\nu|<3(8^{1+\delta}/P)^\nu y_\nu^{-1-(1/n)-\delta}\quad(\nu\geqslant k).$$

Therefore $\underline{\underline{x}}_\nu$ is a (δ,M)-approximation to $\underline{\underline{\xi}}$ if $P>P_0(\delta,M)$. We now choose P, and then Q, to satisfy all our conditions. Then P, Q depend only on n,δ,M.

How many values of ν are there with $\nu\geqslant k$ and $y_\nu\leqslant B$? Observe that

$$y_\nu\leqslant|\underline{\underline{X}}_\nu|\leqslant 8^\nu C_{\nu-1}\cdots C_{\nu-n}<8^\nu e^{nE\rho^\nu}<e^{E_1\rho^\nu}$$

by Lemma 3(a), where E and E_1 ultimately depend only on n, δ, M. The number of integers ν in question is therefore at least given by (6).

It only remains for us to show that $1, \xi_1, \cdots, \xi_n$ are linearly independent over $\mathscr{R}$. Otherwise, there is a nontrivial relation $a_0 + a_1\xi_1 + \cdots + a_n\xi_n = 0$ with integer coefficients. Since x_ν is a (δ, M)-approximation we have

$$|a_0 + a_1 \frac{x_{\nu 1}}{y_\nu} + \cdots + a_n \frac{x_{\nu n}}{y_\nu}| \leqslant |a_1| |\xi_1 - \frac{x_{\nu 1}}{y_\nu}| + \cdots + |a_n| |\xi_n - \frac{x_{\nu n}}{y_\nu}| < c y_\nu^{-1-(1/n)-\delta}$$

where $c = c(M, a_1, \cdots, a_n)$. Thus for large ν we have

$$a_0 + a_1 \frac{x_{\nu 1}}{y_\nu} + \cdots + a_n \frac{x_{\nu n}}{y_\nu} = 0,$$

and x_ν lies on a fixed hyperplane. But since the sequence $\{x_\nu\}$ is independent, this cannot happen for all large values of ν.

References

[1] E. Bombieri and A. J. Van der Poorten. Some quantitative results related to Roth's Theorem, *J. Austral. Math. Soc.* (to appear).

[2] J. Mueller and W. M. Schmidt. On the number of good rational approximations to algebraic numbers, *Proc. A.M.S.* (submitted).

[3] W. M. Schmidt. Simultaneous approximation to algebraic numbers by rationals, *Acta Math.* 125(1970), 189—201.

[4] W. M. Schmidt. The Subspace Theorem in Diophantine Approximation, *Compositio Math.* (to appear).

[5] W. M. Schmidt. The number of rational approximations to algebraic numbers and the number of solutions of norm form equations, *Proc. Tata Conference in honor of Ramanujan*, Bombay 1988.

[6] P. Szüsz. Über die metrische Theorie der diophantischen Approximationen, II., *Acta Arith.* 8(1962/63), 225—241.

ON A GENERALIZED WARING'S PROBLEM IN ALGEBRAIC NUMBER FIELDS

M.V. SUBBARAO* AND WANG YUAN

Department of Mathematics
University of Alberta
Edmonton, Alberta T6G 2G1

Institute of Mathematics
Academia Sinica
Beijing, China

1. Introduction.

Let K be a totally real agebraic number field of degree n. Let $K^{(i)}\,(1 \le i \le n)$ be the conjugate fields of K. For $\gamma \in K$, we denote by $\gamma^{(i)}$ $(1 \le i \le n)$ the conjugates of γ and $N(\gamma) = \prod_{i=1}^{n} \gamma^{(i)}$ the norm of γ. Let γ_j $(1 \le j \le n)$ be numbers of K and x_i $(1 \le i \le n)$ be real numbers. We set $\xi = \sum_{j=1}^{n} x_j \gamma_j$ and define $\xi^{(i)} = \sum_{j=1}^{n} x_j \gamma_j^{(i)}$ $(1 \le i \le n)$. We use the notations

$$\|\xi\| = \max_i |\xi^{(i)}|,\ S(\xi) = \sum_{i=1}^{n} \xi^{(i)} \text{ and } E(\xi) = \exp(2\pi i S(\xi)).$$

where $\exp(x) = e^x$. A number γ of K is called totally nonnegative if $\gamma^{(i)} \ge 0\ (1 \le i \le n)$.

It was Siegel [5,6] who succeeded in dealing with Waring's problem in an arbitrary algebraic number field by his generalized circle method, and obtained the result corresponding to Hardy-Littlewood's estimation on $G(k)$.

Ayoub [1] gave an extension of Siegel's theorem, namely to replace the kth powers by polynomial summands for totally real algebraic number fields. Let ν, α, α_i $(1 \le i \le k-1)$ be nonzero totally nonnegative integers of K and $k \ge 2$. Consider the polynomial

$$\phi(\xi) = \alpha\xi^k + \alpha_{k-1}\,\xi^{k-1} + \cdots + \alpha_1\xi$$

and the equation

$$\nu = \phi(\xi_1) + \cdots + \phi(\xi_s). \tag{1}$$

*Partly supported by a grant from Natural Sciences and Engineering Research Council of Canada.

Let $A(\nu)$ be the number of solutions of (1) in totally nonnegative integers $\xi_1, \ldots, \xi_s$ satisfying $N(\xi_i) \le N(\nu)^{1/k}$ $(1 \le i \le s)$. Ayoub established that if $s \ge n(2^k + n) + 1$, then

$$(2) \qquad A(\nu) = D^{1/2(1-s)}\, \mathfrak{S}(\nu) \Big(\frac{\Gamma(1+\frac{1}{k})}{\Gamma(s/k)}\Big)^n N(\alpha)^{-s/k}\, N(\nu)^{-1+s/k}\, (1 + o(1))$$

where D is the absolute value of the discrimant of K and $\mathfrak{S}(\nu)$ the singular series of (1).

It is our object to give a natural extension of Ayoub's theorem, namely to replace the $\phi(\xi)$ by different polynomials of degree k, and to give a slight improvement on the lower bound for s to $\max(4kn, 2^k + 1)$. In addition, it seems that there is a gap in his proof of (2) (See Remark in §6).

Consider the polynomials

$$(3) \qquad \phi_i(\lambda) = \alpha_{ki}\lambda^{k-1} + \alpha_{k-1,i}\,\lambda^{k-1} + \cdots + \alpha_{1i}\lambda, \ 1 \le i \le s$$

and the equation

$$(4) \qquad \nu = \phi_1(\lambda_1) + \cdots + \phi_s(\lambda_s),$$

where ν and $\alpha_{ki}\,(1 \le i \le s)$ are given nonzero totally nonnegative integers and $\alpha_{ji}\,(1 \le j \le k-1,\ 1 \le i \le s)$ are integers. Let $B(\nu)$ be the number of solutions of (4) in totally nonnegative integers $\lambda_i\,(1 \le i \le s)$ satisfying $N(\lambda_i) \le N(\nu)^{1/k}$ $(1 \le i \le s)$.

THEOREM. *If $s \ge \max(4kn, 2^k + 1)$, then*

$$B(\nu) = \mathfrak{S}' J\, N(\nu)^{-1+s/k}\, (1 + o(1))$$

where $\mathfrak{S}'$ and J denote the singular series and singular integral of (4) respectively. (See §5).

2. The generalized circle method.

Let $\omega_1, \ldots, \omega_n$ be an integral basis of K and δ the different of K. We can choose a basis $\rho_1, \ldots, \rho_n$ of δ^{-1} such that

$$S(\rho_i, \omega_j) = \begin{cases} 1, & \text{if } i = j, \\ 0, & \text{if } i \neq j. \end{cases}$$

Set $\xi = x_, \rho_1 + \cdots + x_n \rho_n$ and $\eta = y_1 \omega_1 + \cdots + y_n \omega_n$ where x_i and y_i $(i \le i \le n)$ are real numbers. We denote by $dx = \prod_{i=1}^{n} dx_i$, $dy = \prod_{i=1}^{n} dy_i$ $P(T)$ the set of $\underline{\underline{y}} = (y_1, \ldots, y_n)$ satisfying $0 \le \eta^{(i)} \le T$ $(1 \le i \le n)$ and $\sum_{\lambda \in P(T)}$ a sum where λ runs over all integers such that $0 \le \lambda^{(i)} \le T$ $(1 \le i \le n)$.

Let $\mathbb{Q}$ denote the rational number field and U_n the n-dimensional unit cube $\{\underline{\underline{x}} = (x_1, \ldots, x_n) : 0 \le x_i < 1 (1 \le i \le n)\}$. Let h and t be real numbers satisfying $h > 2Dt$ and $t > 1$. For any $\gamma \in K$, we can determine uniquely two integral ideals $\mathcal{A}_0$, $\mathcal{L}$ such that

$$\gamma \delta = \mathcal{L} / \mathcal{A}, \ (\mathcal{A}, \mathcal{L}) = 1.$$

We write $\gamma \to \mathcal{A}$. Let $\Gamma(t)$ be the set consisting of $\gamma = x_1 \rho_1 + \cdots + x_n \rho_n$ satisfying

$$\underline{\underline{x}} \in U_n : x_i \in \mathbb{Q}(1 \le i \le n), \ \gamma \to \mathcal{A} \text{ and } N(\mathcal{A}) \le t^n.$$

For every $\gamma \in \Gamma(t)$ subject to $\gamma \to \mathcal{A}$, we define the basis domain B_γ by

$$\underline{\underline{x}} : \underline{\underline{x}} \in U_n, \ \prod_{i=1}^{n} \max(h|\xi^{(i)} - \gamma_0^{(i)}|, t^{-1}) \le N(\mathcal{A})^{-1} \text{ for some } \gamma_0 \equiv \gamma (\mathrm{mod}\ \delta^{-1}).$$

We can show that if γ_1 and γ_2 belong to $\Gamma(t)$ and $\gamma_1 \neq \gamma_2$, then

$$B_{\gamma_1} \cap B_{\gamma_2} = \phi.$$

(See, Siegel [5,6] or Wang [9]). We set

$$B = \bigcup_{\gamma \in \Gamma(t)} B_r$$

and define the supplementary domain S of B with respect to U_n by

$$S = U_n - B.$$

This division of U_n into B and S depends on (h, t). We shall call this division the Farey division of U_n with respect is (h, t).

Let

$$S_i(\xi,T) = S_i(\xi) = \sum_{\lambda\in P(T)} E(\phi_i(\gamma)\xi) \text{ and } S(\xi,T) = \prod_{i=1}^{s} S_i(\xi).$$

Since for any integer α

$$\int_{U_n} E(\alpha\xi)dx = \begin{cases} 1, & \text{if } \alpha = 0, \\ 0, & \text{if } \alpha \neq 0, \end{cases}$$

the number of solutions of (4) in totally nonnegative integers $\lambda_i (1 \le i \le s)$ satisfying $\|\lambda_i\| \le T (1 \le i \le s)$ is equal to

$$Z(\nu) = \int_{U_n} S(\xi,T)E(-\nu\xi)dx = \int_B S(\xi,T)E(-\nu\xi)dx + \int_S S(\xi,T)E(-\nu\xi)dx$$

$$= Z_1 + Z_2, \tag{5}$$

say.

3. Asymptotic expansion for $S_\ell(\xi)$.

Denote by

$$a = \frac{1}{4} + \frac{1}{4kn}, \quad t = T^{1-a} \quad \text{and } h = T^{k-1+a}$$

where $T > 1$. For any $\gamma \in \Gamma(t)$, let $\mathcal{L}_\ell = (\alpha_{k\ell}\gamma, \dots, \alpha_{1\ell}\gamma)$ be the ideal generated by $\alpha_{k\ell}\gamma, \dots, \alpha_{1\ell}\gamma$ and $\mathcal{A}_\ell$ denote the denominator of $\mathcal{L}_\ell\delta$. We use the notations

$$\xi - \gamma = \varsigma, \; G_\ell(\gamma) = N(\mathcal{A}_\ell)^{-1} \sum_{\lambda(\mathcal{A}_\ell)} E(\phi_\ell(\lambda)\gamma) \text{ and } I_\ell(\varsigma,T) = \int_{P(T)} E(\phi_\ell(\eta)\varsigma)dy,$$

where $\sum_{\lambda(\mathcal{A}_\ell)}$ denotes a sum in which λ runs over a complete residue system mod $\mathcal{A}_\ell$. We use E_n to denote the whole n-dimensional Euclidean space.

LEMMA 1. *(Hua [3]). For any given $\varepsilon > 0$,*

$$G_\ell(\gamma) \ll N(\mathcal{A}_\ell)^{\varepsilon - 1/k},$$

hereafter the constants implicit in "$\ll$" or "O" may depend on ε and K. (Note that α_{ij}'s are given integers of K).

LEMMA 2. *(Vinogradov [8]). Let $f(x) = \gamma_k x^k + \cdots + \gamma_1 x$ be a polynomial with real coefficients. Then*

$$\int_0^1 exp(2\pi i f(x))dx \ll min(1, |\gamma_1|^{-1/k}, \ldots, |\gamma_k|^{-1/k}).$$

LEMMA 3. $I_\ell(\varsigma, T) \ll \prod_{i=1}^{n} min(T, |\varsigma^{(i)}|^{-1/k})$.

PROOF: Let $\eta^{(j)} = Tu_j$, $1 \le j \le n$. Then the Jacobian of $y_1, \ldots, y_n$ with respect to $u_1, \ldots, u_n$ is equal to $D^{-1/2}T^n$. Therefore by Lemma 2,

$$I_\ell(\varsigma, T) = D^{-1/2}T^n \prod_{j=1}^{n} \int_0^1 \exp(2\pi i \varphi_\ell(u_j T)\varsigma^{(j)})du_j \ll \prod_{j=1}^{n} \min(T, |\varsigma^{(j)}|^{-1/k}),$$

and the lemma is proved.

LEMMA 4. *If $s < k$, then*

$$\int_{E_n} \prod_{i=1}^{n} min(T, |\varsigma^{(i)}|^{-1/k})^s dx \ll T^{(s-k)n}.$$

PROOF: Let $\varsigma^{(j)} = v_j, 1 \le j \le n$. Then the Jacobian of $x_1, \ldots, x_n$ with respect to $v_1, \ldots, v_n$ is $D^{1/2}$. Therefore

$$\int_{E_n} \prod_{j=1}^{n} \min(T, |\varsigma^{(j)}|^{-1/k})^s dx \ll D^{1/2} \prod_{j=1}^{n} \int_0^\infty \min(T^s, v_j^{-s/k})dv_j$$

$$\ll \left(\int_0^{T^{-k}} T^s dv + \int_{T^{-k}}^\infty v^{-s/k} dv\right)^n \ll T^{(s-k)n}$$

and the lemma is proved.

LEMMA 5. *Let μ be an integer. Then*

$$\sum_{\substack{\gamma+\mu \in P(T) \\ A|\lambda}} E(\varphi_\ell(\lambda+\mu)\varsigma) + N(A_\ell)^{-1} I_\ell(\varsigma, T) + O(N(A_\ell)^{-1}T^{n-a}).$$

PROOF: The proof is similar to the proof for the sum of $E(\alpha(\lambda+\mu)^k\varsigma)$ (See, Siegel [5,6] or Wang [9]). We give the proof here, for the sake of completeness.

Determine positive numbers $\theta^{(i)}$ $(1 \le i \le n)$ such that

$$\theta^{(i)} \max(h|\varsigma^{(i)}|, t) = D^{1/2n} \prod_{i=1}^{n} \max(h|\varsigma^{(i)}|, t^{-1})^{1/n} N(A_\ell)^{1/n} \tag{6}$$

then

$$\prod_{i=1}^{n} \theta^{(i)} = D^{1/2} N(\mathcal{A}_\ell),$$

and it follows by Minkowski's linear form theorem that there exists $\sigma \in \mathcal{A}_\ell$ such that

$$(7) \qquad 0 < |\sigma^{(i)}| \leq \theta^{(i)}, \ 1 \leq i \leq n.$$

Hence $\sigma \mathcal{A}_\ell^{-1} = \mathcal{L}_\ell$ is an integral ideal and

$$N(\mathcal{L}_\ell) = |N(\sigma)| N(\mathcal{A}_\ell)^{-1} \leq \prod_{i=1}^{n} \theta^{(i)} N(\mathcal{A}_\ell)^{-1} = \sqrt{D}.$$

Therefore $\mathcal{L}_\ell$ belongs to a finite set depending on K only. Let $\sigma_1, \ldots, \sigma_n$ be a basis of $\mathcal{L}_\ell^{-1}$. Then $\mathcal{A}_\ell = \sigma \mathcal{L}_\ell^{-1}$ has a basis

$$\tau_i = \sigma \sigma_i, \ 1 \leq i \leq n.$$

By (6), (7) we obtain

$$\|\tau_i\| = O(\|\sigma\|) = O(\|\theta\|) = O(t).$$

Expressing λ in terms of τ_i's we have

$$\lambda = g_1 \tau_1 + \cdots + g_n \tau_n,$$

where g_i's are rational integers. Let $U(\lambda)$ denote the unit cube

$$\underline{\underline{s}} : \tau = s_1 \tau_1 + \cdots + s_n \tau_n, \ g_i \leq s_i < g_i + 1 (1 \leq i \leq n)$$

and $ds = \prod_{i=1}^{n} ds_i$. Then by (6) we have

$$\begin{aligned} &E(\varphi_\ell(\lambda + \mu)\varsigma) - \int_{U(\lambda)} E(\varphi_\ell(\tau + \mu)\varsigma) ds \\ &\ll \|(\lambda - \tau)\varsigma\| (\|\tau + \mu\|^{k-1} + \|\lambda + \mu\|^{k-1} + 1) \\ &\ll \|\theta\varsigma\| T^{k-1} \ll h^{-1} T^{k-1} \ll T^{-a}. \end{aligned}$$

Since the number of integers λ with $\mathcal{A}_\ell \mid \lambda$ and $\lambda + \mu \in P(T)$ is $O(N(\mathcal{A}_\ell)^{-1} T^n)$, we have

$$\begin{aligned} \sum_{\substack{\lambda+\mu \in P(T) \\ \mathcal{A}_\ell \mid \lambda}} E(\varphi_\ell(\lambda + \mu)\varsigma) = &\sum_{\substack{\lambda+\mu \in P(T) \\ \mathcal{A}_\ell \mid \lambda}} \int_{U(\lambda)} E(\varphi_\ell(\tau + \mu)\varsigma) ds \\ &+ O(N(\mathcal{A}_\ell)^{-1} T^{n-a}) \end{aligned}$$

$$= \int_{0\le\mu^{(i)}+\tau^{(i)}\le T} \cdots \int E(\varphi_\ell(\tau+\mu)\varsigma)ds + O(N(\mathcal{A}_\ell)^{-1}T^{n-a}).$$

Let $\tau+\mu=\eta$. Since the Jacobian of s_i's with respect to y_i's is $D^{1/2}|\det(\tau_j^{(i)})|^{-1} = N(\mathcal{A}_\ell)^{-1}$, the lemma follows.

LEMMA 6. $S_\ell(\xi) = G_\ell(\gamma)I_\ell(\varsigma,T) + O(T^{n-a})$.

PROOF: By Lemma 5,

$$\begin{aligned} S_\ell(\xi) &= \sum_{\mu(\mathcal{A}_\ell)} E(\phi_\ell(\mu)\gamma) \sum_{\substack{\lambda+\mu\in P(T)\\ \mathcal{A}_\ell|\lambda}} E(\phi(\lambda+\mu)\varsigma) \\ &= \sum_{\mu(\mathcal{A}_\ell)} E(\phi_\ell(\mu)\gamma)(N(\mathcal{A}_\ell)^{-1}I_\ell(\varsigma,T) + O(N(\mathcal{A}_\ell)^{-1}T^{n-a})) \\ &= G_\ell(\gamma)I_\ell(\varsigma,T) + O(T^{n-a}). \end{aligned}$$

The lemma is proved.

4. Basic domains.

Set $G(\gamma) = \prod_{i=1}^{s} G_i(\gamma)$, $S(\xi,T) = \prod_{i=1}^{s} S_i(\xi)$ and $I(\varsigma,T) = \prod_{i=1}^{s} I_i(\varsigma,T)$.

LEMMA 7. *If $s \ge 4kn$, then*

$$\int_B S(\xi,T)E(-\nu\xi)dx = \sum_{\gamma\in\Gamma(t)} G(\gamma)E(-\nu\gamma)\int_{B_\gamma} I(\varsigma,T)E(-\nu\varsigma)dx + 0(T^{(s-k)n-a}).$$

PROOF: Let $\gamma\to\mathcal{A}$. Since $\mathcal{A}_i|\mathcal{A}|\alpha_{k1}\ldots\alpha_{1i}\mathcal{A}_i$, we have

$$N(\mathcal{A}_i)^{-1} \ll N(\mathcal{A})^{-1} \ll N(\mathcal{A}_i)^{-1},\ 1\le i\le s.$$

By Lemma 1, 3 and 6, we have

$$\begin{aligned} S(\xi,T) &= G(\gamma)I(\varsigma,T) + O\Big(T^{n-a}\max_j (T^{n-a}, |G_j(\gamma)I_j(\varsigma,T)|)^{s-1}\Big) \\ &= G(\gamma)I(\varsigma,T) + O(T^{(n-a)s}) + O(T^{n-a}(N(\mathcal{A})^{1/k+\varepsilon/s}\prod_{i=1}^{n}\min(T,|\varsigma^{(i)}|^{-1/k})^{s-1}). \end{aligned}$$

Since the number of $\gamma\in\Gamma(t)$ such that $\gamma\to\mathcal{A}$ is $O(N(\mathcal{A}))$ and the number $\mathcal{A}$ such that $N(\mathcal{A})=m$ is $O(m^\varepsilon)$, we have

$$(8) \qquad \sum_{\gamma\in\Gamma(t)} N(\mathcal{A})^{-\frac{s-1}{k}+\varepsilon} \ll \sum_{m\le t^n} m^{-\frac{4kn-1}{k}+1+2\varepsilon} \ll 1,$$

and by Lemma 4 and $a = \frac{1}{4} + \frac{1}{4k^n}$,

$$\int_B S(\xi,T)E(-\nu\xi)dx = \sum_{\gamma\in\Gamma(t)} G(\gamma)E(-\nu\gamma)\int_{B_\gamma} I(\varsigma,T)E(-\nu\varsigma)dx$$

$$+ O(T^{(n-a)s}) + O\Big(T^{n-a}\sum_{\gamma\in\Gamma(t)} N(\mathcal{A})^{-\frac{s-1}{k}+\varepsilon}\int_{E_n}\prod_{i=1}^{n}\min(T,|\varsigma^{(i)}|^{-1/k})^{s-1}dx\Big)$$

$$= \sum_{\gamma\in\Gamma(t)} G(\gamma)E(-\nu\gamma)\int_{B_\gamma} I(\varsigma,T)E(-\nu\varsigma)dx + O(T^{(s-k)n-a})$$

The lemma is proved.

Lemma 8. *If $s \ge 4kn$, then*

$$\int_B S(\xi,T)E(-\nu\xi)dx = \sum_{\gamma\in\Gamma(t)} G(\gamma)E(-\nu\gamma)\int_{E_n} I(\varsigma,T)E(-\nu\varsigma)dx + O(T^{(s-k)n-a}).$$

Proof: By Lemma 7, it suffices to show that

$$W = \sum_{\gamma\in\Gamma(t)} G(\gamma)E(-\nu\gamma)\int_{E_n-B_\gamma} I(\varsigma,T)E(-\nu\varsigma)dx = O(T^{(s-k)n-a}).$$

If $\underline{\underline{x}} \in E_n - B_\gamma$, then there is at least one index i such that $h|\varsigma^{(i)}| > N(\mathcal{A})^{1/n}$. Hence by Lemma 3

$$\int_{E_n-B_\gamma} I(\varsigma,T)E(-\nu\varsigma)dx \ll \int_{E_n-B_\gamma}\prod_{i=1}^{n}\min(T^s,|\varsigma^{(i)}|^{-s/k})dx$$

$$\ll \int_{h^{-1}N(\mathcal{A})^{-1/n}}^{\infty} u^{s/k}du\Big(\int_0^\infty \min(T^s,v^{-s/k})dv\Big)^{n-1}$$

$$\ll h^{s/k-1}N(\mathcal{A})^{\frac{1}{n}(\frac{s}{k}-1)}T^{(s-k)(n-1)}.$$

By Lemma 1, we have

$$\sum_{\gamma\in\Gamma(t)} G(\gamma)N(\mathcal{A})^{\frac{1}{n}(\frac{s}{k}-1)} \ll \sum_{\gamma\in\Gamma(t)} N(\mathcal{A})^{\varepsilon-\frac{s}{k}+\frac{s}{k}-1} \ll t^{2n}.$$

Therefore

$$W \ll T^{(s-k)n-(\frac{s}{k}-1)(1-a)+2(1-a)n} \ll T^{(s-k)n-a}.$$

The lemma is proved.

5. Some more lemmas.

Set

$$J_i(\varsigma,T)=\int_{P(T)}E(\alpha_{ki}\eta^k\varsigma)dy(1\le i\le s)\quad\text{and}\quad J(\varsigma,T)=\prod_{i=1}^{s}J_i(s,T).$$

Since

$$E(\phi_i(\eta)\varsigma)-E(\alpha_{ki}\eta^k\varsigma)\ll S(|\varsigma|)T^{k-1},$$

we have for $\theta=-k+\frac{1}{n+1}$,

$$\int_{E_n}(I(\varsigma,T)-J(\varsigma,T))E(-\nu\varsigma)dx\ll\int_{\|\varsigma\|\le T^\theta}|I(\varsigma,T)-J(\varsigma,T)|dx$$

$$+\sum_{i=1}^{n}\int_{|\varsigma^{(i)}|>T^\theta}|I(\varsigma,T)-J(\varsigma,T)|dv=\Sigma_1+\Sigma_2,\text{ say.}$$

By Lemma 3, we have

$$\sum_1\ll\int_0^{T^\theta}\cdots\int_0^{T^\theta}T^{k-1}\sum_{j=1}^{n}v_j\prod_{i=1}^{n}\min(T,v_i^{1/k})^{s-1}\,dv_1\ldots dv_n$$

$$\ll T^{(s-1)n+k-1+(n+1)\theta}\quad\ll T^{(s-k)n-n}$$

and for $s\ge 4kn$,

$$\Sigma_2\ll\int_{T^\theta}^{\infty}r^{-s/k}\,dv\int_{E_{n-1}}\prod_{j=1}^{n-1}\min(T,v_j^{-1/k})^s\,dv_1\ldots dv_{n-1}$$

$$\ll T^{(-s/k+1)\theta+(s-k)(n-1)}\quad\ll T^{(s-k)n-(s/k-1)\frac{1}{n+1}}\quad\ll T^{(s-k)n-1}.$$

By (8) and Lemmas 1 and 8, we have

LEMMA 9. *If $s\ge 4kn$, then*

$$\int_B S(\xi,T)E(-\nu\xi)dx=\sum_{\gamma\in\Gamma(t)}G(\gamma)E(-\nu\gamma)\int_{E_n}J(\varsigma,T)E(-\nu\varsigma)dx$$

$$+0(T^{(s-k)n-a}).$$

Let $\eta'=y_1'\omega_1+\cdots+y_n'\omega_n$, $\varsigma'=x_1'\rho_1+\cdots+x_n'\rho_n$, $dy'=dy_1'\ldots dy_n'$, $dx'=dx_1'\ldots dx_n'$, $\eta=T\eta'$ and $\varsigma=T^{-k}\varsigma'$. The Jacobian of $y_1,\ldots,y_n$ and $x_1,\ldots,x_n$ with respect to $y'_n, \ldots, y'_n$ and $x'_1, \ldots, x'_n$ are T^n and T^{-kn} respectively. Then

$$\alpha_{ki}\eta^k\varsigma = \alpha_{ki}\eta'^k\varsigma'.$$

Write η' and ς' by η and ς again. Then

$$\int_{E_n} J(\varsigma, T)E(-\nu\varsigma)dx = T^{(s-k)n} J(\mu), \tag{9}$$

where $\mu = \nu T^{-k}$ and

$$J(\mu) = J = \int_{E_n} \Big(\prod_{i=1}^{s} J_i(\varsigma, 1)\Big)E(-\mu\varsigma)dx.$$

J is called the singular integral which is absolutely convergent by Lemma 3 if $s > k$. Similar to Siegal [5,6] or Tatuzawa [7], Wang [9], we have

$$J = D^{\frac{1}{2}(1-s)}\, k^{-sn}\, N(\alpha_{k1} \ldots \alpha_{ks})^{1/k} \prod_{\ell=1}^{n} F_\ell,$$

where

$$F_\ell = \int_{W_\ell} \prod_{i=1}^{s} w^{\frac{1}{s}-1}\, dw$$

in which $dw = \prod_{i=1}^{s-1} dw_i$ and W_ℓ denotes the domain

$$0 \le \omega_i \le \alpha_{ki}^{(\ell)} \ (1 \le i \le s),\ \mu^{(\ell)} - \omega_1 - \cdots - \omega_s = 0.$$

By Lemma 1 we have for $s \ge 4kn$

$$\sum_{\substack{\gamma \to A \\ N(A) > t^n}} G(\gamma)E(-\nu\gamma) \ll \sum_{\substack{\gamma \to A \\ N(A) > t^n}} N(A)^{-\frac{s}{k}+\varepsilon} \ll T^{-(1-a)n}\ .$$

and thus

$$\sum_{\gamma \in \Gamma(t)} G(\gamma)E(-\nu\gamma) = \mathfrak{S}' + O(T^{-(1-a)n}) \tag{10}$$

where

$$\mathfrak{S}' = \sum_{\gamma} G(\gamma)E(-\nu\gamma)$$

in which γ runs over a complete residue system of $(A\delta)^{-1}$, mod δ^{-1}, $\mathfrak{S}'$ is called the singular series.

By Lemma 9. (9), (10) and (11), we have

LEMMA 10. *If $s \ge 4k^n$, then*

$$\int_B \delta(\xi, T)E(-\nu\xi)dx = \mathfrak{S}' J\ T^{(s-k)n}\ (1 + o(1)).$$

6. Proof of the theorem.

Let θ be a number satisfying $0 < \theta < (1-a)2^{1-k}$.

LEMMA 11. *If $\underline{\underline{x}} \in S$, then*

$$S_i(\xi) \ll T^{n-\theta}$$

where the constant implicit in $\ll$ depends on θ.

This can be proved a a Siegel's lemma [5,6] on the diophantine approximation for $\xi(\underline{\underline{x}} \in S)$ and a theorem of Mitsui [5]. See, Tatuzawa [7], p. 54.

LEMMA 12. *If $1 \leq j \leq k$, then*

$$\int_{U_n} |S_i(\xi)|^{2^j} dx \ll T^{(2^j-j)n+\varepsilon} .$$

This is a generalization of Hua's inequality [2] in algebraic number fields. See, Ayoub [1], p. 447.

LEMMA 13. *If $s \geq 2^k + 1$, then*

$$\int_S S(\xi, T)E(-\nu\xi)dx \ll T^{(s-k)-\theta} .$$

PROOF: Put

$$\theta_1 = \frac{\theta}{2} + \frac{1-a}{2^k}.$$

Then

$$\theta < \theta_1 < (1-a)2^{1-k}.$$

By Lemmas 11 and 12 we have

$$\int_S |S_i(\xi)|^s dx \ll T^{(s-2^k)(n-\theta_1)} \int_{U_n} |S_i(\xi)|^{2^k} dx$$

$$\ll T^{(s-2^k)(n-\theta_1)+(2^k-k)n+\theta_1-\theta} \quad \ll T^{(s-k)n-\theta} .$$

Therefore by Hölder's inequality,

$$\int_S S(\xi,T)E(-\nu\xi)dx \ll \int_S |S(\xi,T)|dx \ll \prod_{i=1}^{s}\Big(\int_S |S_i(\xi)|^s dx\Big)^{1/s}$$

$$\ll T^{(s-k)n-\theta} \ .$$

The lemma is proved.

Proof of Theorem. Set $T = N(\nu)^{1/kn}$. Out main theorem follows from Lemmas 10 and 13 and (5).

Remarks. 1.) We have not discussed the singular series.

2.) In Ayoub's proof of his Theorem 5.3 ([1], p. 443), the estimation

$$\sum_{\substack{\lambda+\mu\in\mathcal{F}\\ \mathcal{A}|\lambda}} S(|\varsigma|T^{k-1}) \ll T^{k+n-1} N(\mathcal{A})^{-1} h^{-1} N(\mathcal{A})^{-1/n}$$

is used. Since from the definition of ς ([1], p. 441), i.e.

$$N(\max(h|\varsigma|, t^{-1}) \leq N(\mathcal{A})^{-1}$$

or

$$\prod_{i=1}^{n} \max(h|\varsigma^{(i)}|, t^{-1}) \leq N(\mathcal{A})^{-1},$$

if $n > 1$, i.e. if K is not rational field it seems that we cannot derive that for all i,

$$h|\varsigma^{(i)}| \ll N(\mathcal{A})^{-1/n}$$

or

$$S(|\varsigma|) \ll h^{-1} N(\mathcal{A})^{-1/n} .$$

Another thing he used is that $A(\nu) = A(\nu\eta^k)$ ([1], p. 449). This seems not trivial when $\phi(\xi) \neq \alpha\xi^k$.

REFERENCES

[1] R.G. Ayoub, *On the Waring-Siegel Theorem*, Can. J. Math. **5**, 1958, 439–450.

[2] L.K. Hua, *On Waring's problem*, Quart. J. Math. **9**, 1938, 199-201.

[3] L.K. Hua, *On Exponential sums over an algebraic number field*, Can. J. Math. **3**, 1951, 44–45.

[4] T. Mitsui, *On the Goldback problem in an algebraic number field*, I. J. Math. Soc. Japan **12**, 1960, 290–324.

[5] C.L. Siegel, *Generalization of Waring's problem to algebraic number fields*, Amer. J. Math. **66**, 1944, 122–136.

[6] C.L. Siegel, *Sums of m-th powers of algebraic integers*, Ann. of Math. **46**, 1945, 313–339.

[7] T. Tautzawa, *On Waring's problem in algebraic number fields*, Acta Arith. **24**, 1973, 37–60.

[8] I.M. Vinogradov, Selected works, Springer Verlag, 1985, p. 199.

[9] Y. Wang, *Bounds for solutions of additive equations in an algebraic number field I*, Acta Arith. **48**, 1987, 21–48.

Solutions of $x+y=z$ in Numbers Which Are Almost $\mathcal{S}$-Units*

Wang Lianxiang

Institute of Mathematics, Academia Sinica

Abstract

We discuss the finiteness of the numbers of the solutions to the equation $x+y=z$ and the inequality $|x-y|<x^{\delta}$ in positive integers x, y, z if the major parts of x, y, z are composed of prime numbers from a finite set under some condition which is the weakest possible.

Part I

Let $\mathcal{T}=\{p_1,\cdots,p_s\}$ be a set of prime numbers with $s\geqslant 3$ and $p_1<p_2<\cdots<p_s=P$. Let $\mathcal{S}$ be the set of all integers composed of prime numbers from $\mathcal{T}$. Consider the equation.

$$x+y=z \tag{1}$$

in x, y, z $\in \mathcal{S}$ with g. c. d. (x, y, z) = 1. Without the loss of generality we may assume that x, y and z are positive. It follows from a result of Mahler [6] on the greatest prime factor of a binary form that max (x, y, z) is bounded by a number depending only on P. Coates [1, 2] proved that this number is effectively computable in terms of P. De Weger [12] showed how this method can be used to determine all solutions of an equation (1) in practice. He dealt with the case $\mathcal{T}=\{2, 3, 5, 7, 11, 13\}$. For more data, see Shorey and Tijdeman [9] ch. 1.

It is well known that similar results can be obtained in case xyz is not entirely composed of primes from $\mathcal{T}$, but only for the major part. Some result in this direction can be found in Stewart and Tijdeman [10] in connection with the Oesterlé-Masser conjecture (= abc conjecture). In the present paper we give explicit bounds for the solutions of Equation (1) in x, y, z $\in \mathbb{Z}$ if the major part of xyz is composed of primes from $\mathcal{T}$. Put $x=x_1x_2$, $y=y_1y_2$, $z=z_1z_2$ where $x_1,y_1,z_1 \in \mathcal{S}$, and x_2,y_2,z_2

* The research was partly done at the University of Leiden in the academic year 1986/1987 and supported by a grant from the Ministry of Education and Science of the Netherlands.

have no prime divisors from $\mathcal{T}$. Obviously $x_1, x_2, y_1, y_2, z_1, z_2$ are pairwise coprime. We define

$$m(a)=\max_{1\leqslant k\leqslant S}\operatorname{ord}_{pk}(a)\ \textit{for}\ a\in\mathbb{Z},\ a=0,$$
$$x_i=\|x_i, y_i, z_i\|=\max(|x_i|, |y_i|, |z_i|, e)\ \text{for}\ i=1,2.$$

In this notation we have the following results.

Theorem 1. *Let* γ, ε *be any real numbers with* $y\geqslant 2, 0<\varepsilon<1$ *Put*

$$C_1=\varepsilon^{-1}\max(\gamma, \log p_1)2^{3\varepsilon+2S}\,3^{2s}\,(s+1)^{2s+4}.$$
$$P^2(\log P/\log p_1)^{s-1}\log(2^{14}s^2\log P). \tag{2}$$

Then the solutions to Equation (1), *under the condition*

$$\log X_2\leqslant\gamma\,(\log X_1)^{1-\varepsilon}, \tag{3}$$

satisfy

$$X_1\leqslant\exp\{(2C_1\log C_1)^{1/\varepsilon}\}. \tag{4}$$

Here and elsewhere log x denotes the natural logarithm of x.

It follows from Theorem 1 that Equation (1) subject to the condition (3) has only finitely many solutions (x, y, z) in $\mathbb{Z}$. Theorem 2 below provides an upper bound for $m(x_1y_1z_1)$ which implies a better result than Theorem 1 with respect to certain parameters.

Theorem 2. *Let* γ, ε *be any real numbers with* $\gamma\geqslant 2, 0<\varepsilon<1$. *Put* $u=[2s/3]$ *and*

$$C_2=\gamma s2^{8u+60}(u+1)^{2u+2}(\log P)^{u+1}(\log p_1)^{-1}\log(\gamma se\log P)\log(e\log P),$$
$$C_3=\gamma s\varepsilon^{-1}2^{3u+30}(u+1)^{2u+2}(\log P/\log p_1)^{u+1}$$
$$[s3^{2u+2}(u+1)^3P^2(\log P/\log p_1)\log(2^{14}(u+1)^2\log P)$$
$$+2^{5u+30}(\log P_1)^{u}\log(e\log P)].$$

Then the solutions to Equation (1), *under the condition* (3), *satisfy*

$$m(x_1y_1z_1)\leqslant(2C_2+2C_3\log C_3)^{1/\varepsilon}. \tag{5}$$

In the proofs of these theorems we shall use the following lemmas. For any rational number $\alpha=a/b$ with $a, b\in\mathbb{Z}, b>0$, *g. c. d.* $(a, b)=1$, we put

$$h(\alpha)=\log\max(|a|, b),\qquad n^+(\alpha)=\max(1, n(\alpha)).$$

Lemma 1. *Let* $p_1,\cdots, p_r$ $(r\geqslant 2)$ *be distincl prime numbers from* $\mathcal{T}$. *Let* α *be a rational number such that* $\operatorname{ord}_{p_i}\alpha=0$ *for* $j=1,\cdots, r$. *Let* $b_1,\cdots, b_r\in\mathbb{Z}$. *Take* $B\geqslant\max_{1\leqslant j\leqslant r}$

$|b_j|$ *and*

$$C_4=2^{8r+59}(r+1)^{2r+2}(\log P)^r\log(e\log P). \qquad (6)$$

If $p_1^{b_1}\cdots p_r^{b_r}\ \alpha\neq 1$, *then*

$|p_1^{b_1}\cdots p_r^{b_r}\alpha-1|>$

$$\frac{1}{2}\exp\{-C_4\, n^+(\alpha)[\log B+\log(e\max(h^+(\alpha),\log P))]\}.$$

Proof. Obviously we may assume $\frac{1}{2}<P_1^{b_1}\cdots P_r^{b_r}\alpha<\frac{3}{2}$. Hence

$$|p_1^{b_1}\cdots p_r^{b_r}\alpha-1|>\frac{1}{2}|b_1\log p_1+\cdots+b_r\log p_r+\log\alpha|.$$

Then we apply Theorem of Waldschmidt [11] with $K=\mathbb{Q}$, $n=r+1$, $\beta_0=0$, $\beta_j=b_j$, $\alpha_j=p_j$ for $j=1,\cdots,r$ and $(\beta_n,\alpha_n)=(1,\alpha)$, and

$$C(n)=C(r+1)<2^{8r+59}(r+1)^{2r+2}.$$

Note that $P\geqslant 5$. We can take $D=1$, $V_1\cdots V_n=h^+(\alpha)(\log P)^r$, $W=\log B$, $V_n^+=\max(h(\alpha),\log P)$, $V_{n-1}=\log P$ and $E=e$. On substituting these values in Waldschmidt's lower bound we obtain Lemma 1.

Write

$$\zeta_3=e^{2\pi i/3}, \qquad \zeta_4=e^{\pi i/2}.$$

Let p be a prime number, and write

$$K=\begin{cases}\mathbb{Q}(\zeta_3), & \text{if } p=2,\\ \mathbb{Q}(\zeta_4), & \text{if } p>2.\end{cases}$$

Then

$$D=[K:\mathbb{Q}]=2. \qquad (7)$$

Let $\mathscr{P}$ be a prime ideal of the ring of intergers in K, lying above p. Denote by $e_{\mathscr{P}}$ and $f_{\mathscr{P}}$ the ramification index and residue class degree, respectively. It is easy to see from the formula $e_{\mathscr{P}}f_{\mathscr{P}}\leqslant D$ that

$$I\leqslant f_{\mathscr{P}}\leqslant 2. \qquad (8)$$

For $\xi\in K$, $\xi\ x\ 0$, denote by $\operatorname{ord}_{\mathscr{P}}\xi$ the order to which p divides the fractional ideal (ξ) generated by ξ, set $\operatorname{ord}_{\mathscr{P}}0=\infty$.

Let $\alpha_1,\cdots,\alpha_n$ $(n\geqslant 2)$ be non-zero rational numbers. Let $V_1,\cdots,V_n$ be real numbers such that

$$V_3\geqslant\max\left(h(\alpha_j),\ \frac{|\log\alpha_j|}{2\pi D},\ \frac{f_{\mathscr{P}}\log P}{D}\right) \textit{for } j=1,\cdots,n,$$

$$V_1\leqslant V_2\leqslant\cdots\leqslant V_n,$$

where $\log\alpha_j=\log|\alpha_j|+i\arg\alpha_j$ with $-\pi<\arg\alpha_j\leqslant\pi$. Let $\beta_1,\cdots,\beta_n\in\mathbb{Z}$, not all zero, and put

$$V=\begin{cases} 2^{14}n^2V_n, & if\ \mathrm{ord}_p\,\beta_n=\min\limits_{1\leqslant j\leqslant n}\mathrm{ord}_{\mathscr{P}}\,b_j, \\ 2^{14}n^2V_{n-1}, & \text{otherwise.}\end{cases} \tag{9}$$

$$B=\max\left(\max_{1\leqslant j\leqslant n}(|\beta_j|),\ 2^{12}n\right). \tag{10}$$

Then from (7)—(10) and a recent Theorem without the Kummen condition given by Yu [14] (Yu has proved some theorems under the Kummer condition in [13]) we obtain the following.

Lemma 2. *Suppose that*
$\mathrm{ord}_{\mathscr{P}}\,\alpha_j=0$ for $j=1,\cdots,n$ and $\alpha_1^{\beta_1}\cdots\alpha_n^{\beta_n}\neq 1.$
Then we have

$$\mathrm{ord}_{\mathscr{P}}(\alpha_1^{\beta_1}\cdots\alpha_n^{\beta_n}-1)$$
$$<2^{3n+27}3^{2n}(n+1)^{2n+3}P^2(\log P)^{-n}V_1\cdots V_n\log V\log B.$$

Furthermore we obtain

Lemma 3. *Let $p_1,\cdots,p_r\ (r\geqslant 2)$ be distinct prime numbers from $\mathscr{T}$. Let α be a rational number such that $\mathrm{ord}_{p_j}\alpha=0$ for $j=1,\cdots,r$. Let $b_1,\cdots,b_r\in\mathbb{Z}$. Take $B=max(\max\limits_{1\leqslant j\leqslant r}|b_j|,2^{12}r)$. Let p be a prime number from $\mathscr{T}$ such that*

$$\mathrm{ord}_{\mathscr{P}}\,\alpha=0\ \textit{and}\ \mathrm{ord}_{\mathscr{P}}\,p_j=0\ \textit{for}\ j=1,\cdots,r.$$

Put

$$C_5=2^{3r+30}3^{2r+2}(r+2)^{2r+5}P^2(\log P)^{-r-1}(\log P)^r\log(2^{14}(r+1)^2\log P).$$

If $p_1^{b_1}\cdots p_r^{b_r}\alpha\neq 1$, then

$$\mathrm{ord}_{\mathscr{P}}(p_1^{b_1}\cdots p_r^{b_r}\alpha-1)\leqslant C_5\log B\max(1,h(\alpha),\log p).$$

Proof. If $\alpha=1$, we apply Lemma 2 with $n=r\geqslant 2$, $\alpha_j=p_j$, $\beta_j=b_j$ for $j=1,\cdots,r$. Without the loss of generality we may assume that $p_1<P_2<\cdots<P_r$. Put

$$V_j=\max(\log p_j,\log p),\ \text{for}\ j=1,\cdots,r,$$

$$V=2^{14}n^2V_r.$$

Clearly we have $V_1\leqslant V_2\leqslant\cdots\leqslant V_r$, and so

$$V_1\cdots V_n\log V\leqslant(\log P)^r\log(2^{14}r^2\log P)$$

which yields the estimate in the case $\alpha=1$.

It is easy to verify that

$$|\mathrm{ord}_{\mathscr{P}}(p_1^{b_1}\cdots p_r^{b_r}+1)|\leqslant|\mathrm{ord}_{\mathscr{P}}(p_1^{2b_1}\cdots p_r^{2b_r}-1)|$$

Using the estimate for the case $\alpha=1$ we see

$$\operatorname{ord}_{\mathscr{P}}(-p_1^{b_1}\cdots p_r^{b_r}-1)\leqslant C_5\log B\max(1, h(\alpha), \log p).$$

This provides the required estimate in the case $\alpha=-1$.

When $\alpha\neq\pm1$, in order to apply Lemma 2 we assume $p_1<p_2<\cdots<p_r$, and write

$$U_j=\max(\log p_j, \log p),\quad \text{for } j=1, \cdots, r,$$

$$U=\max(1, h(\alpha), \log p).$$

Then

$$U_1\leqslant U_2\leqslant\cdots\leqslant U_r.$$

If $U\geqslant U_r$, then we apply Lemma 3 with $n=r+1\geqslant3$, $\beta_j=b_j$, $\alpha_j=p_j$ for $j=1,\cdots,r$, $(\beta_n, \alpha_n)=(1, \alpha)$, and with

$$V_n=U,\quad V_j=U_j,\quad \text{for}\quad j=1,\cdots,r.$$

Clearly, $\operatorname{ord}_p\beta_n=0=\min_{1\leqslant j\leqslant n}\operatorname{ord}_p\beta_j$. Then we put

$$V=2^{14}n^2V_{n-1}=2^{14}n^2V_r.$$

Otherwise, there exists a non-negative integer k such that

$$U_1\leqslant\cdots\leqslant U_k\leqslant U\leqslant U_{k+1}\leqslant\cdots\leqslant U_r$$

(if $k=0$, we take $U_0=0$). In this case we put

$$(V_1,\cdots,V_r)=(U_1,\cdots,U_k, U, U_{k+1},\cdots,U_{r-1})$$

and

$$V_n=U_r\qquad\text{and}\qquad V=2^{14}n^2V_n.$$

Thus we have

$$V_1\cdots V_n\log V\leqslant(\log P)^r\max(1, h(\alpha),\log p)\log(2^{14}(r+1)^2\log P).$$

This completes the proof of Lemma 3.

Lemma 4. *Let $a\geqslant0$, $b>e^2$ and let $x\in\mathbb{R}$ satisfy $x<a+b\log x$. Then*

$$x<2(a+b\log b).$$

Proof. This is Lemma 2.2 with $h=1$ in a paper by Pethő and de Weger [7].

Proof of Theorem 1

Without the loss of generality, we may assume that

$$z_1=X_1\geqslant\exp(2^{12}s)$$

When we allow x, y and z to be negative. Then x_1 and y_1 are composed of prime numbers from a proper non-empty subset $\{p_{i_1},\cdots,p_{i_t}\}$ of $\mathcal{S}$ with $2\leqslant t\leqslant s-1$ and

$p_{i_1} < p_{i_2} < \cdots < p_{i_t} \leqslant P$. Let

$$x_1 = P_{i_1}^{a_1} \cdots P_{i_t}^{a_t}, \quad y_1 = P_{i_1}^{b_1} \cdots P_{i_t}^{b_t},$$

where a_j and b_j are non-negative integers with $a_j b_j = 0$ for $j = 1, \cdots, t$. Then

$$\frac{x}{y} = \left(\frac{x_1}{y_1}\right)\left(\frac{x_2}{y_2}\right) = p_{i_1}^{a_1 - b_1} \cdots P_{i_t}^{a_t - b_t}\left(\frac{x_2}{y_2}\right).$$

Observe that

$$\max_{1 \leqslant j \leqslant t} |a_j - b_j| \leqslant \max_{1 \leqslant j \leqslant t} \max(a_j, b_j) \leqslant 2 \log \|x_1, y_1\| \leqslant 2 \log X_1.$$

For any prime p dividing z_1 we have

$$p \leqslant P, \ \mathrm{ord}_{\mathfrak{p}}\left(-\frac{x_2}{y_2}\right) = 0 \text{ and } \mathrm{ord}_{\mathfrak{p}} P_{i_j} = 0 \text{ for } j = 1, \cdots, t,$$

and

$$\mathrm{ord}_p z_1 = \mathrm{ord}_p z$$

$$= \mathrm{ord}_{\mathfrak{p}}\left(-\frac{x}{y} - 1\right) = \mathrm{ord}_{\mathfrak{p}}\left(p_{i_1}^{a_1 - b_1} \cdots p_{i_t}^{a_t - b_t}\left[-\frac{x_2}{y_2}\right] - 1\right).$$

We apply Lemma 3 to the right hand side of the above identity with $r = t$, $\alpha = -x_2/y_2$, $B = 2 \log X_1$, $h(\alpha) = \log \max(|x_2|, |y_2|)$ and obtain

$$\mathrm{ord}_{\mathfrak{p}} z_1 \leqslant 2C_6 (\log P)^{-t-1} \log \log X_1 \max(\log \|x_2, y_2\|, \log p), \tag{11}$$

where

$$C_6 = 2^{3t+30} 3^{2t+2} (t+1)^{2t+5} P^2 (\log P)^t \log(2^{14}(t+1)^2 \log P). \tag{12}$$

We consider two cases.

Case a. $\|x_2, y_2\| \geqslant p_1$. It follows from (11) that

$$\log z_1 = \sum_{p | z_1} \mathrm{ord}_{\mathfrak{p}} z_1 \log p$$

$$\leqslant 2C_6 \log \log X_1 \Bigg[\sum_{\substack{p | z_1 \\ p \leqslant \|x_2, y_2\|}}{}_1 (\log p)^{-t} \log \|x_2, y_2\| + \sum_{\substack{p | z_1 \\ p > \|x_2, y_2\|}}{}_2 (\log p)^{-t+1}\Bigg]$$

$$\leqslant 2C_6 \log \log X_1 \Bigg[\sum{}_1 (\log p_1)^{-t} \log \|x_2, y_2\| + \sum{}_2 (\log \|x_2, y_2\|)^{-t+1}\Bigg]$$

$$\leqslant 2C_6 \log \log X_1 (\log p_1)^{-t} \log \|x_2, y_2\| \sum_{p | z_1} 1$$

$$\leqslant 2sC_6 (\log p_1)^{-t} \log \log X_1 \log X_2.$$

Hence

$$\log X_1 \leqslant 2sC_6 (\log p_1)^{-t} \log\log X_1 \log X_2 . \tag{13}$$

By virtue of (3), (12) and (13) we have

$$\log X_1 \leqslant 2s\gamma C_6 (\log p_1)^{-t} \log\log X_1 (\log X_1)^{1-\varepsilon}$$

$$\leqslant s\gamma 2^{3t+31} 3^{2t+2} (t+2)^{2t+5} P^2 (\log P/\log P_1)^t .$$

$$\log (2^{14}(t+1)^2 \log P) \log\log X_1 (\log X_1)^{1-\varepsilon}.$$

Hence

$$\log^{\varepsilon} X_1 \leqslant C_1 \log (\log^{\varepsilon} X_1)$$

by noting (2) and $t+1 \leqslant s$. This implies (4). According to Lemma 4 with $a=0$, $b=C_1$, $x=\log^{\varepsilon} X_1$, we deduce

$$\log X_1 \leqslant (2C_1 \log C_1)^{1/\varepsilon}. \tag{14}$$

Case b. $\|x_2, y_2\| \leqslant p_1$. In this case it follows from (11) that

$$\log X_1 = \log Z_1 \leqslant \sum_{p|z_1} 2C_6 (\log p)^{-t-1} \log\log X_1 (\log p)^2$$

$$\leqslant 2sC_6 (\log p_1)^{-t+1} \log\log X_1 .$$

Since $0<\varepsilon<1$, $\gamma>1$ we also obtain (14) in this case.

Proof of Theorem 2

Instead of Equation (1) we consider the equivalent equation

$$x+y=z$$

in $x, y, z \in \mathbb{N}$ with *g. c. d.* $(x, y, z)=1$ and $x=x_1x_2$, $y=y_1y_2$, $z=z_1z_2$ where x_1, y_1, $z_1 \in \mathscr{S}$ and x_2, y_2, z_2 have no prime divisors from $\mathscr{T}$. By permutation (if necessary) we may assume that x_1y_1 has u different prime divisors $p_1 \leqslant p_{i_1} < p_{i_2} < \cdots < p_{i_u} \leqslant p$ where $u \leqslant [2s/3]$. Let p be any prime dividing z_1. Clearly, we have $\mathrm{ord}_{\mathscr{P}}(x_1y_1) = \mathrm{ord}_{\mathscr{P}}(x_1/y_1) = 0$. We can apply Lemma 3 with $r=u$, $\alpha = \mp x_2/y_2$ and $B=\max(m(x_1y_1), 2^{12}s)$ to obtain

$$\mathrm{ord}_{\mathscr{P}} z_1 = \mathrm{ord}_{\mathscr{P}} z = \mathrm{ord}_{\mathscr{P}} (\mp x/y - 1)$$

$$= \mathrm{ord}_{\mathscr{P}} \left[p_{i_1}{}^{a_1-b_1} \cdots p_{i_u}{}^{a_u-b_u} \left(+ \frac{x_2}{y_2} \right) - 1 \right] \tag{15}$$

$$\leqslant C_6'(\log p)^{-u-1} \log B \max (\log \|x_2, y_2\|, \log p)$$

where C_6' is defined as C_6 in (12) with $t=u$. Thus

$$m(z_1) \leqslant C_6{}'(\log p_1)^{-u-1}\log B\ \max(\log\|x_2, y_2\|, \log P). \tag{16}$$

Note that the condition (3) implies that

$$\log X_2 < \gamma s \log P\,(m(x_1y_1z_1))^{1-\varepsilon}. \tag{17}$$

First we deal with the trivial case $m(x_1y_1) \leqslant 2^{12}s$. We have by (17)

$$p_1^{m(z_1)} < z < 2\max(x, y) \leqslant 2\left(p_{i_1}\cdots p_{i_u}\right)^{2^{12}s}\|x_2, y_2\| < 2P^{2^{12}us}X_2.$$

Hence

$$m(x_1y_1z_1) < 1 + 2^{12}u\,s\,(\log P/\log P_1) + \gamma s\,(\log p_1/\log P_1)(m(x_1y_1z_1))^{1-\varepsilon}$$
$$< 2^{12}\gamma s^2(\log P/\log p_1)(m(x_1y_1z_1))^{1-\varepsilon}$$

and so

$$m(x_1y_1z_1) < (2^{12}\gamma s^2\log P/\log p_1)^{1/\varepsilon}.$$

Therefore (5) holds in this case. Henceforth we may assume that $m(x_1y_1) > 2^{12}s$. We apply Lemma 1 with $r=u$, $\alpha = \pm y_2/x_2$, $B=m(x_1y_1)$ to obtain

$$\frac{(p_2\cdots p_s)^{m(z_1)}z_2}{p_1^{m(x_1y_1)}} = \frac{z}{\max(x, y)} = \frac{z}{x} = \left|\pm\frac{y}{x} - 1\right|$$

$$= \left|P_{i_1}^{a_1-b_1}\cdots P_{i_u}^{a_u-b_u}\left(\pm\frac{y_2}{x_2}\right) - 1\right|$$

$$> \frac{1}{2}\exp\{-C_4{}'\log\|x_2, y_2\|$$
$$[\log m(x_1y_1) + \log(e\max(\log\|x_2, y_2\|, \log P))]\}$$

$$=: (1/2)\exp\{-A\},$$

where $C_4{}'$ is defined as C_4 in (6) with $r=u$. Hence

$$m(x_1y_1)\log p_1 < sm(z_1)\log P + A + \log(2z_2).$$

This, together with (16), implies that

$$\begin{aligned} m(x_1y_1z_1) &< sC_6{}'\log P\,(\log p_1)^{-u-2} \\ &\times\log m(x_1y_1)\max(\log\|x_2, y_2\|, \log P) \\ &+ C_4{}'(\log p_1)^{-1}\log\|x_2, y_2\|\log m(x_1y_1) \\ &+ C_4{}'(\log p_1)^{-1}\log\|x_2, y_2\|\log(e\max(\log\|x_2, y_2\|, \log P)) \\ &+ \log z_2(\log p_1)^{-1} + 1. \end{aligned} \tag{18}$$

We consider two subcases.

i) $\|x_2, y_2\| \geqslant P$. By (17) and (18) we have

$$m(x_1y_1z_1) < sC_6' \log P (\log p_1)^{-u-2} \log m(x_1y_1z_1) \log X_2$$
$$+ C_4' (\log p_1)^{-1} \log m(x_1y_1z_1) \log X_2$$
$$+ C_4' (\log p_1)^{-1} \log X_2 + C_4' (\log p_1)^{-1} \log X_2 \log\log X_2$$
$$+ (\log P_1)^{-1} \log X_2 + 1$$
$$< \gamma s^2 C_6' \log^2 P (\log p_1)^{-u-2} \log m(x_1y_1z_1) (m(x_1y_1z_1))^{1-\varepsilon}$$
$$+ \gamma s C_4' (\log P/\log p_1) \log m(x_1y_1z_1)(m(x_1y_1z_1))^{1-\varepsilon}$$
$$+ \gamma s C_4' (\log P/\log p_1)(m(x_1y_1z_1))^{1-\varepsilon}$$
$$+ \gamma s C_4' (\log P/\log p_1) \log m(x_1y_1z_1)(m(x_1y_1z_1))^{1-\varepsilon}$$
$$+ \gamma s C_4' (\log P/\log p_1) \log (rs \log P)(m(x_1y_1z_1))^{1-\varepsilon}$$
$$+ \gamma s (\log P/\log p_1)(m(x_1y_1z_1))^{1-\varepsilon} + 1,$$

where C_6' is defined as C_6 in (12) with $t=u$. Hence

$$(m(x_1y_1z_1))^{\varepsilon} < \gamma s (\log P/\log P_1)[C_4' + C_4' \log(\gamma s \log P) + 2]$$
$$+ \gamma s \varepsilon^{-1} (\log P/\log p_1)[sC_6' \log P \log^{-u-1} p_1 + 2C_4'] \log((m(x_1y_1z_1)^{\varepsilon}).$$

ii) $\|x_2, y_2\| < P$. It follows from (17) and (18) that

$$m(x_1y_1z_1) < sC_6' \log^2 P (\log p_1)^{-u-2} \log m(x_1y_1z_1)(m(x_1y_1z_1))^{1-\varepsilon}$$
$$+ \gamma s C_4' (\log P/\log p_1) \log m(x_1y_1z_1)(m(x_1y_1z_1))^{1-\varepsilon}$$
$$+ \gamma s C_4' (\log P/\log p_1) \log (e \log P)(m(x_1y_1z_1))^{1-\varepsilon}$$
$$+ \gamma s (\log P/\log p_1)(m(x_1y_1z_1))^{1-\varepsilon} + 1$$
$$< \gamma s (\log P/\log p_1)[C_4' \log(e \log P) + 2] (m(x_1y_1z_1))^{1-\varepsilon}$$
$$+ \gamma s (\log P/\log p_1)[\gamma^{-1} C_6' \log P (\log p_1)^{-u-1} + C_4']$$
$$\times \log m(x_1y_1z_1)(m(x_1y_1z_1))^{1-\varepsilon}$$

and so

$$(m(x_1y_1z_1))^{\varepsilon} < \gamma s (\log P/\log p_1)[C_4' \log(e \log P) + 2]$$
$$+ \gamma s \varepsilon^{-1} (\log P/\log p_1)[\gamma^{-1} C_6' \log P (\log p_1)^{-u-1} + C_4'] \log((m(x_1y_1z_1))^{\varepsilon}).$$

Now put

$$C_7=\gamma s\,(\log P/\log p_1)[C_4'\log(\gamma se\log P)+2],$$

$$C_8=\gamma s\varepsilon^{-1}(\log P/\log p_1)[sC_6'\log P\,(\log p_1)^{-u-1}+2C_4']$$

and apply Lemma 4 with $\alpha=C_7$, $b=C_8$. Then in both subcases we have

$$m\,(x_1y_1z_1)<(2C_7+2C_8\log C_8)^{1/\varepsilon}. \qquad \square$$

Remark 1

The exponent $1-\varepsilon$ in (3) is the best possible in the sense that we cannot take $\varepsilon=0$. This is clear from the following trivial example. Choose $\mathcal{T}=\{2,3\}$ and put

$$x_1=2^k,\qquad y_1=3^m,\qquad z_1=1,$$

$$x_2=1,\qquad y_2=1,\qquad z_2=2^k+3^m=z.$$

Since the inequality

$$2^k+3^m\leqslant\gamma\ max\,(2^k,3^m)\text{ with }\gamma\geqslant2$$

has infinitely many solutions (k,m) in positive integers, we cannot find an upper bound for the values of x, y, z in (1).

Part II

Ellison [3] proved that if $p\leqslant q$ and δ is a given positive real number then for all $x\geqslant X_0\,(\delta,a,b,p,q)$ we have either $ap^r-bq^t=0$ or

$$|ap^r-bq^t|\geqslant p^{\delta r} \qquad (19)$$

where X_0 is given explicitly.

De Weger [12] studied the inequality

$$0<x-y<y^\delta$$

in $x,y\in\mathcal{S}$ consisting of all positive integers composed of prime numbers from a finite set $\mathcal{T}=\{p_1,\cdots,p_r\}$ with $r\geqslant2$. Put

$$X=\max_{1\leqslant i\leqslant r}\ ord_{p_i}(xy).$$

Then there is a constant C_9 depending only on $\mathcal{T}$, δ such that $X<C_9$.

Now we generalize these results. We consider the following diophantine inequality

$$|x-y|<x^\delta, \qquad (20)$$

where $x=x_1x_2$, $y=y_1y_2$, g. c. d. $(x,y)=1$ with $x_1,y_1\in\mathcal{S}$ and $x_2,y_2\in\mathbb{N}$ having no prime divisors from $\mathscr{T}$. Write

$$X_i=\max(x_i, y_i, e) \quad \text{for} \quad i=1, 2.$$

We obtain the following :

Theorem 3. *Let $\gamma, \varepsilon, \delta$ be any real numbers with $\gamma\geqslant 2$, $0<\varepsilon<1$, $0<\delta<1$. Put*

$$C_{10}=\gamma 2^{8r+59}(r+1)^{2r+2}(\log P)^{\gamma}\log(e\log P)\log(e\gamma\log P)/(1-\delta),$$

$$C_{11}=\gamma\varepsilon^{-1}2^{8r+61}(r+1)^{2r+2}(\log P)^{\gamma}\log(e\log P)/(1-\delta).$$

Then the solutions to inequality (20), *under the condition*

$$\log X_2\leqslant\gamma(\log X_1)^{1-\varepsilon}, \tag{21}$$

satisty

$$X_1\leqslant\exp\{(2C_{10}+2C_{11}\log C_{11})^{1/\varepsilon}\}.$$

Furthemore, we have

$$X\leqslant(2C_{10}+2C_{11}\log C_{11})^{1/\varepsilon}/\log p_1.$$

Proof. Without the loss of generality, we may assume $\max(x, y)\geqslant 3$. First, we show that

$$|y/x-1|<2^{1-\delta}X_1^{-(1-\delta)}. \tag{22}$$

Clearly, $\max(x, y)\geqslant\max(x_1, y_1)=X_1$. If $x>y$, then $x\geqslant X_1$, and hence

$$|y/x-1|<X_1^{-(1-\delta)}$$

by (20).

In the case $x<y$, we have $y\geqslant X_1$. If $x\leqslant y/2$, then

$$x^{\delta}>y-x\geqslant y-y/2=y/2\geqslant x$$

which contradicts $x\geqslant 1$. Therefore $x>y/2$. Thus it follows from (20) that

$$|y/x-1|<x^{-(1-\delta)}<2^{1-\delta}y^{-(1-\delta)}\leqslant 2^{1-\delta}X_1^{-(1-\delta)}.$$

Hence (22) holds for any case.

By a reason similar to that in Part I, we may apply Lemma 1 with $B=2\log X_1$, $\alpha=y_2/x_2$ and $h^+(\alpha)=\log X_2$ to obtain that

$$|y/x-1|> \tag{23}$$

$$>\frac{1}{2}\exp\{-C_4\log X_2[2\log\log X_1+\log(e\max(\log X_2, \log P))]\}.$$

In virtue of (21)–(23) we see

$$(1-\delta)\log X_1<(2-\delta)\log 2+C_4\gamma(\log X_1)^{1-\varepsilon}\cdot$$
$$\cdot[2\log\log X_1+\log(e\max(\log X_2,\log P))].$$

Without the loss of generality, we assume $X_1>e^{\theta}$. Then we have

$$(\log X_1)^{\varepsilon}<(3C_4\gamma/(1-\delta))\log\log X_1+C_4\gamma/(1-\delta)+$$
$$+(C_4\gamma/(1-\delta))\log(\mathit{max}(\log X_2,\log P)).$$

If $X_2\geqslant P$, then

$$(\log X_1)^{\varepsilon}<(3C_4\gamma/(1-\delta))\log\log X_1+C_4\gamma/(1-\delta)+$$
$$+(C_4\gamma/(1-\delta))\log\gamma+(C_4\gamma(1-\varepsilon)/(1-\delta))\log\log X_1$$
$$<(4C_4\gamma/(1-\delta))\log\log X_1+(C_4\gamma/(1-\delta))\log(e\gamma).$$

If $X_2<P$, then we have

$$(\log X_1)^{\varepsilon}<(3C_4\gamma/(1-\delta))\log\log X_1+(C_4\gamma/(1-\delta))\log(e\log P).$$

Put

$$C_{12}=(C_4\gamma/(1-\delta))\log(e\gamma\log P),\qquad C_{13}=4C_4\gamma\varepsilon^{-1}/(1-\delta).$$

Then we obtain

$$(\log X_1)^{\varepsilon}<C_{12}+C_{13}\log(\log X_1)^{\varepsilon}.$$

By means of Lemma 4 we complete the proof of Theorem 3. □

Remark 2

The exponent $1-\varepsilon$ in (21) is the best possible in the sense that we cannot take $\varepsilon=0$. This is clear from the following trivial example. Choose $\mathcal{T}=\{2,3\}$ and put

$$x_1=2^k,\qquad y_1=2^k3^m,\qquad x_2=2^k+3^m,\qquad y_2=1.$$

Since the inequality

$$2^k+3^m\leqslant\gamma 2^k3^m,$$

with $\gamma\geqslant 2$ and $m>k((2-\delta)/\delta)(\log 2/\log 3)$, has infinitely many solutions $(k,m)\in\mathbb{N}$, we have

$$|x-y|<|2^k(2^k+3^m)-2^k3^m|=2^{2k}<2^{\delta k}3^{\delta m}<[2^k(2^k+3^m)]^{\delta}<x^{\delta}$$

and therefore we cannot find an upper bound for the values of x, y in (20).

Acknowledgement The author wishes to thank Professor R. Tijdeman for his helpful comments.

References

[1] Coates, P., An effective P-adic analogue of a theorem of Thue, *Acta Arith.*, **15**(1969), 279—305.

[2] ———, An effective p-adic analogue of a theorem of Thue II: The greatest prime factor of a binary form, *Acta Arith.*, **16** (1970), 399—412.

[3] Ellison, W. J., On a theorem of S. Sivasankaranarayana Pillai, *Sem. Th. Nombr.*, 1970—1971, Exp. No. 12, Lab. Theorie Nombres, C. N. R. S., Talence, 1971.

[4] Gurak, S., The Hasse norm principle in a compositum of radical extensions, *J. London Math. Soc.*, 2-nd Ser., **22**(1980), No. 3, 385—397.

[5] Hua Lookeng, *Introduction to Number Theory*, in Chinese, Science Press, Beijing, 1975; in English, Springer-Verlag, Berlin, 1982.

[6] Mahler, K., Zur Approximation algebraischer Zahlen I: Ueber den grössten primteiler binärer Formen, Math. Ann., **107**(1933), 691—730.

[7] Pethö, A. and de Weger, B. M. M., Products of prime powers in binary recurrence sequences, Part I: The hyperbolic case, with an application to the generalized Ramanujan-Nagell equation, *Math. Comp.*, **47**(1986), No. 176, 713—727.

[8] Schinzel, A., On the linear dependence of roots, *Acta Arith.*, **28**(1975), 161—175.

[9] Shorey, T. N. and Tijdeman, R., *Exponential diophantine equations*, Cambridge Univ. Press, London, 1986.

[10] Stewart, C. L. and Tijdeman, R., On the Oesterlé-Masser conjecture, *Monatsh. Math.*, **102**(1986), 251—257.

[11] Waldschmidt, M., A lower bound for linear forms in logarithms, *Acta Arith.*, **37**(1980), 257—283.

[12] De Weger, B. M. M., Solving exponential diophantine equations using lattice basis reduction algorithm, *J. Number Theory*, **26**(1987), No. 3, 325—367.

[13] Yu Kunrui, Linear forms in logarithms in the p-adic case, New Advance in Transcendence Theory, *Proc. Conf. Durham*, 1986, ed. by A. Baker, Cambridge Univ. Press, London, 1988.

[14] ————, Linear forms in p-adic logarithms II, to appear.

Fourth Power Mean Value of Dirichlet's L-Functions*

Wang Wei

Department of Mathematics, Shandong University

§1. Introduction

Let $q \geqslant 2$ be an integer. In this paper we shall consider the fourth power mean value of Dirichlet's L-functions of the following type:

$$\sum_{\chi \bmod q}^{*} \int_0^T \left| L\left(\frac{1}{2}+it, \chi\right)\right|^4 dt$$

where Σ^* indicates the summation over primitive characters modulo q. For $q=1$, i.e. in the case of Riemann zeta-function, Ingham [2] first proved

$$\int_0^T \cdots\cdots \int_0^T \left|\zeta\left(\frac{1}{2}+it\right)\right|^4 dt = (1/(2\pi^2))T\log T + O(T\log T), \qquad (1)$$

which is of fundamental importance in the study of zero density of zeta-function.

Heath-Brown[1] proved the following sharpening of (1):

$$\int_0^T \left|\zeta\left(\frac{1}{2}+it\right)\right|^4 dt = TP_4(\log T) + O(T^{\frac{7}{8}+\varepsilon}), \qquad (2)$$

where $P_4(x)$ represents a polynomial of x of degree 4. For the general case $q \geqslant 3$, the best kown result is duc to Rane[3]. He proved the following result:

$$(1/N)\sum_{\chi \bmod q}^{*} \int_0^T \left|L\left(\frac{1}{2}+it, \chi\right)\right|^4 dt = (1/(2\pi^2))\left\{\prod_{p|q}(1-1/p)^4(1-1/p^2)\right\}$$
$$T\log^4(qT) + O(2^r T\log^3(qT)), \qquad (3)$$

where N denotes the number of primitive characlers modulo q, and r is the number of distinet prime factors of q.

The aim of the present paper is to prove the following theorem, which is a generalization of (2) and an inprovement of (3).

** The project is supported by Natrional Natural Science Foundation of China.

Theorem. *There exist constants a_4, a_3, a_2, a_1, a_0, all depending on q, such that for arbitrary real numbers T and ε satisfying $T \geqslant 2, \varepsilon > 0, 1 < q \leqslant T$.*

$$\sum_{\chi \bmod q}^{*} \int_0^T \left| L\left(\frac{1}{2}+it, \chi\right)\right|^4 dt = TP_4(\log (qT/2\pi)) + O(\min\{q^{\frac{9}{8}}T^{\frac{7}{9}}q, T^{\frac{11}{12}}\}(qT)^{\varepsilon}),$$

where

$$P_4(x) = a_4 x^4 + a_3 x^3 + a_2 x^2 + a_1 x + a_0,$$

$$a_4 = (1/(2\pi^2))\, \varphi_1(q) \prod_{p|q} (1-1/p)^4 (1-1/p^2)^{-1},$$

$$a_i = O(q^{1+\varepsilon}) \text{ for } 0 \leqslant i \leqslant 3,$$

and $\varphi_1(q)$ denotes the number of primitive characters modulo q. The constant implied in the error term depends on ε only.

Throughout this paper, we shall denote by ε an arbitrary small constant, which, however, is not necessarily the same everywhere. As usual, we shall use the notations $e(x)$ for $e^{2\pi i x}$ and r for Euler's constant.

§ 2. An Asymptotic Functional Equation

Let us first prove the following lemma.

Lemma 1. *Let m, n be integers such that $(mn, q) = 1$. Define*

$$F(m, n) = \sum_{\chi \bmod q}^{*} \chi(m) \overline{\chi}(n).$$

Then, $F(m, n)$ can be regarded as a function of $m-n$. Hence we can write it as $F(m-n)$. Moreover, we have

$$F(r) = F((r, q)). \tag{4}$$

Proof. Let us set $F(m, n, q) = F(m, n)$. It is easy to see that $F(m, n, q)$ is a multiplicative function of q. A further calculation shows that for $q = p^{\alpha}$, p prime, we have

$$F(m, n, p^{\alpha}) = \begin{cases} p^{\alpha-2}(p-1)^2, & \text{for } p^{\alpha} \mid m-n, \alpha \geqslant 2, \\ -p^{\alpha-2}(p-1), & \text{for } p^{\alpha-1} \| m-n, \alpha \geqslant 2, \\ p-2, & \text{for } p \mid m-n, \alpha = 1, \\ -1, & \text{for } p \nmid m-n, \alpha = 1, \\ 0, & \text{otherwise}. \end{cases}$$

From this, We can easily see that $F(m, n)$ depends only on the divisibility of $m-n$ by some divisors of q. Thus it can be regarded as a function of $m-n$.

(4) follows immediately from the above formula. This completes the proof of Lemma 1.

The main result in this section is

Lemma 2. *There exist constants* c, $\alpha(u,v)$, $\beta(u,v)$ *such that* $c\geqslant 1$, *and*

$$\sum_{\chi \bmod q}^{*} \left|L\left(\frac{1}{2}+it\right)\right|^4=\sum_{\substack{mn\leqslant c(qT)^2\\(mn,q)=1}} d(m)d(n)(mn)^{-\frac{1}{2}} F(m-n)(m/n)^{it} K_1(mn,t)$$
$$+\sum_{\substack{mn\leqslant c(qT)^2\\(mn,q)=1}} d(m)d(n)(mn)^{-\frac{1}{2}} F(m+n)(m/n)^{it} K_2(mn,t)$$
$$+O(q^{1+\varepsilon}T^{-2}) \qquad (5)$$

holds uniformly for $T\leqslant t\leqslant 2T$, *where it is defined that*

$$K_1(\chi,t)=\frac{1}{\pi i}\int_{1-i\infty}^{1+i\infty}\left\{(qt/(2\pi))^2\chi^{-1}\right\}^z\left(1+\sum_{u=1}^{30}\sum_{v} a(u,v)z^u t^{-2v}\right)e^{z^2/T}\,dz/z,$$

$$K_2(x,t)=\frac{1}{\pi i}\int_{1-i\infty}^{1+i\infty}\left\{(qt/(2\pi))^2\chi^{-1}\right\}^z\sum_{u=1}^{30}\sum_{v}\beta(u,v)z^u t^{-2v}e^{z^2/T}\,dz/z,$$

where Σ_v *denotes summation over the range* $\max\{1,u/3\}\leqslant v\leqslant 30$.

Proof. Define

$$f(w,\chi)=\left\{(\pi/q)^{-w}\,\Gamma\left(\frac{1}{2}(w+it+\delta)\right)\Gamma\left(\frac{1}{2}(w-it+\delta)\right)L(w+it,\chi)L(w-it,\overline{\chi})\right\}^2$$

where $\delta=(1-\chi(-1))/2$. Then it follows from the functional equation of $L(s,\chi)$ that for a primitive character χ modulo q, $f(w,\chi)=f(1-w,\chi)$, and $f(w,\chi)$ is an entire function, and is bounded in any strip $a\leqslant \mathrm{Re}(w)\leqslant b$. Consider

$$1/(2\pi i)\int_{1-i\infty}^{1+i\infty} f\left(\frac{1}{2}+z,\chi\right)e^{z^2/T}\,dz/z.$$

Shift the line of integration to $\mathrm{Re}(z)=-1$, and substitute $z=-w$. Since $f\left(\frac{1}{2}+z,\chi\right)=f\left(\frac{1}{2}-z,\chi\right)$, this yields

$$-1/(2\pi i)\int_{1-i\infty}^{1+i\infty} f\left(\frac{1}{2}+w,\chi\right)e^{z^2/T}\,dw/w.$$

The residue at the pole $z=0$ is obviously $f\left(\frac{1}{2},\chi\right)$. Hence

$$f(\frac{1}{2},\chi)=\frac{1}{\pi i}\int_{1-i\infty}^{1+i\infty} f(\frac{1}{2}+z,\chi)e^{z^2/T}\,dz/z.$$

Replace $(L(\frac{1}{2}+z\pm it,\chi))^2$ by $\sum_{n=1}^{\infty} d(n)\chi(n)n^{-\frac{1}{2}-z\mp it}$ and multiply both sides by $\{(\pi/q)^{-\frac{1}{2}}|\Gamma(\frac{1}{2}(\frac{1}{2}+it+\delta))|^2\}^{-2}$ in the above equation. Then we obtain

$$|L(\frac{1}{2}+it,\chi)|^4=\sum_{m,n=1}^{\infty} d(m)d(n)\overline{\chi}(m)\chi(n)(mn)^{-\frac{1}{2}}(m/n)^{it}I(mn,t). \quad (6)$$

Here we define

$$I(\chi,t)=\frac{1}{\pi i}\int_{1-i\infty}^{1+i\infty}((\pi/q)^2\chi)^{-z}G^2(z,t,\delta)e^{z^2/T}\,dz/z,$$

$$G(z,t,\delta)=\Gamma(\frac{1}{2}(\frac{1}{2}+it+z+\delta))\Gamma(\frac{1}{2}(\frac{1}{2}-it+z+\delta))|\Gamma(\frac{1}{2}(\frac{1}{2}+it+\delta))|^{-2}$$

Since δ equals 0 or 1, we have

$$\begin{aligned}G^2(z,t,\delta)&=G^2(z,t,0)+\delta(G^2(z,t,1)-G^2(z,t,0))\\&=\frac{1}{2}(G^2(z,t,0)+G^2(z,t,1))+\frac{1}{2}\chi(-1)(G^2(z,t,0)\\&\quad-G^2(z,t,1)).\end{aligned}$$

Substituting into (6) and summing it over all primitive characters modulo q. we conclude that

$$\sum_{\chi \bmod q}^{*}|L(\frac{1}{2}+it,\chi)|^4=\sum_{m,n=1}^{\infty} d(m)d(n)(mn)^{-\frac{1}{2}}(\frac{m}{n})^{it}\{F(m-n)I_1(mn,t)$$
$$+F(m+n)I_2(mn,t)\}. \quad (7)$$

Here we define for $k=1$ and 2,

$$I_k(\chi,t)=1/(2\pi i)\int_{1-i\infty}^{1+i\infty}((\pi/q)^2\chi)^{-2}e^{z^2/T}(G^2(z,t,0)+(-1)^{k-1}G^2(z,t,1))dz/z$$

Let $z=c_0+iy$, $\delta=0$ or 1. We shall give an asymptotic estimate of $G^2(z,t,\delta)$. Let us assume that $|y|\leqslant T^{\frac{1}{2}}\log T$. From Stirling's formula we have

$$\log\Gamma(z)=(z-\frac{1}{2})\log z-z+\log(2\pi)+\sum_{r=1}^{N}c_r z^{1-2r}+0(|z|^{-1-2N}),$$

where c_r $(1\leqslant r\leqslant N)$ are absolute constants. Write $\alpha=(\frac{1}{2}+it+\delta)$, $\beta=z/2$, and set $N=2$. For $|y|\leqslant T^{\frac{1}{2}}\log T$, we obtain

$$\log\Gamma(\alpha+\beta)-\log\Gamma(\alpha)=\beta\log\alpha+(\alpha-\beta-\frac{1}{2})\log(1+\beta/\alpha)-\beta$$
$$+\sum_{r=1}^{2}c_r((\alpha+\beta)^{1-2r}-\alpha^{1-2r})+O(T^{-5})$$
$$=\beta\log\alpha+\sum_{u,v}c(u,v)\beta^u\alpha^{-v}+O(T^{-5}),$$

where $c(u,v)$ are absolute constants, and $\sum_{u,v}$ denotes the summation over the range $1\leqslant u\leqslant 11$, $\max\{1,u-1\}\leqslant v\leqslant(u+10)/2$. The other terms are absorbed in the error term. We now express α and β in terms of it and z respectively. This yields

$$\frac{1}{2}z\log(\frac{1}{2}it)+\sum_{u,v}c_\delta'(u,v)z^u(it)^{-v}+O(T^{-5}),$$

where the summation is over the same range as before. Thus

$$\log G^2(z,t,\delta)=2z\log(t/2)+\sum_{u,v}c''(u,v)z^ut^{-2v}+O(T^{-5}),$$

where the summation with respect to v is over

$$\max\{1,(u-1)/2\}\leqslant v\leqslant(u+10)/4.$$

Replacing it by the weaker condition

$$\max\{1,u/3\}\leqslant v<10,$$

we obtain

$$G^2(z,t,\delta)=(t/2)^{2z}(1+\sum_{u,v}c_\delta'''(u,v)z^u t^{-2v}+O(T^{-5})), \tag{8}$$

where the summation is over the range

$$1\leqslant u\leqslant 30,\ \max\{1,u/3\}\leqslant v\leqslant 30.$$

We now have for $|y|\leqslant T^{\frac{1}{2}}\log T$,

$$((\pi/q)^2\chi)^{-z}G^2(z,t,\delta)e^{z^2/T}/z$$
$$=\{(qt/(2\pi))^2\chi^{-1}\}^z(1+\sum_{u,v}c_\delta'''(u,v)z^u t^{-2v})e^{z^2/T}/z$$
$$+0(q^{2C_0}T^{2C_0-5}\chi^{-C_0}e^{-y^2/T}C_0^{-1}). \tag{9}$$

For $|y|>T^{\frac{1}{2}}\log T$, we have

$$G^2(z,t,\delta) \ll (T+|y|)^A \exp\{-\frac{1}{2}\pi(|t+y|+|t-y|-2t)\}$$
$$\ll (T+|y|)^A.$$

Thus we have

$$((\pi/q)^2\chi)^{-z} G^2(z,t,\delta) e^{z^2/T}/z \ll q^{2C_0}\chi^{-C_0}C_0^{-1}(T+|y|)^A e^{-y^2/T}$$
$$\ll q^{2C_0}\chi^{-C_0}C_0^{-1}e^{-y^2/2T}T^{2C_0-5}. \qquad (10)$$

Moreover, we have trivially for $|y|>T^{\frac{1}{2}}\log T$,

$$\{(qt/(2/\pi))^2\chi^{-1}\}^{-z} e^{z^2/T}(1+\sum_{u,v} C_\delta'''(u,v)z^u t^{-2v}) \ll q^{2C_0}\chi^{-C_0}T^{2C_0-5}e^{-y^2/2T}. \quad (11)$$

Define

$$\alpha(u,v)=\frac{1}{2}(C_\delta'''(u,v)+C_\delta'''(u,v)),$$
$$\beta(u,v)=\frac{1}{2}(C_\delta'''(u,v)-C_\delta'''(u,v)).$$

Since the abscissae of the lines of integration in the integrals $I_k(x,t)$ and $K_k(x,t)$ may be-chosen arbitrarily in the range Re $(z)>0$, we obtain from (9)—(11), for $c_0\geqslant\varepsilon>0$,

$$I_k(x,t)-K_k(x,t) \ll q^{2C_0}\chi^{-C_0}T^{2C_0-5}\int_{-\infty}^{\infty} e^{-y^2/2T}dy$$
$$\ll q^{2C_0}\chi^{-C_0}T^{2C_0-\frac{9}{2}}, \text{ for } k=1,2. \qquad (12)$$

Replace $I_k(mn,t)$ by $K_k(mn,t)$ in (7), put $C_0=\varepsilon$ for $mn\leqslant(qT)^2$ and $C_0=1$ otherwise. Then the respective contributions of (12) to (7) are at most

$$\sum_{mn\leqslant(qT)^2} d(m)d(n)(mn)^{-\frac{1}{2}}|F(m\pm n)|q^\varepsilon T^{-4}$$

$$\ll \sum_{d|q} d \sum_{\substack{mn\leqslant(qT)^2\\ m\equiv n \pmod d}} (mn)^{-\frac{1}{2}}(qT)^\varepsilon T^{-4} \ll q^{1+\varepsilon}T^{-3+\varepsilon};$$

$$\sum_{mn>(qT)^2} d(m)d(n)(mn)^{-\frac{1}{2}}|F(m\pm n)|(mn)^{-1}q^2T^{-\frac{5}{2}}$$

$$\ll \sum_{d|q} d \sum_{\substack{mn>(qT)^2\\ m\equiv n \pmod d}} (mn)^{\varepsilon-\frac{3}{2}}q^2T^{-\frac{5}{2}} \ll q^{1+\varepsilon}T^{-3}.$$

Hence

$$\sum_{\chi \bmod q}^{*} |L(\frac{1}{2}+it,\chi)|^4 = \sum_{m,n=1}^{\infty} d(m)d(n)(mn)^{-\frac{1}{2}}(\frac{m}{n})^{it}$$
$$\cdot\{F(m-n)K_1(mn,t)+F(m+n)K_2(mn,t)\}$$
$$+(q^{1+\varepsilon}T^{-2}). \qquad (13)$$

In order to complete the proof of the lemma, we shall give some estimates of $K_k(\chi,t)$, $k=1,2$. Write $A=(qt/(2\pi))^2\chi^{-1}$ and define

$$K_{u,v}(\chi,t)=\frac{1}{\pi i}\int_{1-i\infty}^{1+i\infty} A^z e^{z^2/T} z^{u-1} t^{-2v}\,dz$$

$$K_{0,0}(\chi,t)= J(\chi,t).$$

Then

$$K_1(\chi,t)=J(\chi,t)+\sum_{u=1}^{30}\sum_{v}\alpha(u,v)K_{u,v}(\chi,t),$$

$$K_2(\chi,t)=\sum_{u=1}^{30}\sum_{v}\beta(u,v)K_{u,v}(\chi,t)$$

For $u\geqslant 1$, we shift the line of integration to $\mathrm{Re}(z)=-(T\log A)/2$. Then the integrand is

$$A^z C^{z^2/T} z^{u-1} t^{-2v} \ll T^{-2v}(|y|^{u-1}+|T\log A|^{u-1})\exp\{-\frac{1}{4}T(\log A)^2-y^2/T\}.$$

Hence

$$K_{u,v}(\chi,t) \ll T^{\frac{1}{2}-2v}(T^{\frac{u-1}{2}}+|T\log A|^{u-1})\exp\{-\frac{1}{4}T(\log A)^2\}$$
$$\ll T^{\frac{u}{2}-2v}\exp\{-\frac{1}{8}T(\log A)^2\}.$$

For $J(\chi,t)$ we have

$$J(\chi,t)=H(\chi,t)+\frac{1}{\pi i}\int_{-\frac{T}{2}\log A-i\infty}^{-\frac{T}{2}\log A+i\infty} A^z e^{z^2/T}\,dz/z,$$

where we have defined $H(\chi,t)=0$ for $A<1$ and 2 for $A>1$. The above integral is

$$1/(2\pi)\int_{-\infty}^{\infty} e^{-y^2/T}\left\{\frac{1}{-\frac{T}{2}\log A+iy}+\frac{1}{-\frac{T}{2}\log A-iy}\right\}dy\cdot\exp\left\{-\frac{1}{4}T(\log A)^2\right\}$$

$$=-1/(2\pi)\int_{-\infty}^{\infty} e^{-y^2/T}(T\log A)/(T\log A)^2/4+y^2)\,dy\cdot\exp\left\{-\frac{1}{4}T(\log A)^2\right\}.$$

From the above discussion we see that for $A \leqslant e^{-1}$

$$K_k(\chi, t) \ll \exp\left\{-\frac{1}{8}T(\log A)^2\right\}, \quad \text{for } k=1,2.$$

Put $c=e$. Then for $\chi > c(qT)^2$ we have $A \leqslant e^{-1}$, whence

$$\exp\left(-\frac{1}{8}T(\log A)^2\right) \ll \chi^{-1}T^{-2}.$$

This yields immediately the result of the lemma.

We state here two more estimates for $K_{u,v}(\chi, t)$ and $J(\chi, t)$, which were given by Heath-Brown [1].

$$\frac{d}{dt}K_{u,v}(\chi, t) \ll T^{\frac{u}{2}-2v-\frac{1}{2}}\exp\left\{-\frac{1}{8}T(\log A)^2\right\}. \tag{14}$$

$$\frac{d}{dt}J(\chi, t) = (2/t)\cdot(T/\pi)^{\frac{1}{2}}\exp\left\{-\frac{1}{4}T(\log A)^2\right\}. \tag{15}$$

§ 3. A Formula of the Fourth Power Mean

In this section, we shall prove mainly the following lemma.

Lemma 3. *Assume $T \ll T_1 < T_2 \ll T$. Then we have*

$$\sum_{\chi \bmod q}^{*}\int_{T_1}^{T_2}\left|L\left(\frac{1}{2}+it, \chi\right)\right|^4 dt$$

$$= \Phi_1(q)\sum_{\substack{m \leqslant \frac{qT_2}{2\pi} \\ (mn, q)=1}} d^2(m)m^{-1}\int_{T_1}^{T_2} H(m^2, t)\,dt$$

$$+2\left\{\sum_{\substack{mn \leqslant \left(\frac{qt}{2\pi}\right)^2 \\ (mn, q)=1}} d(m)d(n)(mn)^{-\frac{1}{2}}F(m-n)\left(\frac{m}{n}\right)^{it}\left(i\log\frac{m}{n}\right)^{-1}\right\}$$

$$-2\sum_{\substack{T_1 < \tau(mn) \leqslant T_2 \\ (mn, q)=1}} d(m)d(n)(mn)^{-\frac{1}{2}}F(m-n)\left(\frac{m}{n}\right)^{i\tau}\left(i\log\frac{m}{n}\right)^{-1}$$

$$\cdot\exp\left\{-\frac{1}{4}\tau^2\left(\log\frac{m}{n}\right)^2/T\right\} + O(q^{1+\varepsilon}T^{\frac{1}{2}+\varepsilon}) \tag{16}$$

uniformly in T_1 and T_2. Here we have defined $\tau=\tau(mn)=2\pi(mn)^{\frac{1}{2}}/q$, and $\Phi_1(q)$ denotes the number of primitive characters modulo q.

Proof. The proof is based on termwise integration of (5), where $J(\chi, t)$ gives rise to the main terms and $K_{u,v}(\chi, t)$, *for* $u\geq 1$, contributes the error term. We have

$$\sum_{\chi \bmod q}{}^{*}\int_{T_1}^{T_2}|L(\frac{1}{2}+it,\chi)|^4\,dt$$

$$=\sum_{\substack{mn\leq c(qT)^2\\(mn,q)=1}} d(m)d(n)(mn)^{-\frac{1}{2}}F(m-n)\int_{T_1}^{T_2}\left(\frac{m}{n}\right)^{i\tau}J(mn,t)\,dt$$

$$+\sum_{u,v}\sum_{\substack{mn\leq C(qT)^2\\(mn,q)=1}} d(m)d(n)(mn)^{-\frac{1}{2}}(\alpha(u,v)F(m-n)+\beta(u,v)F(m+n))$$

$$\cdot\int_{T_1}^{T_2}\left(\frac{m}{n}\right)^{i\tau}K_{u,v}(mn,t)\,dt+0(q^{1+\varepsilon}T^{-1}). \tag{17}$$

First, we shall show that for $u\geq 1$, $K_{u,v}(mn, t)$ contributes to (17) at most $q^{1+\varepsilon}T^{\frac{1}{2}+\varepsilon}$. As before, we write $A=(qt/(2\pi))^2(mn)^{-1}$. Then, for $m=n$, we have

$$\int_{T_1}^{T_2}K_{u,v}(m^2,t)\,dt \ll T^{\frac{u}{2}-2v}\int_{T_1}^{T_2}\exp\left\{-\frac{1}{8}T(\log A)^2\right\}dt.$$

For $m\neq n$, the integration by parts together with (14) yields

$$\int_{T_1}^{T_2}\left(\frac{m}{n}\right)^{i\tau}K_{u,v}(mn,t)\,dt$$

$$=\left(i\log\frac{m}{n}\right)^{-1}\left(\frac{m}{n}\right)^{i\tau}K_{u,v}(mn,t)\Big|_{T_1}^{T_2}-\left(i\log\frac{m}{n}\right)^{-1}\int_{T_1}^{T_2}\left(\frac{m}{n}\right)^{i\tau}\frac{d}{dt}K_{u,v}(mn,t)\,dt$$

$$\ll\left|\log\frac{m}{n}\right|^{-1}T^{\frac{u}{2}-2v}\left(1+T^{-\frac{1}{2}}\int_{T_1}^{T_2}\exp\left(-\frac{1}{8}T(\log A)^2\right)dt.\right.$$

It is easy to see that

$$\int_{T_1}^{T_2}\exp\left(-\frac{1}{8}T(\log A)^2\right)dt \ll T^{\frac{1}{2}}\log T.$$

Therefore the contribution of $K_{u,v}(mn, t)$ to (17) is

$$\Phi_1(q) \sum_{\substack{mn \leqslant c(qT)^2 \\ (mn, q)=1}} d^2(m) m^{-1} T^{\frac{u}{2}-2v+\frac{1}{2}} \log T$$

$$+ \sum_{\substack{mn \leqslant c(qT)^2 \\ (mn, q)=1 \\ m \neq n}} d(m) d(n) (mn)^{-\frac{1}{2}} |F(m \pm n)| \cdot \left|\log \frac{m}{n}\right|^{-1} T^{\frac{u}{2}-2v+\varepsilon}$$

$$\ll q^{1+\varepsilon} T^{\frac{1}{2}+\varepsilon}.$$

By the normal method of estimating, since for $u, v \geqslant 1$, $v \geqslant u/3$, we have $\frac{1}{2}u - 2v + \frac{1}{2} \leqslant 0$.

Dealing with $J(mn, t)$, we consider the two cases $m=n$ and $m \neq n$, seperately. For $m=n$, we have

$$\int_{T_1}^{T_2} J(m^2, t)\, dt = \int_{T_1}^{T_2} H(m^2, t)\, dt + O\left(\int_{T_1}^{T_2} \exp\left\{-\frac{1}{4} T (\log A)^2\right\}\right)$$

$$= \int_{T_1}^{T_2} H(m^2, t)\, dt + O\left(T^{\frac{1}{2}} \log T\right),$$

which gives rise to the first sum in (16) with an error $O(q^{1+\varepsilon} T^{\frac{1}{2}+\varepsilon})$. For $m \neq n$ we have, on integrating by parts,

$$\int_{T_1}^{T_2} J(mn, t) \left(\frac{m}{n}\right)^{i\tau} dt = \left(i \log \frac{m}{n}\right)^{-1} \left\{ J(mn, t) \left(\frac{m}{n}\right)^{i\tau} \Big|_{T_1}^{T_2} - N(m, n) \right\}$$

$$N(m, n) = \int_{T_1}^{T_2} \left(\frac{m}{n}\right)^{i\tau} \frac{d}{dt} J(mn, t)\, dt.$$

We shall show that $N(m, n)$ contributes to (16) the third sum with a total error $\ll q^{1+\varepsilon} T^{\frac{1}{2}+\varepsilon}$. By (16) we have

$$N(m, n) = 2(T/\pi)^{\frac{1}{2}} \int_{T_1}^{T_2} \left(\frac{m}{n}\right)^{i\tau} t^{-1} \exp\left\{-\frac{1}{4} T (\log A)^2\right\} dt.$$

In the case where $|t - \tau| > T^{\frac{1}{2}} \log T$, we have

$$\exp\left\{-\frac{1}{4} T (\log A)^2\right\} \ll T^{-1}.$$

Therefore if $\tau < T_1 - T^{\frac{1}{2}} \log T$ or $\tau > T_2 + T^{\frac{1}{2}} \log T$, then

$$N(m,n) \ll T^{-4}. \tag{18}$$

Further, for $|\tau - T_j| \leqslant T^{\frac{1}{2}} \log T$ we have

$$N(m,n) \ll \log T. \tag{19}$$

(18) contributes to (17) at most

$$T^{\frac{1}{2}}(qT)^{\varepsilon} \sum_{mn \leqslant c(qT)^2} (mn)^{-\frac{1}{2}} |F(m-n)| \left|\log \frac{m}{n}\right|^{-1} \ll q^{1+\varepsilon} T^{\frac{1}{2}+\varepsilon}.$$

Now we assume that $T \ + T^{\frac{1}{2}} \log T < \tau < T_2 - T^{\frac{1}{2}} \log T$. Then

$$\int_{T_1}^{T_2} \left(\frac{m}{n}\right)^{it} t^{-1} \exp\left\{-\frac{1}{4} T (\log A)^2\right\} dt$$

$$= \int_{\tau - T^{\frac{1}{2}} \log T}^{\tau + T^{\frac{1}{2}} \log T} \left(\frac{m}{n}\right)^{it} t^{-1} \exp\left\{-\frac{1}{4} T (\log A)^2\right\} dt + O(T^{-1}). \tag{20}$$

Setting $t = \tau + n$, and noticing that for $|n| \leqslant T^{\frac{1}{2}} \log T$.

$$-\frac{1}{4} T (\log A)^2 = -Tn^2 \tau^{-2} + O(T^{\varepsilon - \frac{1}{2}})$$

$$t^{-1} = \tau^{-1} + O(T^{\varepsilon - \frac{3}{2}}),$$

we obtain that (20) is equivalent to

$$\tau^{-1}\left(\frac{m}{n}\right)^{i\tau} \int_{-T^{\frac{1}{2}} \log T}^{T^{\frac{1}{2}} \log T} \left(\frac{m}{n}\right)^{i\eta} \exp\{-T\tau^{-2} n^2\}\, dn + O(T^{\varepsilon - 1})$$

$$= \tau^{-1}\left(\frac{m}{n}\right)^{i\tau} \int_{-\infty}^{\infty} \left(\frac{m}{n}\right)^{i\eta} \exp\{-T\tau^{-2} n^2\} dn + O(T^{\varepsilon - 1})$$

$$= (\pi/T)^{\frac{1}{2}} \left(\frac{m}{n}\right)^{i\tau} \exp\left\{-\frac{1}{4}\tau^2 \left(\log \frac{m}{n}\right)^2 / T\right\} + O(T^{\varepsilon - 1}).$$

From the above discussion we see that $N(m,n)$ contributes to (17)

$$-2 \cdot \sum_{\substack{T_1 + T^{\frac{1}{2}} \log T < \tau(mn) \leqslant T_2 - T^{\frac{1}{2}} \log T \\ (mn,\, q) = 1,\, m \neq n}} d(m) d(n) (mn) F(m-n) \left(\frac{m}{n}\right)^{i\tau} \left(i \log \frac{m}{n}\right)^{-1}$$

$$\cdot \exp\left\{-\frac{1}{4}\tau^2 \left(\log \frac{m}{n}\right)^2 / T\right\}$$

$$+ O\left(\sum_{\substack{|\tau(mn)-Tj|\leqslant T^{\frac{1}{2}}\log T \\ m\neq n}} d(m)\,d(n)\,(mn)^{-\frac{1}{2}}|F(m-n)||\log\frac{m}{n}|^{-1}\right)$$

$$+ O(q^{1+\varepsilon}T^{\frac{1}{2}+\varepsilon}).$$

Replacing the above range of summation by $T_1 < \tau(mn) \leqslant T_2$, we obtain the third sum in (16) and the error given can be absorbed in the error term above. We also have

$$\sum_{\substack{|\tau(mn)-Tj|\leqslant T^{\frac{1}{2}}\log T \\ m\neq n}} d(m)\,d(n)\,(mn)^{-\frac{1}{2}}|F(m-n)||\log\frac{m}{n}|^{-1}$$

$$\ll (qT)^{\varepsilon}\sum_{d|q} d\cdot \sum_{\substack{|\tau(mn)-Tj|\leqslant T^{\frac{1}{2}}\log T \\ m\equiv n\,(\mathrm{mod}\,d)\, m<n\leqslant 2m}} (mn)^{-\frac{1}{2}}\, n/(n-m)$$

$$+ \sum_{d|q} d \sum_{\substack{|\tau(mn)-T_3|\leqslant T^{\frac{1}{2}}\log T \\ m\equiv n\,(\mathrm{mod}\,d)}} (mn)^{-\frac{1}{2}}(qT)^{\varepsilon}. \tag{21}$$

In the first sum on the right-hand side of (21), set $n=m+dk$, then $dk\leqslant qT$. For fixed d and k, the condition $|\tau(m(m+dk))-T|\leqslant T^{\frac{1}{2}}\log T$ implies

$$|m-\frac{1}{2}\{((qT/\pi)^2+(dk)^2)^{\frac{1}{2}}-dk\}|\leqslant cqT^{\frac{1}{2}}\log T,$$

where c is a suitable constant. But

$$qT \ll \frac{1}{2}\{((qT/\pi)^2+(dk)^2)^{\frac{1}{2}}-dk\} \ll qT.$$

Thus the first sum on the right-hand side is

$$\ll \sum_{d|q} d \sum_{k\leqslant\frac{qT}{d}} (qT)^{\varepsilon}\, qT^{\frac{1}{2}}/(dk) \ll q^{1+\varepsilon}T^{\frac{1}{2}+\varepsilon}.$$

The second sum can be treated strictly and it is

$$\ll \sum_{d|q} d \sum_{m^2\leqslant c(qT)^2} m^{-\frac{1}{2}} \sum_{\substack{m<n\leqslant c(qT)^2m^{-1} \\ |\tau(mn)-Tj|\ll T^{\frac{1}{2}}\log T \\ n\equiv m\,(\mathrm{mod}\,d)}} n^{-\frac{1}{2}}(qT)^{-1+\varepsilon}$$

$$\ll \sum_{d|q} d \sum_{m^2\leqslant c(qT)^2} (qT)^{-1+\varepsilon}q^2T^{\frac{3}{2}}/(dk) \ll q^{1+\varepsilon}T^{\frac{1}{2}+\varepsilon}.$$

Now for the proof of the lemma, it is sufficient to show that if we replace $(\frac{m}{n})^{it}$ $J(mn,t)|_{T_1}^{T_2}$ by $(\frac{m}{n})^{it}H(mn,t)|_{T_1}^{T_2}$, then the error caused contributes to (17) at most $q^{1+\varepsilon}T^{\frac{1}{2}+\varepsilon}$. In fact, this contribution is

$$\ll (qT)^{\varepsilon} \sum_{mn \leqslant c(qT)^2} (mn)^{-\frac{1}{2}} | F(m-n) | \, | \log \frac{m}{n} |^{-1} \exp \left\{ -\frac{1}{4} T (\log A)^2 \right\},$$

where $A=(qT_j/2\pi)^2 (mn)^{-1}, j=1$ or 2. Using

$$\exp \left\{ -\frac{1}{4} T (\log A)^2 \right\} \ll \begin{cases} 1, & \text{for } |t-\tau| \leqslant T^{\frac{1}{2}} \log T, \\ T^{-1}, & \text{for } |t-\tau| > T^{\frac{1}{2}} \log T, \end{cases}$$

and treating it in the same way as in the estimation of (21), we conclude that the above sum is $O(q^{1+\varepsilon} T^{\frac{1}{2}+\varepsilon})$. This completes the proof of Lemma 3.

§ 4. A Weighted Mean Value Estimate

Since (16) contains too many terms for us to estimate, following Heath-Brown, we define the following weight-function and consider a weighted mean value estimate. Let

$$T^{\frac{1}{2}} \leqslant \Delta \leqslant T(\log T)^{-2}.$$

$$3T/2 \leqslant T_1 \leqslant T_2 \leqslant 9T/2.$$

We define the weight-function as follows:

$$w(t) = \pi^{-\frac{1}{2}} \Delta^{-1} \int_{T_1}^{T_2} \exp \{ -(t-u)^2 \Delta^{-2} \} \, du.$$

We state here several simple properties of this function:

$$0 \leqslant w(t) \leqslant 1, \quad \text{for } all\ t. \tag{22}$$

$$w(t) \ll T^{-1}, \quad \text{for } t \leqslant T_1 - \Delta \log T \text{ or } t \geqslant T_2 + \Delta \log T, \tag{23}$$

$$w(t) = 1 + 0(T^{-1}), \text{ for } T_1 + \Delta \log T < t < T_2 - \Delta \log T, \tag{24}$$

$$w'(t) = \pi^{-\frac{1}{2}} \Delta^{-1} (\exp\{-(t-T_1)^2 \Delta^{-2}\} - \exp\{-(t-T_2)^2 \Delta^{-2}\}). \tag{25}$$

We have the following:

Lemma 4. *Define*

$$f(t) = 2\varphi_1(q) \sum_{\substack{m \leqslant qt/2\pi \\ (m,q)=1}} d^2(m) m^{-1} (t - \tau(m^2))$$

$$+ 2 \sum_{\substack{\tau(mn) \leqslant t, (mn,q)=1 \\ 0 < |\log \frac{m}{n}| \leqslant \Delta^{-1} \log T}} d(m) d(n) (mn)^{-\frac{1}{2}} F(m-n) \left(\frac{m}{n} \right)^{i\tau}$$

$$\cdot \left(i \log \frac{m}{n}\right)^{-1} \exp\left\{-\frac{1}{4}\Delta^2\left(\log\frac{m}{n}\right)^2\right\}$$

$$-2 \sum_{\substack{\tau(mn) \leqslant t \\ (mn, q)=1 \\ 0<|\log \frac{m}{n}| \leqslant T^{-\frac{1}{2}} \log T}} d(m)\, d(n)\, (mn)^{-\frac{1}{2}} F(m-n)\left(\frac{m}{n}\right)^{i\tau}$$

$$\cdot \left(i \log \frac{m}{n}\right)^{-1} \exp\left\{-\frac{1}{4}\tau^2\left(\log\frac{m}{n}\right)^2 / T\right\},$$

then for $3T/2 \leqslant T_1 \leqslant T_2 \leqslant 9T/2$, *we have, uniformly in* T_1 *and* T_2.

$$\int_T^{5T} w(t) \sum_{\chi \bmod q}{}^{*} \left|L\left(\frac{1}{2}+it, \chi\right)\right|^4 dt = f(T_2) - f(T_1) + O(q\,\Delta\,(qT)^{\varepsilon}).$$

Proof. On integrating by parts we obtain

$$\int_T^{5T} w(t) \sum_{\chi \bmod q}{}^{*} \left|L\left(\frac{1}{2}+it, \chi\right)\right|^4 dt$$

$$= w(5T)\int_T^{5T} \sum_{\chi \bmod q}{}^{*} \left|L\left(\frac{1}{2}+it, \chi\right)\right|^4 dt - \int_T^{5T} w'(t) \int_T^{t} \sum_{\chi \bmod q}{}^{*} \left|L\left(\frac{1}{2}+iy, \chi\right)\right|^4 dy dt. \quad (26)$$

It is well known that

$$\int_0^{t} \sum_{\chi \bmod q}{}^{*} \left|L\left(\frac{1}{2}+it, \chi\right)\right|^4 dt \ll qT \log^4 (qT).$$

Hence from (23) we see that the first term on the right hand side of (26) is $\ll q^{1+\varepsilon} T^{\varepsilon}$. We define

$$I_1 = \int_T^{5T} w'(t) \sum_{\substack{\tau(m^2) \leqslant t \\ (m, q)=1}} d^2(m) m^{-1} \int_T^{t} H(m^2, y)\, dy dt,$$

$$I_2 = \int_T^{5T} w'(t) \sum_{\substack{\tau(mn) \leqslant y \\ (mn, q)=1,\, m \neq n}} d(m)\, d(n)\, (mn)^{-\frac{1}{2}} F(m-n)\left(\frac{m}{n}\right)^{iy}\left(i \log\frac{m}{n}\right)^{-1} \Big|_T^t dt,$$

$$I_3 = \int_T^{5T} w'(t) \sum_{\substack{T < \tau(mn) \leqslant t \\ (mn, q)=1,\, m \neq n}} d(m) d(n) (mn)^{-\frac{1}{2}} F(m-n)\left(\frac{m}{n}\right)^{i\tau}$$

$$\exp\left\{-\frac{1}{4}\tau^2\left(\log\frac{m}{n}\right)^2/\mathrm{T}\right\}\left(i\log\frac{m}{n}\right)^{-1}.$$

Then from the result of Lemma 3 we see immediately that

$$\int_T^{5T} w(t)\sum_{\chi \bmod q}^{*}\left|L\left(\frac{1}{2}+it,\chi\right)\right|^4 dt = -\varphi_1(q)I_1-2I_2+2I_3+O\left(q^{1+\varepsilon}T^{\frac{1}{2}+\varepsilon}\right).$$

So we need to estimate I_1, I_2 and I_3, separately. For I_1, we have

$$I_1=\sum_{\substack{m\leqslant \frac{qT}{2\pi}\\ (m,q)=1}} d^2(m)m^{-1}\int_T^{5T} w'(t)\cdot 2(t-T)\,dt$$

$$=\sum_{\substack{\frac{qT}{2\pi}<m\leqslant\frac{5qT}{2\pi}\\ (m,q)=1}} d^2(m)m^{-1}\int_\tau^{5T} w'(t)\cdot 2(t-\tau)\,dt.$$

For $\tau(m^2)\leqslant T_j-\Delta\log T$, we have

$$\pi^{-\frac{1}{2}}\Delta^{-1}\int_\tau^{5T}(t-\tau)\exp\{-(t-T_j)^2\Delta^{-2}\}dt$$

$$=\pi^{-\frac{1}{2}}\Delta^{-1}\int_{-\infty}^{\infty}(t-\tau)\exp\{-(t-T_j)^2\Delta^{-2}\}dt+O(T^{-1})$$

$$=T_j\setminus\tau+O(T^{-1}),$$

For $\tau(m^2)\geqslant T_j+\Delta\log T$, we have

$$\pi^{-\frac{1}{2}}\Delta^{-1}\int_\tau^{5T}(t-\tau)\exp\{-(t-T_j)^2\Delta^{-2}\}dt=O(T^{-1}).$$

For $|\tau-T_j|\leqslant\Delta\log T$, we have

$$\pi^{-\frac{1}{2}}\Delta^{-1}\int_\tau^{5T}(t-\tau)\exp\{-(t-T_j)^2\Delta^{-2}\}dt$$

$$=\pi^{-\frac{1}{2}}\Delta^{-1}\int_{\substack{\tau\\ |t-T_j|\leqslant\Delta\log T}}^{5T}(t-\tau)\exp\{-(t-T_j)^2\Delta^{-2}\}dt+O(T^{-1})$$

$$\ll\Delta\log T.$$

Moreover,

$$\sum_{|\tau(m^2)-T_j|\leqslant\Delta\log T} d^2(m)m^{-1}\leqslant(qT)^{\varepsilon}\Delta T^{-1}.$$

Thus from the above discussion we obtain

$$I_1 = \sum_{\substack{\tau(m^2) \leqslant T \\ (m,q)=1}} d^2(m) m^{-1} 2(T_1 - T_2) + 2 \sum_{\substack{T < \tau(m^2) \leqslant T_1 - \Delta \log T \\ (m,q)=1}} d^2(m) m^{-1} (T_1 - \tau)$$

$$-2 \sum_{\substack{T < \tau(m^2) \leqslant T_2 - \Delta \log T \\ (m,q)=1}} d^2(m) m^{-1} (T_2 - \tau) + O((qT)^\varepsilon \Delta).$$

For I_2 we have

$$I_2 = \sum_{\substack{\tau(mn) \leqslant 5T \\ (mn,q)=1,\ m \neq n}} d(m) d(n) (mn)^{-\frac{1}{2}} F(m-n) \left(i \log \frac{m}{n}\right)^{-1} \int_{\max(T,\tau)}^{5T} w'(t) \left(\frac{m}{n}\right)^{it} dt$$

$$-\sum_{\substack{\tau(mn) \leqslant T \\ (mn,q)=1,\ m \neq n}} d(m) d(n) (mn)^{-\frac{1}{2}} F(m-n) \left(i \log \frac{m}{n}\right)^{-1} \int_T^{5T} w'(t)\, dt \left(\frac{m}{n}\right)^{iT}. \quad (27)$$

From (23), we obtain

$$\int_T^{5T} w'(t)\, dt = w(5T) - w(T) \ll T^{-1}.$$

Therefore the second sum on the right hand side of (27) is $\ll q^{1+\varepsilon}$ by standard argument. For $\tau(mn) \leqslant T_j - \Delta \log T$, we have

$$\pi^{-\frac{1}{2}} \Delta^{-1} \int_{\max(\tau, T)}^{5T} \exp\{-(t - T_j)^2 \Delta^{-2}\} \left(\frac{m}{n}\right)^{it} dt$$

$$= \left(\frac{m}{n}\right) \exp\left\{-\frac{1}{4} \Delta^2 \left(\log \frac{m}{n}\right)^2\right\} + O(T^{-1}),$$

For $\tau(mn) \geqslant T_j + \Delta \log T$, we have

$$\pi^{-\frac{1}{2}} \Delta^{-1} \int_\tau^{5T} \exp\{-(t - T_j)^2 \Delta^{-2}\} \left(\frac{m}{n}\right)^{it} dt \ll T^{-1}.$$

But by the same argument as in the estimation of (21) we can prove

$$\sum_{\substack{|\tau(mn) - T_j| \leqslant \Delta \log T \\ m \neq n}} (mn)^{-\frac{1}{2}} |F(m-n)| \left|\log \frac{m}{n}\right|^{-1} \ll (q\Delta)(qT)^\varepsilon.$$

Then the above estimates together yield

$$I_2 = \sum_{\substack{\tau(mn) \leqslant t \\ (mn,q)=1,\ m \neq n}} d(m) d(n) (mn)^{-\frac{1}{2}} F(m-n) \left(i \log \frac{m}{n}\right)^{-1}$$

$$\cdot\left(\frac{m}{n}\right)^{it}\exp\left\{-\frac{1}{4}\Delta^2\left(\log\frac{m}{n}\right)^2\right\}| + O\left(q\,\Delta\,(qT)^{\varepsilon}\right).$$

We can treat I_3, in the same way as in the treatment of I_2, so we conclude

$$I_3=\sum_{\substack{T_1<\tau(mn)\leqslant T_2\\(mn,q)=1,\,m\neq n}} d(m)\,d(n)\,(mn)^{-\frac{1}{2}}F(m-n)\left(i\log\frac{m}{n}\right)^{-1}$$

$$\cdot\left(\frac{m}{n}\right)^{i\tau}\exp\left\{-\frac{1}{4}\tau^2\left(\log\frac{m}{n}\right)^2/T\right\}+O\left(q\Delta\,(qT)^{\varepsilon}\right).$$

This completes the proof of Lemma 4.

In order to obtain the mean value estimates of the required form from the weighted one, we need the following:

Lemma 5. *Assume that there exists a function $g(t)\in C'(0,\infty)$ such that, uniformly in T_1 and T_2,*

$$\int_T^{5T} w(t)\sum_{\chi \bmod q}{}^{*}\left|L\left(\frac{1}{2}+it,\chi\right)\right|^4 dt=g(T_2)-g(T_1)+O\left(q\Delta\,(qT)^{\varepsilon}\right)$$

for $3T/2\leqslant T_1<T_2\leqslant 9T/2$, and $g'(t)=O(q^{1+\varepsilon}T^{\varepsilon})$. Then

$$\int_{2T}^{4T}\sum_{\chi \bmod q}{}^{*}\left|L\left(\frac{1}{2}+it,\chi\right)\right|^4 dt=g(4T)-g(2T)+O\left(q\Delta\,(qT)^{\varepsilon}\right).$$

Proof. First set $T_1=2T-\Delta\log T$, $T_2=4T+\Delta\log T$. Then (23) and (24) yield

$$\int_{2T}^{4T}\sum_{\chi \bmod q}{}^{*}\left|L\left(\frac{1}{2}+it,\chi\right)\right|^4 dt\leqslant\int_T^{5T} w(t)\sum_{\chi \bmod q}{}^{*}\left|L\left(\frac{1}{2}+it,\chi\right)\right|^4 dt+O(q^{1+\varepsilon})$$

$$=g(T_2)-g(T_1)+O\left(q\Delta\,(qT)^{\varepsilon}\right)$$

$$=g(4T)-g(2T)+O\left(q\Delta\,(qT)^{\varepsilon}\right).$$

Next, set $T_1=2T+\Delta\log T$, $T_2=4T-\Delta\log T$. Then

$$\int_{2T}^{4T}\sum_{\chi \bmod q}{}^{*}\left|L\left(\frac{1}{2}+it,\chi\right)\right|^4 dt\geqslant\int_T^{5T} w(t)\sum_{\chi \bmod q}{}^{*}\left|L\left(\frac{1}{2}+it,\chi\right)\right|^4 dt+O(q^{1+\varepsilon})$$

$$=g(T_2)-g(T_1)+O\left(q\Delta(qT)^{\varepsilon}\right)$$

$$=g(4T)-g(2T)+O\left(q\Delta\,(qT)^{\varepsilon}\right).$$

This completes the proof of Lemma 5.

§5. A Divisor Sum

The aim of this section is to give an asymptotic estimate for the following divisor sum:

$$S(x;r,q)=\sum_{\substack{n\leqslant \chi\\(n(n+r),\,q)=1}} d(n)(n+r) \tag{28}$$

Lemma 6. *There exist functions* $h^{j};(d,q)$, *for* $j=0,1,2$ *such that*

$$S(x;r,q)=\sum_{j=0}^{2} x(\log x)^{j}\sum_{d\mid r} h_{j}(d,q)+E(x;r,q), \tag{29}$$

where the functions $h_j(d,q)$ *satisfy*

$$h_{j}(d,q)\ll d^{-1}q^{\varepsilon},\ \textit{for}\ j=0,1,2. \tag{30}$$

The following estimates for the remainder term $E(\chi;r,q)$ *hold uniformly in the given ranges*:

$$E(x;r,q)\ll q^{\varepsilon}x^{\frac{5}{6}+\varepsilon} \qquad \textit{for}\ \ r\leqslant x^{\frac{5}{6}}, \tag{31}$$

$$\int_{x}^{2x} E^{2}(y;r,q)\,dy\ll q^{1+\varepsilon}x^{\frac{5}{2}+\varepsilon} \qquad \textit{for}\ \ r\leqslant x. \tag{32}$$

Proof. It is easy to see from (28) that

$$S(x;r,q)=\sum_{v\mid q}\mu(v)\sum_{\substack{e(l-r)\equiv 0\,(v)\\0\leqslant l<v}}\ \sum_{\substack{n\leqslant x\\n\equiv l-r\,(v)}} d(n)d(n+r).$$

For $n\leqslant x$, using

$$d(n)=2\sum_{\substack{k\mid n\\k\leqslant x^{\frac{1}{2}}}}1-\sum_{\substack{k\mid n\\nx^{-\frac{1}{2}}\leqslant k\leqslant x^{\frac{1}{2}}}}1,$$

we obtain

$$S(x;r,q)=\sum_{v\mid q}\mu(v)\sum_{\substack{l(l-r)\equiv 0\,(\gamma)\\0\leqslant l\leqslant v}}\ \sum_{\substack{k\leqslant x^{\frac{1}{2}}\\(k,v)\mid r-l}}$$

$$\cdot\Big\{2\sum_{\substack{T<n\leqslant x+r\\n\equiv l\,(v)\\n\equiv r\,(k)}} d(n)-\sum_{\substack{r<n\leqslant kx^{\frac{1}{2}}+r\\n\equiv l\,(v)\\n\equiv r\,(k)}} d(n)\Big\}. \tag{33}$$

Let $(k,v)=\rho$, $v=\rho v_1$, $\lambda=(v_1,1)$, $v_1=\lambda v_2$. Then, for $\mu(v)\neq 0$, v_2, ρ and λ are pairwise coprime. Here we can assume

$$l \equiv l_1 v' \lambda \rho + l_2 \rho' \lambda v_2 + l_3 \lambda' \rho v_2 \pmod{v},$$

where v', ρ' and λ' are integers satisfying

$$v' \lambda \rho \equiv 1 \pmod{v_2},\quad v_2 \lambda \rho' \equiv 1 \pmod{\rho},\quad v \lambda' \rho \equiv 1 \pmod{\lambda},$$

and as l_1, l_2, l_3 run through the complete residue classes modulo v_2, ρ_1, λ, respectively and independently, l runs through the complete residue class modulo v. Hence we can replace the conditions

$$l(l-r) \equiv 0 \pmod{v},\quad \rho \mid r-l,\ 0 \leqslant l \leqslant v-1$$

by the following,

$$\begin{aligned} &l_1(l_1-r) \equiv 0 \pmod{v_2}, && 0 \leqslant l_1 \leqslant v_2-1,\ (l_1, v_2) \equiv 1,\\ &l_2 \equiv r \pmod{\rho}, && 0 \leqslant l_2 < \rho,\\ &l_3 \equiv 0 \pmod{\lambda}, && 0 \leqslant l_3 < \lambda. \end{aligned}$$

But these congruences have exactly one group of simultaneous solutions, therefore we deduce from (33) that

$$S(x; r, q) = \sum_{\substack{v\rho\lambda \mid q \\ (v, r)=1}} \mu(v\rho\lambda) \sum_{\substack{k \leqslant x^{\frac{1}{2}} \\ k \equiv 0 \pmod{l} \\ (k, v\lambda)=1}} \{2 \sum_{\substack{n \leqslant x \\ n \equiv 0\ (\lambda) \\ n \equiv r\ (kv)}} d(n)$$

$$- \sum_{\substack{n \leqslant kx^{\frac{1}{2}} \\ n \equiv 0 \pmod{\lambda} \\ n \equiv r \pmod{kv}}} d(n)\} + 0(rx^{\varepsilon})$$

$$= 2S_1 - S_2 + O(rx^{\varepsilon}). \tag{34}$$

Since the congruences

$$n \equiv r \pmod{kv},\quad n \equiv 0 \pmod{\lambda}$$

have exactly one group of simultaneous'solutions, say

$$n \equiv r\lambda\bar{\lambda} \pmod{kv\lambda},$$

where $\bar{\lambda}$ is an integer satisfying $\lambda\bar{\lambda} \equiv 1 \pmod{kv}$, and $(r\lambda\bar{\lambda}, kv\lambda) = \lambda(k, r)$, therefore from Heath-Brown[1] § 6 we see that

$$\sum_{\substack{n \leqslant y \\ n \equiv r \pmod{kv} \\ n \equiv 0 \pmod{\lambda}}} d(n) = (y/(kv\lambda)) \sum_{d \mid \lambda(k,r)} \sum_{\delta \mid \frac{\lambda kv}{d}} \mu(\delta)\delta^{-1}$$

$$\cdot (\log y + 2r - 1 - 2\log(d\delta)) + \Delta(y; kv\lambda, r\lambda\bar{\lambda}).$$

Since k, v and λ are pairwise coprime, the sum over d and δ can be written as

$$\sum_{d_1 \mid (k,r),} \ \sum^{*}_{\delta_1 \mid \frac{k}{d_1},} \ \sum_{d_2 \mid \lambda,} \ \sum_{\delta_2 \mid \frac{\lambda v}{d_2}} .$$

Therefore

$$S_1 = \sum_{\substack{v\rho\lambda \mid q \\ (v,r)=1}} \mu(v\rho\lambda)(v\lambda)^{-1} \sum_{\substack{k \leqslant x^{\frac{1}{2}} \\ k \equiv 0 \ (\mathrm{mod}\ \rho) \\ (k, v\lambda)=1}} xk^{-1} \sum_{d_1 \mid (k,r),} \ \sum_{\delta_1 \mid \frac{k}{d_1}}$$

$$\sum_{d_2 \mid \lambda} \ \sum_{\delta_2 \mid \frac{\lambda v}{d_1}} \mu(\delta_1\delta_2)(\delta_1\delta_2)^{-1}(\log x + 2r - 1 - 2\log(d_1\delta_1 d_2\delta_2) + S_1(\Delta), \quad (35)$$

$$S_2 = \sum_{\substack{v\rho\lambda \mid q \\ (v,r)=1}} (v\rho\lambda)(v\lambda)^{-1} \sum_{\substack{k \leqslant x^{\frac{1}{2}} \\ k \equiv 0 \ (\rho) \\ (k, v\lambda)=1}} x^{\frac{1}{2}} \sum_{d_1 \mid (k,r),} \ \sum_{\delta_1 \mid \frac{k}{d_1},} \ \sum_{d_2 \mid \lambda,} \ \sum_{\delta_2 \mid \frac{\lambda v}{d_2}} \mu(\delta_1\delta_2)(\delta_1\delta_2)^{-1}$$

$$\cdot \left(\frac{1}{2} \log x + \log k + 2r - 1 - 2\log(d\,\delta\,d\,\delta) \right) + S_2(\Delta) \text{ say.} \quad (36)$$

We replace the condition $(k, v\lambda) = 1$ by the sum $\sum_{e \mid (k, \mu l)} \mu(e)$, and change the order of summation in (35) and (36) so that the sum is taken over k first. Then k must satisfy

$$k \equiv 0 \ (\mathrm{mod}\ [e, \rho, d_1, \delta_1]).$$

We need to estimate the following type of sums

$$\sum_{\substack{k \leqslant x^{\frac{1}{2}} \\ k \equiv 0 \ (\mathrm{mod}\ [e, \rho, d_1\delta_1])}} a(k),$$

where $a(k)$ equals to $1/k$, 1 or $\log k$. In fact, we have

$$\sum_{\substack{k \leqslant x^{\frac{1}{2}} \\ k \equiv 0 \ ([e, \rho, d_1\delta_1])}} 1/k = (1/[e, \rho, d_1\delta_1] \ (\log(x^{\frac{1}{2}}/[e, \rho, d_1\delta_1] + r) + O(x^{-\frac{1}{2}}),$$

$$\sum_{\substack{k \leqslant x^{\frac{1}{2}} \\ k \equiv 0 \ ([e, \rho, d_1\delta_1])}} 1 = x/[e, \rho, d_1\delta_1] + O(1),$$

$$\sum_{\substack{k \leqslant x^{\frac{1}{2}} \\ k \equiv 0 \ ([e, \rho, d_1\delta_1])}} \log k = (x^{\frac{1}{2}}/[e, \rho, d\,\delta\,]) \left(\frac{1}{2} \log x - 1 \right) + O(\log x).$$

Substituting the above estimates in (35), (36) and, in turn, in (34), we conclude

$$S(x;r,q)=\sum_{j=0}^{2} x(\log x)^j \sum_{\alpha,\beta=0}^{1} c_j(\alpha,\beta) \sum_{\substack{v\rho\lambda\mid q\\(v,r)=1}} \mu(v\rho\lambda)(v\lambda)^{-1}$$

$$\cdot \sum_{e\mid v\lambda} \mu(e) \sum_{d_1\mid r,\ \delta_1\leqslant x^{\frac{1}{2}}d_1^{-1}} \sum_{d_2\mid\lambda,\ \delta_2\mid\frac{v\lambda}{d_2}} \mu(\delta_1\delta_2)(\delta_1\delta_2)^{-1}$$

$$\cdot\ (\log(d_1\delta_1 d_2\delta_2))^{\alpha}(\log[e,\rho,d_1\delta_1])^{\beta}\Lambda[e,\rho,d_1\delta_1]$$

$$\cdot\ O(q^{\varepsilon}x^{\frac{1}{2}+\varepsilon})+0(rx^{\varepsilon})+2S_1(\Delta)-S_2(\Delta)$$

Replacing the range of summation in δ_1 by $\delta_1<\infty$, the error caused is at most $q^{\varepsilon}x^{\frac{1}{2}+\varepsilon}$. We introduce the sum $\sum_{d_3\mid(v,r)}\mu(d_3)$ to replace the condition $(v,r)=1$. Then the above estimate yields

$$S(x;r,q)=\sum_{j=0}^{2} x(\log x)^j \sum_{d\mid r} h_j(d,q)+E(x;r,q)$$

where we have defined

$$h_j(d,q)=\sum_{\substack{[d_1,d_3]=d\\d_3/q}} \mu(d_3)\sum_{\alpha=0}^{1}\sum_{\beta=0}^{1} c_j(\alpha,\beta)\sum_{\substack{v\rho\lambda/q\\d_3/v}} \mu(v\rho\lambda)(v\lambda)^{-1}$$

$$\cdot \sum_{e\mid\lambda v}\mu(e)\sum_{\delta_1=1}^{\infty}\sum_{d_2\mid\lambda,}\sum_{\delta_2\mid\frac{\lambda v}{d_2}} \mu(\delta_1\delta_2)(\delta_1\delta_2)^{-1}(\log(d_1\delta_1 d_2\delta_2))^{\alpha}$$

$$\cdot\ (\log[e,\rho,d_1\delta_1]\ \Lambda[e,\rho,d_1\delta_1] \ll d^{-1}q^{\varepsilon}$$

$$E(x;r,q)=2S_1(\Delta)-S_2(\Delta)+O(q^{\varepsilon}x^{\frac{1}{2}+\varepsilon})+O(rx^{\varepsilon}).$$

In order to estimate $E(x;r,q)$, we make use of the two corollaries of Theorem 3 in Heath-Brown's paper[1]. Corollary 1 yields

$$E(x;r,q)\ll\sum_{v\rho\lambda\mid q}(v\lambda)^{-1}\sum_{k\leqslant x^{\frac{1}{2}}} x^{\frac{1}{3}+\varepsilon}+O(q^{\varepsilon}x^{\frac{1}{2}+\varepsilon})+O(rx^{\varepsilon})$$

$$\ll q^{\varepsilon}x^{\frac{5}{6}+\varepsilon}+rx^{\varepsilon},$$

while Corollary 2 yields

$$\int_{x}^{2x}E^2(y;r,q)\,dy\ll q^{\varepsilon}x^{\frac{1}{2}}\sum_{v\rho\lambda\mid q,}\sum_{k\leqslant x^{\frac{1}{2}}}(x^{\frac{3}{2}}+qkx)+q^{\varepsilon}x+r^2x^{1+\varepsilon}$$

$$\ll q^{1+\varepsilon}x^{\frac{5}{2}+\varepsilon}+r^2x^{1+\varepsilon}.$$

Hence (31) and (32) follow. This completes the proof of Lemma 6.

§6. Proof of the Theorem

By Lemmas 4 and 5, we can sufficiently show

$$f(t) = g(t) + 0\,(q\,\Delta\,(qT)^{\varepsilon}), \tag{37}$$

where $g(t)$ is a function satisfying the conditions in Lemma 5, such that (37) holds uniformly for $T \ll t \ll T$. First, we have

$$2 \sum_{\substack{m \leqslant qt/2\pi \\ (m,q)=1}} d^2(m)\, m^{-1}(t - 2\pi m/q)$$

$$= 4\pi\, q^{-1}(2\pi i)^{-1} \int_{1-I\infty}^{1+i\infty} L^4(S+1, \chi_0)\, L^{-1}(2S+2, \chi_0)(qt/2\pi)^{s+1}\, ds/(s(s+1)).$$

We shift the line of integration to Re $s = -\dfrac{1}{2} + \varepsilon$, and then we have

$$4\pi\, q^{-1}(2\pi i)^{-1} \int_{-\frac{1}{2}-i\infty}^{-\frac{1}{2}+i\infty} L^4(s+1, \chi_0)\, L^{-1}(2s+2, \chi_0)(qt/2\pi)^{s+1}\, ds/(s\,(s+1))$$

$$\ll q^{\varepsilon}\, t^{\frac{1}{2}+\varepsilon}.$$

The residue at $s = 0$ is

$$\lim_{s \to 0} \frac{1}{24} \cdot \frac{d^4}{ds^4} \left\{ \frac{s^5 L^4(s+1, \chi_0)}{s(s+1) L\,(2s+2, \chi_0)} \cdot \left(\frac{qt}{2\pi} \right)^{s+1} \cdot \frac{4\pi}{q} \right\}$$

$$= tP\,(\log\,(qt/2\pi)),$$

where $P(x)$ is a polynomial of x of degree 4, in which the coefficient of x^4 is

$$\lim_{s \to 0} \frac{s^4 L^4(s+1, \chi_0)}{12\, L\,(2s+1, \chi_0)} = \frac{1}{12} \left(\frac{\varphi(q)}{q} \right)^4 (\zeta(2))^{-1} \prod_{p \mid q} \left(1 - \frac{1}{p^2}\right)^{-1}$$

$$= \frac{1}{2\pi^2} \prod_{p \mid q} \left(1 - \frac{1}{p}\right)^4 \left(1 - \frac{1}{p^2}\right)^{-1},$$

and the other coefficients are all $O\,(q^{\varepsilon})$.

Next, we define

$$f_0(t) = \sum_{\substack{\tau(mn) \leqslant t \\ 0 < |\log \frac{m}{n}| \leqslant \Delta^{-1} \log T \\ (mn, q)=1}} d(m)\, d(n)\,(mn)^{-\frac{1}{2}}(m/n)^{it}$$

$$\left(i\log\frac{m}{n}\right)^{-1}F(m-n)\exp\left\{-\frac{1}{4}\Delta^2\left(\log\frac{m}{n}\right)^2\right\}.$$

Let $m=n+r$. Then for a fixed r, the condition for n to satisfy is

$$r(\exp\{\Delta^{-1}\log T\}-1)^{-1}\leqslant n\leqslant\frac{1}{2}(-r+(r^2+(qt/\pi)^2)^{\frac{1}{2}}),$$

say $A_r\leqslant n\leqslant Br$. Hence r satisfias

$$r\leqslant(qt/2\pi)(\exp\{\Delta^{-1}\log T\}-1)=R \qquad \text{say.}$$

Therefore

$$qT\Delta^{-1}\log T \ll R\ll qT\Delta^{-1}\log T \tag{38}$$

$$r\Delta(\log T)^{-1}\ll A_r\ll r\Delta(\log T)^{-1} \tag{39}$$

$$B_r=qt/2\pi+O(r). \tag{40}$$

Now we have

$$f_0(t)=2\mathrm{Re}\left\{\sum_{r=1}^{R}\sum_{n=A_r}^{B_r}d(n)d(n+r)F(r)h(n,r)\right\},$$

where it is defined

$$h(n)=n^{-1}(1+r/n)^{-\frac{1}{2}+it}(i\log(1+r/n))^{-1}\exp\left\{-\frac{1}{4}\Delta^2(\log(1+r/n))^2\right\}$$

$$=(ir)^{-1}\exp\left(itm^{-1}-\frac{1}{4}\Delta^2r^2n^{-2}\right)+O(n^{-1})+O(Tm^{-2}),$$

in which the "O" terms contribute to $f_0(t)$

$$\ll(qT)^{\varepsilon}\sum_{r=1}^{R}(1+TrA_r^{-1})|F(r)|\ll(qT)^{\varepsilon}qT^2\Delta^{-2}.$$

Hence

$$f_0(t)=2\mathrm{Re}\sum_{r=1}^{R}(ir)^{-1}F(r)S_r+O((qT)^{\varepsilon}qT^2\Delta^{-2}),$$

where it is defined

$$S_r=\sum_{\substack{n=A_r\\(n(n+r),\,q)=1}}^{B_r}d(n)d(n+r)\exp\left(itm^{-1}-\frac{1}{4}\Delta^2r^2n^{-2}\right).$$

We have, from Lemma 6,

$$S_r = \int_{A_r}^{B_r} \exp\left(itrx^{-1} - \frac{1}{4}\Delta^2 r^2 x^{-2}\right) d(m(x;r,q) + E(x;r,q)),$$

where $m(x;r,q)$ is defined by

$$m(x;r,q) = \sum_{j=0}^{2} x(\log x)^j \sum_{d|r} h_j(\mathrm{d},\mathrm{q}).$$

$E(x;r,q)$ contribute to S_r at most

$$|E(x;r,q)|\,|_{x=A_r,B_r} + \int_{A_r}^{B_r} (Trx^{-2} + \Delta^2 r^2 x^{-3})|E(x;r,q)|\,dx. \tag{41}$$

It follows from (31), (39) and (40) that (41)

$$\ll (qT)^{\frac{5}{6}+\varepsilon} + Tr^{\frac{5}{6}} \Delta^{-\frac{1}{6}} (qT)^{\varepsilon},$$

or, combined with (32), that (41)

$$\ll (qT)^{\frac{5}{6}+\varepsilon} + q^{\frac{1}{2}} Tr^{\frac{3}{4}} \Delta^{-\frac{1}{4}} (qT)^{\varepsilon}.$$

Hence we obtain

$$S_r = \int_{A_r}^{B_r} m'(x;r,q) \exp\left\{itrx^{-1} - \frac{1}{4}\Delta^2 r^2 x^{-2}\right\} dx$$
$$+ O\left((qT)^{\varepsilon}\left\{(qT)^{\frac{5}{6}} + \min\left(\mathrm{Tr}^{\frac{5}{6}} \Delta^{-\frac{1}{6}}, q^{\frac{1}{2}} Tr^{\frac{3}{4}} \Delta^{-\frac{1}{4}}\right)\right\}\right). \tag{42}$$

Further, trivially we have

$$m'(x;r,q)\left(\exp\left\{itrx^{-1} - \frac{1}{4}\Delta^2 r^2 x^{-2}\right\} - \exp\{itrx^{-1}\}\right)$$
$$\ll (qT)^{\varepsilon} \Delta^2 r^2 x^{-2},$$

and

$$\int_{X}^{2X} |(m(x;r,q)\left(\exp\left\{-\frac{1}{4}\Delta^2 r^2 x^{-2}\right\} - 1\right)|\,dx \ll (qT)^{\varepsilon} \Delta^2 r^2 X^{-2}.$$

Thus, on integrating by parts,

$$\int_{X}^{Y} e^{itrx^{-1}} m'(x;r,q)\left(\exp\left\{-\frac{1}{4}\Delta^2 r^2 x^{-2}\right\} - 1\right) dx$$

$$\ll (qT)^{\varepsilon}\Delta^2 r^2 X^{-2} \max_{X\leqslant Z\leqslant Y}\left|\int_X^Z e^{itrx^{-1}}\,dx\right|$$

for $X\leqslant Y\leqslant 2X$. By Titshmarsh [4, Lemma 4.3], we see that

$$\int_X^Z e^{itrx^{-1}}\,dx \ll X^2/(Tr).$$

Thus we have

$$\int_{A_r}^{B_r} m'(x;r,q)(\exp\left\{itrx^{-1}-\frac{1}{4}\Delta^2 r^2 x^{-2}\right\}dx$$

$$=\int_{A_r}^{B_r} e^{itrx^{-1}} m'(x;r,q)\,dx+O(T^{\varepsilon-1}\Delta^2 rq^{\varepsilon}). \tag{43}$$

From (40) we trivially have

$$\int_{B_r}^{\frac{qt}{2\pi}} m'(x;r,q)e^{itrx^{-1}}\,dx \ll r(qT)^{\varepsilon}.$$

Moreover, since $x^2(\log x)^j(tr)^{-1}$ is piecewise monotonic, we have by [4] Lemma 4.3,

$$\int_0^{A_r} m'(x;r,q)e^{itrx^{-1}}\,dx \ll r\Delta^2 T^{-1}(qT)^{\varepsilon}.$$

Hence (42) and (43) yield

$$S_r=\int_0^{\frac{qt}{2\pi}} e^{itrx^{-1}} m'(x;r,q)\,dx+E_0,$$

where

$$E_0\ll (qT)^{\varepsilon}\{(qT)^{\frac{5}{6}}+\min(Tr^{\frac{5}{6}}\Delta^{-\frac{1}{6}}, q^{\frac{1}{2}}Tr^{\frac{3}{4}}\Delta^{-\frac{1}{4}})+r\Delta^2 T^{-1},$$

which contributes to $f_0(t)$

$$\ll (qT)^{\varepsilon}\{(qT)^{\frac{5}{6}}+\min(T^{\frac{11}{6}}\Delta^{-1}q^{\frac{5}{6}}, T^{\frac{7}{4}}\Delta^{-1}q^{\frac{5}{4}})+q\Delta\},$$

We see from the definition of $m(x;r,q)$ that

$$\int_0^{\frac{qt}{2\pi}} e^{itrx^{-1}} m'(x;r,q)\,dx=\sum_{j=0}^{2}\sum_{d\mid r} h_j'(d,q)\int_0^{\frac{qt}{2\pi}} e^{itrx^{-1}}\log^j x\,dx.$$

We substitute $x=qt/(2\pi y)$ and define

$$c_k(r,q)=\int_1^{+\infty}(\log y)^k e(ry/q)\,dy/y^2.$$

Then

$$\int_0^{\frac{qt}{2\pi}} m'(t,r,q)e^{itrx^{-1}}\,dx=\sum_{j,k=0}^{2} qt\log^j(qt/2\pi)\sum_{d\mid r} h_{k,j}(d,q)c_k(r).$$

Therefore

$$f_0(t)=\sum_{j,k=0}^{2} qt(\log(qt/2\pi))^j\cdot 2\mathrm{Re}\left\{\sum_{r=1}^{R}(ir)^{-1}F(r)\sum_{d\mid r}h_{k,j}(d,q)c_k(r)\right\}.$$

$$+O((qT)^\varepsilon\{qT^2\Delta^{-2}+(qT)^{\frac{5}{6}}+\min(T^{\frac{11}{6}}\Delta^{-1}q^{\frac{5}{6}},T^{\frac{9}{4}}\Delta^{-1}q^{\frac{5}{4}})+q\Delta\}).$$

Since $c_k(r)\ll q/r$, we change the range of summation in r into $1\leqslant r<\infty$. Then the terms with $r>R$ contribute at most.

$$(qT)^{1+\varepsilon}\sum_{r>R} qr^{-2}\,|F(r)|\ll q\Delta(qT)^\varepsilon.$$

On taking $\Delta=q^{\frac{1}{8}}T^{\frac{7}{8}}$ and then $\Delta=T^{\frac{11}{12}}$ we obtain

$$f_0(t)=t\,Q_1(\log(qt/2\pi))+O((qT)^\varepsilon\min(q^{\frac{9}{8}}T^{\frac{7}{8}},qT^{\frac{11}{12}})),$$

where $Q_1(x)$ is a polynomial of x of degree 2, with coefficients all $\ll q^{1+\varepsilon}$.

Now we define

$$f(T_1,T_2)=\sum_{\substack{T_1<\tau(mn)\leqslant T_2\\ 0<|\log\frac{m}{n}|\leqslant T^{-\frac{1}{2}}\log T\\ (mn,q)=1}} d(m)d(n)(mn)^{-\frac{1}{2}}$$

$$F(m-n)\left(i\log\frac{m}{n}\right)^{-1}(m/n)^{i\tau}\exp\left\{-\frac{1}{4}\tau^2\left(\log\frac{m}{n}\right)^2/T\right\}.$$

We set $m=n+r$. By an argument similar to that in the treatment of $f_0(t)$ we see that

$$f(T_1,T_2)=2\mathrm{Re}\sum_{\substack{T=1\\(n(n+r),\,q)=1}}^{R}\sum_{n=A_r}^{B_r} d(n)d(n+r)F(r)h(n,r),$$

where

$$h(n,r)=n^{-1}(1+r/n)^{-\frac{1}{2}+i\tau}(i\log(1+r/n))^{-1}\exp\left\{-\frac{1}{4}\tau^2\log^2(1+r/n)/T\right\}$$

$$qT^{\frac{1}{2}}\log T \ll R \ll qT^{\frac{1}{2}}\log T, \tag{44}$$

$$A_r = qT_1/(2\pi) + O(r), \tag{45}$$

$$B_r = qT_2/(2\pi) + O(r), \tag{46}$$

Since $\tau = 2\pi (mn)^{\frac{1}{2}}/q = 2\pi (n(n+r))^{\frac{1}{2}}/q$ we have

$$i\tau \log(1+r/n) = (2\pi i/q)\cdot n(1+r/n)^{\frac{1}{2}}\log(1+r/n) = 2\pi ir/q + O(q^{-1}r^3n^{-2}).$$

Thus

$$h(n,r) = (ir)^{-1} e(r/q)\exp\{-\pi^2 r^2/(q^2T)\} + O(n^{-1}\log T).$$

Combined with (44)—(46), this yields

$$f(T_1,T_2) = 2\operatorname{Re}\sum_{r=1}^{R}(ir)^{-1}F(r)e(r/q)\exp\{-\pi^2r^2/(q^2T)\}$$

$$\sum_{\substack{qT_1<2\pi n\leqslant qT_2 \\ (n(n+r),\,q)=1}} d(n)d(n+r) + O(q^{1+\varepsilon}T^{\frac{1}{2}+\varepsilon}).$$

Using Lemma 6 we obtain

$$f(T_1,T_2) = \sum_{v=0}^{2}\left\{\frac{qT_2}{2\pi}\left(\log\frac{qT_2}{2\pi}\right)^v - \frac{qT_1}{2\pi}\left(\log\frac{qT_1}{2\pi}\right)^v\right\}S_v(q,T) + O(q^{1+\varepsilon}T^{\frac{1}{2}+\varepsilon}),$$

where we have defined

$$S_v(q,T) = \sum_{d\leqslant R} h_v(d,q)\cdot 2\operatorname{Re}\sum_{\substack{r=1 \\ r\equiv O \,(\mathrm{mod}\, d)}}^{R}(ir)^{-1}F(r)e(r/q)\exp\{-\pi^2r^2/(q^2T)\}.$$

Since $F(r) = F(q-r)$, we have, for all integers d.

$$\operatorname{Im}\left\{\sum_{m=1}^{q} e(dm/q)F(dm)\right\} = 0.$$

Further, by the periodic property of $F(r)$ we have

$$\sum_{m=1}^{q} e(dm/q)F(dm) = (d,q)\sum_{m=1}^{q/(d,q)} e(dm/q)F(dm),$$

which leads to the conclusion

$$\sum_{m=1}^{q/(d,q)} \sin(2\pi dm/q)F(dm) = 0.$$

Hence for $x \geqslant 0$, we have

$$|\sum_{m\leqslant x} F(dm)\sin(2\pi\, dm/q)| \leqslant \sum_{m=1}^{q/(d,q)} |F(dm)| \ll q^{1+\varepsilon}. \qquad (47)$$

We divide the sum $S_v(q, T)$ into two parts as

$$S_v(q,T)=2\sum_{dm\leqslant R} h_v(d,q)(dm)^{-1}F(dm)\sin(2\pi\, dm/q)\exp\{-\pi^2(dm)^2/(q^2T)\}$$

$$=2\sum_{dm\leqslant qT^{\frac{1}{3}}} + 2\sum_{qT^{\frac{1}{3}}<dm\leqslant R} = \sum\nolimits_1 + \sum\nolimits_2 \text{ say.}$$

For $dm \leqslant qT^{\frac{1}{3}}$,

$$\exp\{-\pi^2(dm)^2/(q^2\mathrm{T})\}=1+O(T^{-\frac{1}{3}}). \qquad (48)$$

On integrating by parts, (47) yields

$$\sum\nolimits_2=2\sum_{d\leqslant R} h_v(d,q)\sum_{qT^{\frac{1}{3}}<dm\leqslant R}(dm)^{-1}F(dm)\sin(2\pi\, dm/q)$$

$$\exp\{-\pi^2(dm)^2/(q^2T)\} \ll T^{-\frac{1}{3}}(qT)^{\varepsilon};$$

$$\sum_{m<dm\leqslant 2m} d^{-1}h_v(d,q)(dm)^{-1}F(dm)\sin(2\pi\, dm/q)$$

$$=\sum_{d<2m} d^{-1}h_v(d,q)\sum_{m<dm\leqslant 2m} m^{-1}F(dm)\sin(2\pi\, dm/q) \ll q^{1+\varepsilon}M^{-1}.$$

Hence from (48) we obtain

$$\sum\nolimits_1=2\sum_{r=1}^{\infty} r^{-1}F(r)\sin(2\pi\, r/q)\sum_{d\mid r} h_v(d,q)+O(T^{-\frac{1}{3}}(qT)^{\varepsilon}).$$

Thus it follows that

$$F(T_1,T_2)=tQ_2(\log(qt/(2\pi)))\Big|_{T_1}^{T_2} + O(qT^{\frac{2}{3}}(qT)^{\varepsilon}),$$

where Q_2 is a polynomial of x of degree 2 with coefficients all $\ll q^{1+\varepsilon}$. Finally, we define

$$g(t)=t\left\{\varphi_1(q)P\left(\log\frac{qt}{2\pi}\right)+Q_1\left(\log\frac{qt}{2\pi}\right)+Q_2\left(\log\frac{qt}{2\pi}\right)\right\}.$$

Then for both $\Delta=q^{\frac{1}{8}}T^{\frac{7}{8}}$ and $\Delta=T^{\frac{11}{12}}$ we have

$$f(t)=g(t)+O(q\Delta(qT)^{\varepsilon}).$$

This completes the proof of the Theorem.

References

[1] Heath-Brown, D. R.: The fourth power moment of the Riemann zeta-function, *Proc. London Math. Soc.*, (3)**38**(1979), 385—422.

[2] Ingham, A. E.: Mean value theorem in the theory of Riemann zeta-function, *Proc. London Math Soc.*, (2) **27**(1926), 273—300.

[3] Rane, V. V.: A note on the mean value of L-functions, *Proc Indian Acad. Sci. Math. Sci.*, (3)**90**(1981), 273—286.

[4] Titchmarch, E. C.: The theory of the Riemann zeta-function, Oxford, 1951.

FROM BAKER TO MORDELL

G. WÜSTHOLZ
ETH ZÜRICH
ETH-ZENTRUM
CH - 8092 ZÜRICH

1. LINEAR FORMS IN LOGARITHMS

The theory of linear forms in logarithms goes back to the seventh of Hilbert's famous problems which Hilbert stated in 1900. This problem was solved in 1934 independently by Gelfond and Schneider. They proved namely that for algebraic α, β with $\alpha \neq 0,1$ and β irrational the number $\gamma = \alpha^{\beta}$ is transcendental. This statement is equivalent to the following qualitative statement. Suppose that α, β, γ are algebraic numbers and that $\log \alpha$, $\log \gamma$ are defined and not zero. Then if $\Lambda = \beta \log \alpha - \log \gamma$ satisfies $\Lambda = 0$ we have $\beta \in \mathbb{Q}$.

The first quantitative result in that direction is due to Gelfond in 1935. He gives a lower bound for Λ in terms of the height of β, with an effective constant depending on the remaining quantities. Here the height of an algebraic number is defined in modern terms as

$$h(\beta) = \sum_{v} \max(0, \log|\beta|_v)$$

where v runs through all places of the field $\mathbb{Q}(\beta)$ and $|\ |_v$ is the normalized absolute value attached to v. In a sequence of papers Gelfond improved his first lower bound. However he could not succeed to go beyond the case of linear forms in two logarithms.

In 1966 A. Baker managed to overcome the difficulties by a new method of introducing several variables. He got qualitative as well as quantitative results which he improved in the time between 1966 and 1977 in several directions. Also he applied his results in the theory of diophantine equations, e.g. Thue's equation, elliptic and superelliptic equations. Furthermore he succeeded to give a definite solution of the class-number one problem and together with

Stark of the class-number two problem. In all these applications it turned out that the so-called rational case is the crucial one and we restrict ourselves now completely to this case.
We state now the final results of Baker [B] in this situation. For this let $\alpha_1,\ldots,\alpha_n$ be algebraic numbers not 0 or 1 generating a field K of degree at most d. If L is a linear form in the variables $z_1,\ldots,z_n$ with integer coefficients with absolute values at most B then the following holds.

Theorem 1. *If* $\Lambda = L(\log \alpha_1,\ldots,\log \alpha_n) \neq 0$ *then*

$$\log |\Lambda| \geq -C \log B \; \Omega \; \log \Omega'$$

where $\Omega = \bar{h}(\alpha_1) \ldots \bar{h}(\alpha_n)$, $\Omega' \bar{h}(\alpha_n) = \Omega$, *and* $\bar{h}(\alpha) = \max(1, h(\alpha), |\log \alpha|)$.

The constant C can be determined explicitly and Baker gives for it the value $C = (16nd)^{200n}$. Since Baker's work has appeared quite an effort was made to eliminate the additional factor $\log \Omega'$ which occurs very unnaturally and makes the result not symmetrical in the different $\log \alpha_i$ for instance. A careful analysis of Baker's proof shows that this false factor is due to the so-called Kummer descent developed by Baker in order to get his sharp result.

In the period of 1980-1983 a new technique was developed by D. W. Masser and the author first and pushed then further by the author in order to unify the second part, the so-called deconstruction, of a standard transcendence proof. Here one usually shows that a polynomial in several variables which has too many zeroes of a certain type compared with its degree must vanish identically. Such a type of result is called multiplicity estimates and we have proved a general multiplicity estimate on group varieties ([Wü1]). It turned out recently ([Wü3],[Wü4]) that this sort of result which also applies of course in Baker's situation is actually so sharp that Baker's Kummer descent could be replaced by it and thus one was able to eliminate this false factor.

Theorem 2 ([Wü5]). *If* $\Lambda \neq 0$ *then*

$$\log |\Lambda| \geq -C \log B \; \Omega .$$

In the paper quoted we do not give an explicit value for the constant C. In a forthcoming joint work with A. Baker we shall give an extremely precise value for this constant which is very crucial in many applications, especially numerical ones.

After Baker had published his results an obvious problem was to prove the analogue of Baker's

result in the p-adic case. The first result in that direction is due to Sprindzuk. The p-adic analogue of Theorem 1 was finally established by A. v.d. Poorten.
However v.d. Poorten's paper contains several fundamental errors. This was realized by Yu Kunrui. In his thesis [Y] Yu could overcome parts of the difficulties and prove the result stated by v.d. Poorten under a certain technical hypothesis, the so-called Kummer condition. Unfortunately he did not succeed to eliminate this hypothesis.

The author's new approach to the theory of linear forms in logarithms made it then possible to prove the unconditional result. On the other hand Yu recently was then also able to remove the difficulties involved in the Kummer condition and so we have now also in the p-adic case two different approaches available. In a forthcoming paper the author and Yu combine the two methods in the style of the joint work of Baker and the author mentioned above in order to prove a very sharp estimate especially useful in numerical applications. As a nice exotic application of this type of result we mention a recent paper by R. Riley where he applies the lower bounds to a certain problem in knot theory. There it leads to lower bonds for the order of $H_1(M_k,\mathbb{Z})$ where M_k is a k-sheeted cyclic cover of S^3 branched over a tame knot $K \subseteq S^3$.

We end this paragraph with stating Yu's p-adic result. Here we define Ω as $\Omega = h_1 \dots h_n$ where $h_i = \max(h(\alpha_i), \frac{|\log \alpha_i|}{2\pi d}, \log p)$ for $1 \le i \le n$. Suppose that $h_1 \le \dots \le h_n$ and put $\Omega' = h_{n-1}$ if $\mathrm{ord}_p b_n = \min \mathrm{ord}_p b_j$ and $\Omega' = h_n$ otherwise. Let $\wp$ be a prime ideal in K dividing p.

Theorem. *Suppose that* $\alpha_1^{b_1} \dots \alpha_n^{b_n} \neq 1$. *Then*

$$\mathrm{ord}_{\wp}(\alpha_1^{b_1} \dots \alpha_n^{b_n} - 1) < C(n,d,p)\, \Omega \log(4d^2 B) \log\left(2^{15} n^2 d^3 \Omega'\right)$$

where $C(n,d,p) = 19801\,(10(n+1)d)^{2(n+1)} p^{d+1}/(\log p)^{n+1}$.

2. The Analytic Subgroup Theorem

We mentioned in the first part of this report the theory of multiplicity estimates which provided the necessary tool for eliminating the false factor in Baker's result. We try to explain shortly the philosophy behind this theory. It is a theory adapted to group varieties G which in the case of linear forms are just the group varieties of the form $G = \mathbb{G}_a \times \mathbb{G}_m^n$ where $\mathbb{G}_a$ is the additive group and $\mathbb{G}_m$ the multiplicative group.

This theory gives a very precise description (best possible up to a constant) of the relation between degrees of hypersurfaces on G and the number of zeroes of a certain type on this hypersurface. In order to define the degree of a variety one embeds G into $\mathbb{P}^N$ for some N and here one has a degree theory by the Hilbert polynomial which counts the number of independent homogeneous polynomials on projective subvarieties of $\mathbb{P}^N$. So if one tries to construct a hypersurface on G of a given degree D which contains a number of given points with prescribed multiplicity then linear algebra applies if the value of the Hilbert polynomial at D is larger than the number of points counted with multiplicities, i.e. the total number of conditions. The main theorem [Wü1] in the theory gives, up to a constant, the converse, i.e. if the number of conditions is constant multiple larger than the value of the Hilbert polynomial at D then the conditions are not independent and, as a consequence, one obtains explicite relations. Instead of giving a precise statement of the result we give the main applications. This is a result on algebraic points on analytic subgroups of group varieties which covers all known transcendence results in this context and solves several conjectures due to Schneider, Lang, Baker, Waldschmidt, Grothendieck.

Let K be a number field, G a commutative connected algebraic group defined over K, $A \subseteq G(\mathbb{C})$ an analytic subgroup defined over K. By this we mean that the Lie algebra of A is defined as a Lie subalgebra of G by linear forms with coefficients in K. Then we have the following result

Theorem 3 ([Wü4]). *We have* $0 \neq A(\mathbb{Q}) = A \cap G(\mathbb{Q})$ *if and only if there exists an algebraic subgroup* H *of* G *such that*

(i) $\dim H > 0$,

(ii) H is defined over $\overline{\mathbb{Q}}$,

(iii) $H(\mathbb{C}) \subseteq A$.

Here by $G(F)$ respectively $H(F)$ for a field F we mean the set of F-rational points of G and H respectively.

As a consequence we obtain for instance the transcendence of the non-zero periods of rational 1-forms defined over $\overline{\mathbb{Q}}$ on quasiprojective varieties.

In order to demonstrate how to obtain transcendence results from this theorem we prove the transcendence of π. Suppose π is algebraic; hence so is $i\pi$. We put $G = \mathbb{G}_a \times \mathbb{G}_m$. Then $z \longrightarrow (z, e^z)$ defines an analytic subgroup A of G which, as the graph of a transcendental function e^z, is not algebraic nor contains an algebraic subgroup of positive dimension. Thus $0 = A(\mathbb{Q}) \ni (i\pi, -1) \neq 0$. This is a contradiction.

It is similarly easy to deduce the qualitative version of Baker's theorem from this result. That is, in the notation of Theorem 1 either $\Lambda = 0$ or Λ is transcendental.
Furthermore it is possible to prove a quantitative analog of Theorem 3 in the style of Theorem 1 or 2. A very special version of such a quantitative result leads to a transcendence proof of the famous Mordell's conjecture which was proved by Faltings [Fa] in 1983.

3. Mordell's Conjecture

Let K be as above an algebraic number field and X a projective curve defined over K which we may assume to be smooth. Then Mordell conjectured that $X(K)$, the set of K-rational points on X, is finite provided the genus g of X is at least 2. In genus 0 or 1 this is obviously false. Since more than 60 years many attempts were made in order to prove this conjecture before Faltings succeeded to prove it. As a consequence it follows for instance, that the famous Fermat equation

$$x^n + y^n = z^n$$

has only finitely many solutions in coprime integers x, y, z.

The starting point for a proof of Mordell's conjecture is a construction basically due to Kodaira and further developed by Parshin. It attaches to a curve X as above a family of curves $Y \longrightarrow X$. Via Torelli one obtains an embedding

$$\begin{array}{ccc} Y & \hookrightarrow & J(Y) \\ & \searrow \quad \swarrow & \\ & X & \end{array}$$

where $J(Y)$ is the associated family of jacobians. We have thus attached to X a family of abelian varieties. By construction these have good reduction outside a fixed finite set of finite places S. Furthermore by construction X maps to the moduli space of principally polarized abelian varieties and this map is finite. Hence we can regard X as some sort of "moduli space". A K-rational point on X can therefore be interpreted as an abelian variety of some dimension g' which is the genus of the generic curve of the family $Y \longrightarrow X$. The finiteness of $X(K)$ will therefore follow if one proves that there are at most finitely many isomorphism classes of principally polarized abelian varieties with good reduction outside of S. The strategy to prove this is first to prove that there are only finitely many isogeny classes of such objects and then to prove that there are only finitely many isomorphism classes within a fixed isogeny class. Faltings proves this by bounding the Faltings-heights by means of the galois representation associated to such an abelian variety and using heavily the Weil-conjectures.

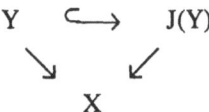

In a joint work D.W. Masser and the author take another approach to the problem. The approach is in some sense a special case of Theorem 3. Namely we prove the following theorem.

Theorem 4. *Let* A *be an abelian variety defined over* K *and* $\Phi : A \longrightarrow A^*$ *be an isogeny defined over* K. *Then there exists an isogeny* Φ_0 *between* A *and* A^* *which has degree bounded by a constant depending only on* A *and the degree of* K.

Remarks. 1) It follows immediately that the number of isomorphism classes in one isogeny class is finite.

2) The constant is effective.

3) If A is an elliptic curve with Weierstraß invariants g_2 and g_3 whose heights are bounded by log G then we can give the following bound:

$$\deg \Phi_0 \leq c\,(\log G)^4$$

where c is an effective constant depending only on d , the degree of the number field generated by g_2 and g_3.

4) One conjectures that the bound of deg Φ_0 should not depend on A , i.e. the exponent of log G should be 0 . This is proved by B. Mazur for the case $d = 1$.

5) Faltings in his paper proves a much weaker result than the one in Theorem 4. Namely he only proves that the difference of the Faltings heights of A and A^* is bounded. This suffices to prove the Mordell conjecture via moduli space theory.

From our Theorem 4 we deduce easily all the standard conjectures which are proved by Faltings. These are the Tate-Conjecture, the semisimplicity of the galois representation on the Tate module and the Shafarevich conjecture.

4. TATE'S CONJECTURE AND THE SEMISIMPLICITY OF THE TATE MODULE

Let again K be a number field and A an abelian variety of dimension n over K which carries a polarization Ψ, i.e. an isogeny $\Psi : A \longrightarrow \check{A}$ where $\check{A}$ is the dual abelian variety. Then we denote by π the Galois group $Gal(\overline{K}/K)$. As A is an algebraic group the multiplication by an integer N on A is defined and a homomorphism. This holds in particular for non-negative powers of a fixed prime number l. We denote the kernel of multiplication by l^m by ${}_{l^m}A$. Thus

$$ {}_{l^m}A := Ker(A \xrightarrow{l^m} A)(\overline{\mathbb{Q}}) $$

The Tate module of A is then defined as

$$ T_l(A) = \lim_{\leftarrow} {}_{l^m}A $$

Each ${}_{l^m}A$ is a π-module and so becomes $T_l(A)$. A homomorphism $\Phi : A^* \longrightarrow A$ induces a homomorphism $T_l(\Phi)$ of Galois modules $T_l(A^*), T_l(A)$ and thus, as can be easily shown, an injection

$$ Hom(A^*, A) \otimes_{\mathbb{Z}} \mathbb{Z}_l \longrightarrow Hom_\pi(T_l(A), T_l(A^*)). $$

Tate conjectured that this injection is an isomorphism and proved that it follows from the following hypothesis:

Hyp (K,A,d,l): *Given an abelian variety* A *of dimension* n *defined over a number field* K *, an integer* $d \geq 1$ *and a prime number* l *there exist only finitely many abelian varieties* A^* *over* K *such that*

(i) *there exists a polarization* ψ *of* A^* *of degree* d^2 *defined over* K,

(ii) *there exists an integer* $m \geq 1$ *and a* K*-isogeny* $\Phi : A^* \longrightarrow A$ *with* $\deg \Phi = l^m$.

Corollary 1 to Theorem 4. Hyp(K,A,d,l) *is true. In particular Tate's conjecture holds.*
Next we regard the Galois module $T_l(A)$. It is a module over $\mathbb{Z}_l[\pi]$ and one of the steps of Faltings' proof of Mordell's conjecture is the semisimplicity of the Galois module $T_l(A)$. Again this is a consequence of our Theorem 4.

Corollary 2 to Theorem 4. *The Galois module* $T_l(A)$ *is semisimple.*

LITERATURE.

[B] A. Baker, The theory of linear forms in logarithms, in *Transcendence Theory: Advances and Applications*, ed. by A. Baker and D.W. Masser, Academic Press, London (1977), 1-27.

[Fa] G. Faltings, Endlichkeitssätze für abelsche Varietäten über Zahlkörpern, Inventiones Math. **73** (1983), 349-366.

[Wü1] G. Wüstholz, Multiplicity estimates on group varieties, Ann. Math., to appear.

[Wü2] G. Wüstholz, A new approach to Baker's theorem on linear forms in logarithms I, SLN 1290, 1987, 189-202.

[Wü3] G. Wüstholz, A new approach to Baker's theorem on linear forms in logarithms II, SLN 1290, 1987, 203-211.

[Wü4] G. Wüstholz, Algebraische Punkte auf analytischen Untergruppen algebraischer Gruppen, Ann. Math., to appear.

[Wü5] G. Wüstholz, A new approach to Baker's theorem on linear forms in logarithms III, in *New Advances in Transcendence Theory*, ed. by A. Baker, Cambridge University Press, Cambridge (1988), 399-410.

[Y] K. Yu, Linear forms in the p-adic logarithms, Thesis, Bonn 1987.

On the Arithmetic Properties of G-Functions

Xu Guangshan

Institute of Mathematics, Academia Sinica

In his fundamental paper in 1929 Siegel[5] developed a method for studying the arithmetic properties of the values of certain classes of analytic functions known as E-and G-functions. He proved the algebraic independence of the values of certain E-functions at algebraic points, and also pointed out that his method could be use to investigate G-functions.

This suggestion of Siegel has been followed by Nurmagomedov, Galochkin, Flicker, Väänänen and Xu, respectively, but the results of these papers use the additional Galochkin's condition on G-functions. This restrictive condition is usually not trivial to verify. In an important paper of Bombieri[1] this condition is replaced by another condition, which is called "Fuchsian operator of arithmetic type".

Using very interesting new ideas, Chudnovsky[2] recently succeeded in considering the arithmetic properties of the values of classsical G-functions without any additional restrictions.

Our aim in the present paper is to give the refined proof of the main theorem in Bombieri's paper[1] without the hypothesis of "Fuchsian operator of arithmetic type", thereby we obtain some applications on the algebraic independence in the values of G-functions, in both the archimedian and the p-adic case, thus it gives a generalization of Chudnovsky[2] in p-adic case, moreover all the relative constants in our paper have been given explicitly. Our proof based on Siegel's method and on the ideas of Chudnovsky[2] regarding the use of Pade approximations of the second kind.

1. Definitions and Notations

Let K be an algebraic number field of degree d over Q, for every place v of K we write $d_v=[K_v:Q_v]$. If the finite place v of K lies over the prime p, we write $v \mid p$, for infinite place v of K, we write $v \mid \infty$. We normalize the absolute value $|\,|_v$ so that

(i) if $v \mid p$, then $|p|_v=p^{-d_v/d}$,

(ii) if $v \mid \infty$, then $|x|_v= |x|^{d_v/d'}$,

here $|\;|$ denotes the ordinary absolute value in R or in C.

The absolute height $h(x)$ of $x \in K$ is defined by the formula

$$h(x)=\prod_{v} \max (1, |x|_v), \tag{1}$$

and for any polynomial $P(z)=\sum_{i=1}^{n} p_i z^i \in K[z]$ we define

$$|P|_v = \max_i |P_i|_v$$

and

$$h(P)=\prod_{v} \max_{i} (1, |P|_v). \tag{2}$$

We write $\log^+ a = \log \max (1, a)$, for all $a \geqslant o$ and denote

$$\alpha_v = \begin{cases} 1, \text{if } v \mid p, \\ 0, \text{if } v \mid \infty, \end{cases} \qquad \beta_v = \begin{cases} 0, \text{if } v \mid p, \\ d_v/d, \text{if } v \mid \infty. \end{cases} \tag{3}$$

We then give the definition of G-function.

Let $y_j(z)=\sum_{m=0}^{\infty} a_{jm} z^m$, $j=1, \ldots, n$ be n Taylor series with coefficients $a_{jm} \in K$. For every place v of K, let $r_v(y_j, o)$ be the v-adic radius of convergence of $y_j(z)$, we denote

$$r_v = \min_j r_v(y_j, o).$$

We suppose that the vector $Y(z)=(y_1(z), \ldots, y_n(z))^t$ is not identically zero. We define the size $\sigma(Y)$ of the vector $Y(z)$ by means of

$$\sigma(Y)=\overline{\lim_{m\to\infty}} \frac{1}{m} \sum_{v} \max_{\substack{1\leqslant j\leqslant n \\ h\leqslant m}} \log |a_{jh}|_v \tag{4}$$

and call $Y(z)$ a G-function vector if $\sigma(Y)<\infty$. We suppose that

$$\sum_{v} \log^+ 1/r_v < \infty \tag{5}$$

and define

$$d(Y)=\overline{\lim_{m\to\infty}} \; 1/m \sum_{v} \max_{\substack{1\leqslant j\leqslant n \\ h<m}} (-h \log^+ 1/r_v + \log |a_{jh}|_v), \tag{6}$$

It is obvious that

$$\sigma(Y) \leqslant d(Y) + \sum_{v} \log^+ 1/r_v$$

and $d(Y) \geqslant 0$.

Let

$$g_1(z), \ldots, g_m(z)$$

denote the power-products

$$Y_1^{k_1}(z)\dots\ y_n^{K_n}(z),\qquad 0\leqslant k_1+\dots+k_n\leqslant N,\qquad m=\binom{n+N}{N},\tag{7}$$

where N is a natural number $\geqslant 3$. As in Lemma 12 of [1] we have

$$\sum_v \log^+ 1/r_v(G)\leqslant \sum_v \log^+ 1/r_v \tag{8}$$

and

$$d(G)\leqslant \left(\sum_{t=1}^{N} 1/t\right)d(Y)\leqslant (1+\log N)d(Y),\tag{9}$$

where $G(z)$ is the vector with components $g_1(z),\dots,\ g_m(z)$.

In the following we suppose that the G-functions $y_1(z),\dots,\ y_n(z)$ satisfy a system of linear differential equations

$$\frac{d}{dz}\ Y = AY+B,\tag{10}$$

where $A=(A_{ij}(z))_{n\times n}$, $B=(b_1(z),\dots,b_n(z))'$, $A_{ij}(z)$ and $b_i(z)\in K(z)$. Let $T(z)\in K[z]$ denote the common denominator of all $A_{ij}(z)$ and $b_i(z)$ and put

$$S=\max\ (\deg\ T(z),\ \max\ (\deg\ T(z)A_{ij}(z),\ \deg\ T(z)b_i(z))).$$

Let $y_0(z)=1$, then $y_0(z),\ y_1(z),\dots,y_n(z)$ satisfy the system of homogeneous linear differential equations

$$\frac{d}{dz}\ Y_1 = A_1 Y_1,\tag{11}$$

where $A_1=\begin{pmatrix}0&0\\B&A\end{pmatrix}$, $Y_1=(y_0(z),\dots,\ y_n(z))'$.

2. Padé Approximation of the Second Kind

In this section we follow the paper of Chudnovsky[2] to give the definition of Padé approximation of the second kind and some lemmas.

Let D and M be natural numbers and let $Q(z)$ be a non-zero polynomial of degree $\leqslant D$ with the coefficients belonging to K. Then for every $i=1,\dots,n$, there exists a unique polynomial

$$P_i(z)=[\ Q\ \cdot\ Y_i\ ]_D$$

of degree $\leqslant\ D$ such that

$$\operatorname{ord}_{z=0}(\ Q(z)y_i(z)-P_i(z)\ )\geqslant D+1,$$

if we now have

$$ord_{z=0}\,(Q(z)\,y_i(z)-P_i(z))\geqslant D+M+1\ ,$$

for every $i=1,\dots,n$, then the system of polynomials $(Q(z), P_1(z),\dots, P_n(z))$ is called a system of Padé approximations of the second kind for G-functions $y_0(z),\dots, y_n(z)$ with parameters (D, M). According to Dirichlet's box principle the system $(Q(z), P_1(z),\dots, P_n(z))$ with parameters (D, M) exists whenever $D \geqslant nM$.

We suppose here that G-functions $y_0(z), y_1(z),\dots, y_n(z)$ are linearly independent over $K(z)$ and satisfy (11).

Lemma 1. (Theorem 1.1 of Chudnovsky [2]) *Let $(Q(z), P_1(z),\dots, P_n(z))$ be a system of Padé approximations of the second kind with parameters (D, M) for G-functions $y_0(z),\dots, y_n(z)$ as the above definition. Let k be a natural number with $M \geqslant k(s+1)$. We define*

$$Q^{<k>}(z) = T^k(z)\left(\frac{d}{dz}\right)^k Q(z)/k!\,,$$

$$P^{<k>}(z) = [Q^{<k>} \cdot Y_i]_{D+ks},\ i=1,\dots,n.$$

Then $(Q^{<k>}(z), P_1^{<k>}(z),\dots, P_n^{<k>}(z))$ is a system of Padé approximations of the second kind with parameters $(D+ks, M-k(s+1))$ for G-functions $y_0(z),\dots, y_n(z)$.

Lemma 2. (Theorem 1.2 of Chudnovsky [2]) *Let δ, $0<\delta<1/n+n^2(s+1)$, be given and $M=[(1/n-\delta)D]$. Let $\theta \in K$ with $\theta\, T(\theta) \neq 0$. There exists a positive constant C_0 depending only on δ, n, the system (11) and G-functions $y_1(z),\dots, y_n(z)$ such that, for all $D > C_0$ there exist integers $0 \leqslant k_0 \leqslant k_1 < \dots < k_n \leqslant J = D - nM + n(n+1)(s+1)/2$, satisfying*

$$\nabla(\theta) = \begin{vmatrix} Q^{<k_0>}(\theta)\, P_1^{<k_0>}(\theta) & \cdots & P_n^{<k_0>}(\theta) \\ \cdots & & \cdots \\ Q^{<k_n>}(\theta)\, P_1^{<k_n>}(\theta) & \cdots & P_n^{<k_n>}(\theta) \end{vmatrix} \neq 0$$

3. A Method of Siegel and Main Theorem

We suppose here that $y_0(z), y_1(z),\dots, y_n(z)$ are *G-functions with parameters* $d(Y) < \infty$ and $\sum_v \log^+ 1/r_v < \infty$.

Lemma 3. *Let δ, $0<\delta<1/n$, be given and let $M=[(1/n-\delta)D]$, for any D there exists a system $(Q(z), P_1(z),\dots, P_n(z))$ of Padé approximations of the second kind with parameters (D, M) for G-functions $y_0(z),\dots, y_n(z)$ such that $Q(z), P_i(z) \in K[z]$ $(i=1,\dots,n)$ and*

$$\log h(Q) \leqslant (1-1/n-\delta)((\delta n)^{-1}-1)D\,(a(Y) + \sum_v \log^+ 1/r_v)$$

$$+(\delta n)^{-1}(\log\Gamma+\log 2(D+1)), \tag{12}$$

where the constant Γ *depends only on* K.

Proof. Let $Q(z)=\sum_{m=0}^{D} q_m z^m$, by the hypothesis of the lemma and the definition of Padé approximation of the second kind, the unknowns q_k must satisfy the system of linear equations

$$\sum_{k=0}^{D} q_k a_{m-k,i}=0,\quad m=D+1,\dots,D+M;\ i=1,\dots,n.$$

The number of equations is nM and the number of unknowns is $D+1$. By Siegel lemma in [1] we can find a non-trivial solutions $q_k \in K$ satisfying

$$\log h(Q)\leqslant(1-1/n-\delta)((\delta n)^{-1}-1)\mathrm{D}\,\sigma(Y)$$
$$+(\delta n)^{-1}(\log\Gamma+\log 2(D+1)),$$

obviously also $P_i(z)=[Q\cdot Y_i]_D\in K(z)$. Thus Lemma 3 is proved.

Lemma 4. *Let* $(Q(z),\ P_1(z),\dots,P_n(z))$ *be the system constructed in Lemma 3. If* $M=[(1/n-\delta)D]\geqslant k(s+1)$ *we have the following estimates*

$$|Q^{<k>}|_v\leqslant c(k,D)^{\beta_v}|T|_v^k|Q|_v,$$
$$|P_i^{<k>}|_v\leqslant c(k,D)^{2\beta_v}|T|_v^k|Q|_v\max_{\substack{1\leqslant i\leqslant n\\ 1\leqslant D+ks}}|a_{li}|_v,\ (1\leqslant i\leqslant n), \tag{13}$$

where $c(k,D)=(s+1)^k(D+1)2^D$.

Proof. We denote $T(z)=\sum_{i=0}^{s} t_i z^i$, then we have

$$|Q^{<k>}(z)|_v=|(\sum_{i=0}^{s} t_i z^i)^k|_v\,|\sum_{j=k}^{D}\binom{j}{k}q_j z^{j-k}|_v.$$

If $v\mid\infty$, we have

$$|Q^{<k>}|_v\leqslant((s+1)^k(D+1)2^D)|Q|_v\,|T|_v^k.$$

On the other hand,

$$P_i^{<k>}(z)=\sum_{j=0}^{\mathrm{D+ks}}(\sum_{m=0}^{j} q_m^{<k>}a_{j-m,i})z^j,$$

thus we obtain

$$|P_i^{<k>}|_v\leqslant(D+ks+1)^{\beta_v}\max|q_m^{<k>}|_v\max_{\substack{1\leqslant i\leqslant n\\ 1\leqslant D+ks}}|a_{li}|_v$$

$$\leqslant (c(k,D)^2)^{\beta_v} |Q|_v |T|_v^k \max_{\substack{1\leqslant i\leqslant n \\ 1\leqslant D+ks}} |a_{li}|_v, 1\leqslant i\leqslant n.$$

If $v \mid p$, we have

$$|Q^{<k>}|_v \leqslant |Q|_v \ |T|_v^k,$$

$$|P_i^{<k>}|_v \leqslant |Q|_v \ |T|_v^k \max_{\substack{1\leqslant i\leqslant n \\ 1\leqslant D+ks}} |a_{li}|_v, \ 1\leqslant i\leqslant n,$$

thus we proved Lemma 4.

We next use Lemma 2 to define rational functions $L_{tj}=L_{tj}(\theta)$, as the solutions of the system of linear equations

$$\sum_{t=0}^{n} L_{tj} Q_t^{<ki>}(\theta)=\delta_{ij}, \ i,j=0,1,\ldots,n,$$

where $Q_0^{<ki>}(\theta)=Q^{<ki>}(\theta)$, $Q_t^{<ki>}(\theta)=p_t^{<ki>}(\theta)$, $t=1,\ldots,n, i=0,1,\ldots,n$.

By Cramer's rule,

$$L_{tj}(\theta)=R_{tj}(\theta)/\nabla(\theta), \ t,j=0,1,\ldots,n,$$

where $R_{tj}(\theta)$ is t, j-cofactor of the matrix corresponding to $\nabla(\theta)$. We now define linear forms $F_j(\theta)$ in $y_0(\theta)$, $y_1(\theta),\ldots,y_n(\theta)$ by the formulae

$$F_j(\theta)=\sum_{t=0}^{m} M_{tj}(\theta)y_t(\theta), \quad j=0,1,\ldots,n,$$

where $M_{tj}(\theta)=R_{tj}(\theta)\theta^{-w}$, $\omega=(n-1)(M+D-J)$.

Using Theorem 4.1 of Chudnovsky [2] we immediately obtain the following important result

Lemma 5. *Let $D>C_0$ and $M\geqslant J(k+1)$. Then the linear forms $F_0(\theta),\ldots,F_n(\theta)$ are linearly independent and have polynomial coefficients $M_{tj}=M_{tj}(\theta)$ satisfying*

$$\deg_\theta M_{tj}\leqslant D-(n-1)M+J(ns+n-1), \quad t,j=0,1,\ldots,n.$$

Further we have

$$\operatorname{ord}_{\theta=0} F_j(\theta)\geqslant D+M-J, \quad j=0,1,\ldots,n.$$

Lemma 6.

$$|M_{tj}|_v \leqslant (n!\,c(J,D)^{3n})^{\beta_v} |Q|_v^n \ |T|_v^{nJ} \ (\max_{\substack{1\leqslant i\leqslant n \\ 1\leqslant D+Js}} |a_{li}|_v)^\sigma,$$

where σ is n or $n-1$.

Proof. The result follows immediately from Lemma 4 and the definition of the polynomials M_{tj}.

Lemma 7. *Let* $\delta, 0<\delta<1/3n^2(s+1)$, *be given, assume that*

$$D > \max\left(\delta^{-1}(1+(n+1)(s+1)/2),\ n/1-3\delta n^2(s+1),\ C_0\right). \quad (14)$$

If $\theta \in K$, *then*

$$|M_{tj}(\theta)|_v \leqslant (D(1/n+\delta(n+2n^2(s+1)))+1)^{\beta_v} |M_{tj}|_v \cdot \max(1, |\theta|_V^{D(1/n+\delta(n+2n^2(s+1))}).$$

Further, if $|\theta|_v < r_v$, *for any* $\varepsilon>0$, *we have*

$$\begin{aligned}\log|F_j(\theta)|_v \leqslant\ & \beta_v \log(n+1)(D(1/n+\delta(n+2n^2(s+1)))+1)\\ &+ (1+1/n-3\delta n)D\log|\theta|_v/r_v + \max_{t,j}\log|M_{tj}|_v\\ &+ D(1/n+\delta(n+2n^2(s+1)))\log^+ r_v + L_1(Y)+\varepsilon,\end{aligned}$$

where the positive constant $L_1(Y)$ *depends only on* $y_0(z), \ldots, y_n(z)$.

Proof. It follows from the hypothesis (14) that the hypotheses of Lemmas 2, 5 and 6 are valid, and $J\leqslant 2\delta nD$. In addition, we obviously have

$$R=D-(n-1)M+J(ns+n-1)\leqslant D(1/n+\delta(n+2n^2(s+1)))$$

and

$$N=D+M-J\geqslant(1+1/n-3\delta n)D.$$

The estimate for $|M_{tj}(\theta)|_v$ now follows from Lemma 5.

By Lemma 5, $\operatorname{ord}_{\theta=0} F_j(\theta)\geqslant N$. Since $z^{-N}F_j(z)$ is regular in $|z|_v< r_v$, by the maximum principle we have for $|\theta|_v< r < r_v$,

$$\begin{aligned}|\theta^{-N}F_j(\theta)|_v &\leqslant \sup_{|z|_v=r}|z^{-N}F_j(z)|_v \leqslant r^{-N}\sup_{|z|_v=r}|F_j(z)|_v,\\ &\leqslant r^{-N}\max_{t,j}|M_{tj}|_v\ \max(1,r^R)((n+1)(R+1))^{\beta_v}\max_t\sup_{|z|_v=r}|y_t(z)|_v,\end{aligned}$$

by choosing r arbitrarily close to r_v, we have

$$|F_j(\theta)|_v \leqslant \left(\frac{|\theta|_v}{r_v}\right)^N |M_{tj}|_v e^{R\log^+ r_v}((n+1)(R+1))^{\beta_v}L(Y)+\varepsilon,$$

where $L(Y)$ depends only on G-functions $y_0(z), \ldots, y_n(z)$.

Thus Lemma 7 is proved.

We shall next prove the main theorem.

Let

$$\sum_{j=0}^{n}\lambda_{ij}y_j(\theta)=0,\quad i=1,\ldots,\rho$$

be ρ linearly independent relations over K, relative to a finite set s of place v *of* K, for which

$$|\theta|_v < r_v,\ \text{all}\ v \in s.$$

By Lemma 5 we have

$$\operatorname{rank}\begin{pmatrix}\lambda\\ \mathrm{M}\end{pmatrix} = \operatorname{rank}\begin{bmatrix}\lambda_{10} & . & . & . & \lambda_{1n}\\ \vdots & & & & \vdots\\ \lambda_{\rho 0} & . & . & . & \lambda_{\rho n}\\ M_{t_1 0}(\theta) & . & . & . & M_{t_1 n}(\theta)\\ \vdots & & & & \vdots\\ M_{t_{n+1-\rho},0}(\theta) & . & . & . & M_{t_{n+1-\rho},n}(\theta)\end{bmatrix} = n+1.$$

Let us consider the linear system of equations

$$\sum_{j=0}^{n} \lambda_{ij}\, y_j(\theta) = 0, \quad i = 1, \ldots, \rho,$$

$$\sum_{j=0}^{n} M_{t_1 j}(\theta)\, y_j(\theta) = F_{t_1}(\theta), \quad i = 1, \ldots, n+1-\rho,$$

the determinant $\Delta(\theta)$ of this linear system is in K and is not zero.
The above system of equations holds v-adically for all $v \in s$. By Cramer's rule we find

$$\Delta(\theta) = \sum_{i=1}^{n+1-\rho} \operatorname{cofactor}_{\rho+i,\,1}\begin{pmatrix}\lambda\\ M\end{pmatrix} F_{t_i}(\theta)$$

where $\operatorname{cofactor}_{\rho+i,\,1}\begin{pmatrix}\lambda\\ M\end{pmatrix}$ means the cofactor relative to the $\rho+i$-th row and 1-th column of the matrix $\begin{pmatrix}\lambda\\ M\end{pmatrix}$. Thus we have

$$\log|\Delta(\theta)|_v \leqslant (n-\rho)\max_{t,j}\log|M_{tj}(\theta)|_v + \max_{1 \leqslant 1 \leqslant n+1-\rho}\log|F_{t_i}(\theta)|_v + L_2(\lambda, n, \rho),$$

where the constnat L_2 depends only on λ, n, ρ. We shall next denote by $L_3(a, b, c, \ldots)$, $L_4(a, b, c, \ldots)$, ... the constants depending only on $a, b, c, \ldots$. By Lemma 7, we have

$$\sum_{v \in s}\log|\Delta(\theta)|_v \leqslant (n-\rho)\sum_{v \in s}\max_{t,j}\log|M_{tj}(\theta)|_v$$

$$+ \sum_{v \in s}\beta_v \log(n+1)(D(1/n+\delta(n+2n^2(s+1)))+1)$$

$$+(1+1/n-3\delta n)D\sum_{v \in s}\log\frac{|\theta|_v}{r_v} + \sum_{v \in s}\max_{t,j}\log|M_{tj}|_v$$

$$+D(1/n+\delta(n+2n^2(s+1)))\sum_{v\in s}\log^+ r_v+L_3(Y(\theta),\lambda,n,\rho,s,\varepsilon).\quad(15)$$

On the other hand

$$\sum_{v\in s}\log|\Delta(\theta)|_v=-\sum_{v\notin s}\log|\Delta(\theta)|_v$$

$$\geqslant -(n+1-\rho)\sum_{v\notin s}\max_{t,j}\log|M_{tj}(\theta)|_v-L_4(\lambda,n,\rho,s).\quad(16)$$

By (15), (16), Lemmas 7 and 6, we have

$$(n+1-\rho)\sum_{v\notin s}\max_{t,j}\log|M_{tj}(\theta)|_v+(n-\rho)\sum_{v\notin s}\max_{t,j}\log|M_{tj}(\theta)|_v$$

$$+D\sum_{v\in s}\left\{(1+1/n-3\delta n)\log\frac{|\theta|_v}{r_v}+(1/n+\delta(n+2n^2(s+1)))\log^+ r_v\right\}$$

$$+\sum_{v\in s}\beta_v\log(n+1)(D(1/n+\delta(n+2n^2(s+1)))+1)+\sum_{v\in s}\max_{t,j}\log|M_{tj}|_v$$

$$\geqslant -L_5(Y(\theta),\lambda,n,\rho,s,\varepsilon),$$

and

$$(n+1-\rho)\Big\{\log(n!\,c(J,D)^{3n})+n\log h(Q)+nJ\log h(T)$$

$$+n(D+Js)(d(Y)+\sum_v\log^+ 1/r_v)+D(1/n+\delta(n+2n^2(s+1)))\log h(\theta)$$

$$+\log(D(1/n+\delta(n+2n^2(s+1)))+1)+\log(m+1)(D(1/n$$

$$+\delta(n+2n^2(s+1)))+1)\Big\}+D(\sum_{v\in s}(1+1/n-3\delta n)\log\frac{|\theta|_v}{r_v}$$

$$+(1/n+\delta(n+2n^2(s+1)))\log^+ r_v$$

$$\geqslant -L_6(Y(\theta),\lambda,n,\rho,s,\varepsilon).$$

We use Lemma 3 and divide both sides of this inequality by D. Finally, we let first $D\to\infty$ and then $\varepsilon\to 0$. This yields

$$(n+1-\rho)\Big\{6n^2\log(s+1)+3n\log 2+2\delta n^2\log h(T)$$

$$+((n+1-\delta n)(1/\delta n-1)+n(1+2\delta ns))(d(Y)+\sum_v\log^+ 1/r_v)$$

$$+(1/n+\delta(n+2n^2(s+1)))\log h(\theta)\Big\}$$

$$+(1+1/n-3\delta n)\sum_{v\in s}\log\frac{|\theta|_v}{r_v}+(1/n+\delta(n+2n^2(s+1)))\log^+ r_v\geqslant 0.$$

We state the result as a theorem.

Theorem 1. (The main theorem) *Let* $y_1(z), \ldots, y_n(z)$ *be G-functions satisfying the system* (10), *with parameters* $d(Y)$ *and* $\sum_v \log^+ 1/r_v$. *Let* $\theta \in K$ *with* $\theta\, T(\theta) \neq 0$ *and let s be a finite set of place v of K such that*

$$|\theta|_v < \min(1, r_v), \qquad \text{if } v \in s$$

and suth that we have ρ *linearly independent relations over K of type*

$$\sum_{j=0}^{n} \lambda_{ij}\, y_j(\theta) = 0, \qquad i = 1, \ldots, \rho, \tag{17}$$

with $\lambda_{ij} \in K$ *and valid in every completion* K_v, $v \in s$.

We further assume that the *G*-functions $y_0(z), \ldots, y_n(z)$ are linearly independent over $K(z)$.

Then for every δ, $0 < \delta < 1/4n + 3n^2(s+1)$, we have

$$(n+1-\rho)(1+\delta(n^2+3n^3(s+1))) \log h(\theta) + (n+1-3\delta n^2) \sum_{v \in s} \log |\theta|_v$$
$$\geqslant -C_1, \tag{18}$$

where

$$C_1 = n(n+1-\rho)\{6\delta n^2 \log(s+1) + 3n \log 2 + 2\delta n^2 \log h(T)$$
$$+ ((n+1-\delta n)((\delta n)^{-1} - 1) + (n(1+2\delta n s)))\,(d(Y) + \sum_v \log^+ 1/r_v)\}$$
$$+ (n+1-3\delta n^2) \sum_v \log^+ 1/r_v \;. \tag{19}$$

4. Applications of the Main Theorem

In this section, we apply the main theorem to obtain that the linear relations (17) can occur only at points θ whose height is bounded in terms of the system (10), of the solutions $y_1(z), \ldots, y_n(z)$, and to deal with algebraic function and the algebraic independence of *G*-functions $y_1(z), \ldots, y_n(z)$ at algebraic points.

Theorem 2. *Under the hypotheses of Theorem* 1, *we have*

$$\log h(\theta) \leqslant L_7(Y(z), \;\; T(z), \; n\,). \tag{20}$$

Proof. Since $|\theta|_v < \min(1, r_v)$, we have

$$\sum_{v \in s} \log |\theta|_v \leqslant -\log h(\theta) + \sum_v \log^+ 1/r_v$$

(see p. 51 in [1]). Using Theorem 1 we have

$$((n+1-\rho)(1+\delta(n^2+3n^3(s+1))-(n+1-3\delta n^2))\log h(\theta)$$

$$\geqslant -C_1-(n+1-3\delta n^2)\sum_v \log^+ 1/r_v \ ,$$

$$(\delta(3n^2+n^2(n+1)+3n^3(n+1)(s+1))-\rho)\log h(\theta)$$

$$\geqslant -C_1-(n+1-3\delta n^2)\sum_v \log^+ 1/r_v \ .$$

Theorem 2 follows by choosing sufficiently small, for example,

$$\delta=1/2\,(3n^2+n^2(n+1)+3n^3(n+1)(s+1)) \ ,$$

and we have

$$L_7=2n(n+1)\{1+3n\log 2+\log h(T)$$
$$+2(n+1)(3n+(n+1)^2+3n^2(n+1)(s+1))\,(d(Y)+\sum_v \log^+ 1/r_v)\}$$
$$+2(n+1)\sum_v \log^+ 1/r_v \ .$$

Analogously to Corollary of Theorem 2 of [1] we have

Corollary 1. *The set of algebraic points θ of bounded degree at which there is some linear relation is a finite set.*

Corollary 2. *Suppose that s consists of only one place and m is a natural number, there exists an effectively computable constant C depending only on $Y(z)$, $T(z)$ and n such that*

(i) if $v \mid p$ and $P^m > C$, then we have that $1, y_1(P^m), \ldots, y_n(P^m)$ are linearly independent over Qp.

(ii) if $v \mid \infty$ and $m > C$, then we have that $1, y_1(1/m), \ldots, y_n(1/m)$ are linearly independent over Q.

We now consider algebraic function, let $u(z)=\sum_{i=0}^{\infty} a_i z^i$ be an algebraic function of one variable regular in a neighborhood of $z=0$, such that we have a polynomial equation

$$P(z, u)=0 \tag{21}$$

with the coefficients in Q. Let $Q(z, u)$ be a function field, if $P(z, u)$ is irreducible over $Q(z)[u]$ and $\deg_u P(z, u)=n$, then $[Q(z, u): Q(z)]=n$. Since

$$\frac{d}{dz}u=-\frac{P_z'(z, u)}{P_u'(z, u)},$$

then the function field $Q(z, u)$ is mapped into itself by the derivation $D=\frac{d}{dz}$ and it is a vector space over $Q(z)$ with basis $1, u, \ldots, u^{n-1}$, it follows that

$$DU = SU, \tag{22}$$

where all the entries $S_{ij}(z)$ of the matrix S belong to the rational function field $Q(z)$ and the vector $U = (1, u, \ldots, u^{n-1})^t$.

Let $K = Q((a_i)_{i \geqslant 0})$, $a_i \in \overline{Q}$ and $[K : \overline{Q}] = d \leqslant \deg_u P(z, u)$. The same equation (21) is satisfied, if we replace u by a conjugate over $K(z)$, now a well-known theorem of Eisenstein asserts that algebraic function $u(z)$ is G-function and $\sum_{2v} \log^+ 1/r_v < \infty$ (see p. 60 in [1]).

Theorem 3. *Let* $\delta = 1/3\,((n-1)^{2v} + 3(n-1)^3(s+1))$, *and let* $\theta \in Q$ *with* $\theta \neq 0$, *let* P_1 *be a divisor of* $P(\theta, u)$ *and irreducible over* $Q[u]$. *Denote by* $s(\theta, P_1)$ *a set of place* v *of* K, *such that*

$$|\theta|_v < \min(1, r_v) \text{ and } P_1(u, (\theta)) = 0.$$

Then we have

$$(4/3(n-1))\,(\deg P_1) \log h(\theta) + \sum_{v \in s(\theta, P_1)} \log |\theta|_v \geqslant -\overline{C}_1,$$

where

$$C_1 = n\left\{1 + 3(n-1)\log 2 + \log h(T) + 3n(n + 3(n-1)^2(s+1))\right.$$
$$\left. \cdot \left(d(Y) + \sum_v \log^+ 1/r_v\right)\right\} + n/n-1 \sum_v \log^+ 1/r_v .$$

Proof. Let $v \in s(\theta, P_1)$, by the irreduction of P_1 we know that $Q(u_v(\theta))$ is a vector space of dimension $\deg P_1$. Then exist $n - \deg P_1$ linearly independent relations for numbers $1, u_v(\theta), \ldots, u_v(\theta)^{n-1}$ over Q of type

$$\sum_{j=1}^{n} \lambda_{ij} u_v(\theta)^{j-1} = 0, \quad i = 1, \cdots, n - \deg P_1 .$$

It follows from Theorem 1 by choosing $\delta = 1/3((n-1)^2 + 3(n-1)^3(s+1))$ that

$$(4/3(n-1))\,(\deg P_1) \log n(\theta) + \sum_{v \in s(\theta, P_1)} \log |\theta|_v \geqslant -\overline{C}_1 .$$

Thus Theorem 3 is proved.

We now consider the algebraic independence of G-functions at algebraic points.

Since the G-functions $y_1(z), \ldots, y_n(z)$ satisfy the system of differential equations (10), the G-functions $g_1(z), \ldots, g_m(z)$ defined by (7) satisfy the system of differential equations

$$\frac{d}{dz} G = WG, \tag{23}$$

where all the entries of matrix W are the linear combinations of $A_{ij}(z)$ and $b_i(z)$ in (10), thus the common denominator of all entries in matrix also W is $T(z)$.

Theorem 4. *Suppose that the G-functions $y_1(z), \ldots, y_n(z)$ are algebraically independent over $K(z)$. There exists a constant C_2 depending only on the system (10) and $Y(z)$, for any place v of K and $\theta \in K$ satisfying*

$$\theta T(\theta) \neq 0, \log h(\theta) \geqslant ((n!/2)^4 + \max(3, \lambda)^{4n}) \log\log h(\theta) \tag{24}$$

and

$$|\theta|_v < \exp(-C_2 \lambda (\log h(\theta))^{(4n-1)/4n} (\log\log h(\theta))^{1/4n}).$$

Then there cannot be any algebraic relation of degree λ among the elements $y_1(\theta), \ldots, y_n(\theta)$ of K_v.

Proof. Suppose that the numbers $y_1(\theta), \ldots, y_n(\theta)$ satisfy a nontrivial algebraic relation $P(y_1(\theta), \ldots, y_n(\theta)) = 0$ of degree λ, $P(z_1, \ldots, z_n) \not\equiv 0 \in K[z_1, \ldots, z_n]$. Let N be natural number with $N \geqslant \lambda$, we consider the power products $g_1(z), \ldots, g_n(z)$ defined by (7) satisfying the system of differential equations (23) and having the properties (8) and (9). Let $t = \binom{n+N-\lambda}{n}$ and $w = m - t$. As in [6], we obtain the estimates

$$N^n/n! \leqslant m \leqslant \gamma_1 N^n, \quad t < m, \quad w \leqslant \gamma_2 \lambda N^{n-1} \tag{25}$$

with positive constants γ_1 and γ_2 depending only on n.

Multiplying $P(y_1(\theta), \ldots, y_n(\theta)) = 0$ by $y_1^{k1}(\theta) \ldots y_n^{kn}(\theta)$, $0 \leqslant k_1 + k_n \leqslant N - \lambda$ we shall have t non-trivial linearly independent forms in $g_1(\theta), \ldots, g_m(\theta)$ with coefficients from K.

We apply Theorem 1 and must replace $n+1$ by m, $d(Y)$ by $(1+\log N)d(Y)$, ρ by t and δ by $1/3((m-1)^2 + 3(m-1)^3(s+1))$, we obtain that

$$4/3\, w \log h(\theta) + (m - 1/1 + 3(m-1)(s+1)) \log |\theta|_v$$
$$\geqslant -m(m-1)^3 w (1+\log N) A,$$

where $A = (12(s+1)+3)(d(Y) + \sum_v \log^+ 1/r_v) + (1 + 3\log 2 + \log h(T))$. We choose

$$N = \left[\left(\frac{\log h(\theta)}{\log\log h(\theta)} \right)^{1/4n} \right].$$

Then we have $N \geqslant \lambda$ and $N^n \geqslant n!/2$ by (24), it follows from (25) that

$$\log |\theta|_v \geqslant -wm(m-1)^2 \ (1+\log N)A - 4/3\, w\, (m-1)^{-1} \log h(\theta)$$
$$\geqslant -\gamma_2 \lambda N^{(n-1)} \gamma_1^3 N^{3n} (1+\log N) A - 2\gamma_2 N^{-1} n!\, \lambda \log h(\theta).$$

Then

$$\log |\theta|_v \geqslant -\gamma_1^3 \gamma_2 \lambda A/2n (\log h(\theta))^{1-1/4n} (\log\log h(\theta))^{1/4n}$$
$$-4\gamma_2 n!\lambda (\log h(\theta))^{1-1/4n} (\log\log h(\theta))^{1/4n},$$
$$\geqslant -C_2 \lambda (\log h(\theta))^{1-1/4n} (\log\log h(\theta))^{1/4n},$$

where $C_2 = \gamma_1^3 \gamma_2 A/2n + 4\gamma_2 n!$.

Thus we complete the proof of Theorem 4.

Theorem 5. *Suppose that the G-functions $y_1(z), \dots, y_n(z)$ are algebraically dependent over $C(z)$ and satisfy the system (10). Let $y_1(z), \dots, y_l(z)$ $(1 \leqslant l \leqslant n-1)$ be algebraically independent over $C(z)$ as well as C. Let $P(z_1, \dots, z_n)$ be any polynomial of degree λ, $P \not\equiv 0$ and $P \in K[z_1, \dots, z_n]$. There exists a constant C_3 depending only on the system (10) and the G-functions $y_1(z), \dots, y_n(z)$, for any place v of K and $\theta \in K$, $\theta T(\theta) \neq 0$ satisfying*

$$\log h(\theta) \geqslant (1 + \max(3, \lambda + N_0)^{4l}) \log\log h(\theta) \tag{26}$$

and

$$|\theta|_v < \exp(-C_3 \lambda (\log h(\theta))^{1-1/4l} (\log\log h(\theta))^{1/4l}),$$

where the positive constant N_0 depends only on the G-functions $y_1(z), \dots, y_n(z)$. Then the numbers $y_1(\theta), \dots, y_n(\theta)$ of K_v have the property

$$P(y_1(\theta), \dots, y_n(\theta)) \neq 0,$$

provided that $P(y_1(z), \dots, y_n(z)) \neq 0$.

Proof. Let

$$N = \left[\left(\frac{\log h(\theta)}{\log\log h(\theta)} \right)^{1/4l} \right],$$

we have from (26) that $N \geqslant \lambda$ and $N \geqslant N_0$. Let L_N be a linear space generated by power-product functions (7) and let

$$g_1(z), \dots, g_r(z) \tag{27}$$

be a basis of the linear space L_N over $C(z)$ as well as C, $r = \varphi(N) < m$, where $\varphi(N)$ is some positive increasing function of N.

We denote the elements of L_N, not contained in the basis (27), by

$$g_{r+1}(z), \dots, g_m(z), \tag{28}$$

any of the functions (28) may be represented in a unique way as a linear combination of the functions (27) with constant coefficients, which by Lemma 10 in [3] may be chosen from K. We have

$$g_i(z)-\sum_{j=1}^{r}\alpha_{ij}\,g_j(z)=0,\quad i=r+1,\ldots,m, \tag{29}$$

one can easily verify that the basis $g_1(z),\ldots,g_r(z)$ satisfies a system of homogeneous linear differential equations

$$\frac{d}{dz}\,g_i(z)=\sum_{j=1}^{r}Q_{ij}(z)\,g_j(z),\quad i=1,\ldots,r,$$

where the coefficients $Q_{ij}(z)$ are some linear combinations of $A_{ij}(z)$ and $b_i(z)$ in (10) with the coefficients α_{ij}. Thus the common denominator of $Q_{ij}(z)$ is also $T(z)$.

Denote a basis of $L_{N-\lambda}$ by $f_1(z),\ldots,f_t(z)$, $t=\varphi(N-\lambda)$, consider the formulae

$$f_j(z)\,P(y_1(z),\ldots,y_n(z))=\sum_{J=1}^{m}\beta_{ji}\,g_i(z),\quad j=1,\ldots,t, \tag{30}$$

we shall prove that the linear forms (29) and (30) of the variables $g_1(z),\ldots,g_m(z)$ with the coefficients from K are linearly independent over K.

Suppose the contrary. Then there are not all zero $a_1,\ldots a_t,\ a_{r+1},\ldots,a_m$ from K such that the equalities

$$\sum_{j=1}^{t}a_j f_j\,P(y_1(z),\ldots,y_n(z))=\sum_{j=1}^{t}a_j\sum_{i=1}^{m}\beta_{ji}\,g_i(z)$$

$$=\sum_{j=r+1}^{m}a_j\,(g_j(z)-\sum_{i=1}^{r}\alpha_{ji}\,g_i(z))$$

hold. As the left-hand side of equations (29) are linearly independent over K regarded as linear forms in the variables $g_1(z),\ldots,g_m(z)$, hence at least one of the $a_1,\ldots,a_t$ is different from zero, this implies that

$$\sum_{j=1}^{t}a_j f_j(z)\,P(y_1(z),\ldots,y_n(z))=0$$

and

$$\sum_{j=1}^{t}a_j f_j(z)=0 \tag{31}$$

by the assumption $P(y_1(z),\ldots,y_n(z))\neq 0$, the equation (31) is imposible, because the functions $f_1(z),\ldots,f_t(z)$ are linearly independent over C. Replacing the functions $g_{r+1}(z),\ldots,g_m(z)$, by the linear combinations of the functions $g_1(z),\ldots,g_r(z)$ in equations (30), we have

$$f_j(z)\,P(y_1(z),\ldots,y_n(z))=\sum_{i=1}^{r}\gamma_{ji}\,g_i(z),$$

$$j=1,\cdots,t,\ \gamma_{ji}\in K\,. \tag{32}$$

Suppose that $P(y_1(\theta), \ldots, y_n(\theta))=0$. Then there exist t nontrivial linearly independent forms in $g_1(\theta), \ldots, g_r(\theta)$ with coefficients from K such that

$$\sum_{i=1}^{r} \gamma_{ji}\, g_i(\theta)=0, \quad j=1, \ldots, t. \tag{33}$$

We use Theorem 1 and must replace $n+1$ by r, $d(Y)$ by $(1+\log N)\, d(Y)$, ρ by t and δ by $1/3\,((r-1)^2+3(r-1)^3(s+1))$. It has been proved for $N \geqslant N_0$ that $r=\varphi(N)$ is polynomial of N and the inequalities

$$\gamma_3 N^l \leqslant r \leqslant \gamma_4 N^l, \quad t < r, \quad w=r-t \leqslant \gamma_5 \lambda N^{l-1}$$

hold, where the positive constants γ_3, γ_4 and γ_5 depend only on the G-functions $y_1(z), \ldots, y_n(z)$ (see *e. g.* [4]).

Analogously to the proof of Theorem 4 we have

$$\log |\theta|_v \geqslant -C_3 \lambda (\log h(\theta))^{1-1/4l} (\log\log h(\theta))^{1/4l},$$

where

$$C_3=\gamma_4^3 \gamma_5 A/2l + 4\gamma_5 \gamma_3^{-1}$$

and

$$A=(3+12\,(s+1))\,(d(Y)+\sum_v \log^+ 1/r_v)+(1+3\log 2+\log h(T)).$$

Thus we complete the proof of Theorem 5.

References

[1] E. Bombieri, On *G-functions*, *Recent Progress in Analytic Number Theory*, Vol. II, H. Halberstam and C. Hooley (eds.), Academic Press, 1981.

[2] G. V. Chudnovsky, On applications of diophantine approximations, *Proc. Natl. Acad. Sci. USA*, Vol. **8**(1984), 7261—7265.

[3] A. B. Shidlovsky, The arithmetic properties of the values of analytic functions, *Trudy Mat. Inst. Steklov*, **132** (1973), 169—205.

[4] A. B. Shidlovsky, Estimates for the moduli of polynomials with algebraic coefficients at the values of E-functions, *Studies in Pure Math.*, 635—657, Birkhäuser, Basel-Boston, Mass., 1983.

[5] C. L. Siegel, Über eininge Anwendungen diophantischer approximationen, Abh, Preuss. Akad. Wiss., Phys. - mat. Kl. No. **1**(1929).

[6] K. Väänänen, On a class of G-functions, Mathematics, University of Oulu 1/81(1981).

Uniform Distribution of Values of Multiplicative Functions

Zhang Wenbin

Department of Mathematics. South China University of Technology Guangzhou, P. R. China

1. Introduction

Let $n \to \varphi(n)$ be a positive valued arithmetic function which tends to infinity as $n \to \infty$. Following [5], we shall say that the values of φ are uniformly distributed in $(0, \infty)$ (briefly, φ is u.d. in $(0, \infty)$) if there exists a positive constant c such that

$$N(x) = N(x; \varphi) := \# \{n \cdot \varphi(n) \leqslant x\} \sim cx$$

as $x \to \infty$. The number c will be called the density of values of φ. Also, we shall say that the values of φ are distributed with zero (respectively infinite) density in $(0, \infty)$ if $N(x)/x$ tends to zero (respectively infinity) as $x \to \infty$.

It is known (*e. g.* [1], [6], [7]) that Euler's phifunction is u.d. in $(0, \infty)$. In [5], Diamond and Erdős gave quite general conditions for uniform distribution in $(0, \infty)$ of values of a multiplicative function (cf. [10]). For certain classes of functions, their conditions are both necessary and sufficient. They also characterized the cases in which the values of a multiplicative function are distributed with zero density or infinite density.

We can show, however, that for more general classes of multiplicative functions the conditions of Diamond and Erdős are still both necessary and sufficient. We can also show more general cases with zero density. For comparison, we shall give examples which our theorems cover but those in [5] do not.

2. Statement of Results

To simplify the statement of results, we will not repeat the hypothesis in theorems and corollaries that φ denotes a positive valued multiplicative function which tends to infinity as $n \to \infty$. For the same reason, we shall say that a function $G(x)$ with support in $[1, \infty)$ satisfies the condition (A) if there exists a nondecreasing function $A(x)$ such that $|G(x)| \leqslant A(x)$ and

$$\int_1^\infty A(x)x^{-2}\,dx < \infty .$$

For φ we define

$$\pi(x)=\pi(x;\varphi)=\#\{\,p:\varphi(p)\leqslant 1 \text{ and } p\leqslant x \text{ or } 1<\varphi(p)\leqslant x\,\}$$

and

$$\Pi(x)=\pi(x)+\frac{1}{2}\pi(x^{1/2})+\frac{1}{3}\pi(x^{1/3})+\dots$$

which are the respective analogue of

$$\pi_0(x)=\#\{\,p:p\leqslant x\,\}$$

and

$$\Pi_0(x)=\pi_0(x)+\frac{1}{2}\pi_0(x^{1/2})+\frac{1}{3}\pi_0(x^{1/3})+\dots$$

of classical prime number theory. Since $\Pi(x)$ has support in $[1,\infty)$, it will be granted that ir a decomposition $\Pi(x)=\Pi_1(x)+\Pi_2(x)$, functions $\Pi_1(x)$ and $\Pi_2(x)$ also have support in $[1,\infty)$.

We shall use multiplicative convolution notations which have been summarized in [4]. The convolution techniques have been described in detail in [2].

Theorem 1. *Suppose that*

$$\sum_{p,\,k\geqslant 2}\varphi(p^k)^{-1}<\infty \tag{1}$$

and that there exists a decomposition $\Pi=\Pi_1+\Pi_2$ *satisfying the following conditions:*

(i) $\Pi_1(x)\uparrow$ *and* $\Pi_1(x)\sim x/\log x$ *as* $x\to\infty$;

(ii) $\int_{1-}^{x}\exp d\Pi_2$ *satisfies the condition* (A);

and

(iii) $\int_{1-}^{\infty}x^{-1}\exp d\Pi_2(x)\neq 0.$

Then φ *is u.d. in* $(0,\infty)$ *if and only if*

$$F(\sigma):=\sum_{n\geqslant 1}\varphi(n)^{-\sigma}\sim c/(\sigma-1) \tag{2}$$

holds for some positive constant c as $\sigma\to 1+$. *In this case, the density of values of* φ *is*

$$c=\lim_{\sigma\to 1+}\prod_p\{(1-p^{-\sigma})\}\sum_{k\geqslant 0}\varphi(p^k)^{-\sigma}\}. \tag{3}$$

Remark. Condition (ii) implies the convergence of $\int_{1-}^{\infty} x^{-1} \exp d\Pi_2(x)$ as we showed in [11].

Corollary 1. (*Diamond-Erdös* [5]) *suppose that* (1) *holds and that*

$$\varphi(p) \sim p \text{ as } p \to \infty.$$

Then φ *is u.d. in* $(0, \infty)$ *if and only if* (2) *holds. Moreover,* (3) *is true.*

Corollary 2. (Diamond-Erdös [5]) *Suppose that the series*

$$\sum_p (\varphi(p)^{-1} - p^{-1})$$

converges and for each $\varepsilon > 0$ *the series*

$$\sum_{|\varphi(p)-p| > \varepsilon p} |\varphi(p)^{-1} - p^{-1}|$$

converges too. If (1) *holds then* φ *is u.d. in* $(0, \infty)$ *and has density*

$$c = \prod_p \{(1-p^{-1}) \sum_{k \geqslant 0} \varphi(p^k)^{-1}\}.$$

The following two theorems involve complex analysis techniques

Theorem 2. *Suppose that* (1) *holds and that there exists a decomposition* $\Pi = \Pi_1 + \Pi_2$ *satisfying the following conditions*:

(i) $\Pi_1(x) \uparrow$ *and* $\Pi_1(x) \ll x/\log x$;

(ii) $\int_{1-}^{x} \exp d\Pi_2$ *satisfies the condition* (A);

and

(iii) $\int_{1-}^{\infty} x^{-1-it} \exp d\Pi_2(x) \neq 0$ *for all* $t \in (-\infty, \infty)$.

Then φ *is u.d. in* $(0, \infty)$ *if and only if*

$$F(s) := \sum_{n \geqslant 1} \varphi(n)^{-s} = \frac{c}{s-1} + o\left(\frac{1}{\sigma - 1}\right) \tag{4}$$

with some positive constant c holds as $\sigma = \operatorname{Re} s \to 1+$ *uniformly for* $-T \leqslant t \leqslant T$ *for each fixed* $T > 0$.

Remark. Condition (ii) implies the uniform convergence of $\int_{1-}^{\infty} x^{-1-it} \exp d\Pi_2(x)$ for all. $t \in (-\infty, \infty)$. (cf. [11]).

Corollary 3. (Diamond-Erdös [5]) *Suppose that* (1) *holds and that*

$$\sum_{\varphi(p)\leqslant x} 1 \leqslant x \leqslant \log x \ .$$

Then φ *is u.d. in* $(0, \infty)$ *if and only if* (4) *holds as* $\sigma = \mathrm{Re}\ s \to 1+$ *uniformly for* $-T \leqslant t \leqslant T$ *for each fixed* $T > 0$.

Condition (4) is the analogue of the hypothesis of Halász's theorem on mean values of multiplicative functions [8]. By analogy with Halász's theory, (4) can be replaced by a condition on the generating function on the vertical line $\sigma = 1$.

Theorem 3. *Suppose that* (1) *holds and that there exists a decomposition* $\Pi = \Pi_1 + \Pi_2$ *satisfying the following conditions*:

(i) $\Pi_1(x)\uparrow$ *and* $x/\log x \ll \Pi_1(x) \ll x/\log x$;

(ii) $\int_{1-}^{x} \exp d\Pi_2$ *satisfies the condition* (A) ;

and

(iii) $\int_{1-}^{\infty} x^{-1-it}\, d\Pi_2(x)$ *converges for all* $t \in (-\infty, \infty)$.

Then φ *is u. d. in* $(0, \infty)$ *if and only if*

$$\int_{1-}^{y} x^{-1}\, d\Pi_0(x) - \mathrm{Re} \int_{1-}^{y} x^{-1-it}\, d\Pi(x) \tag{5}$$

converges to a finite number for $t = 0$ *and diverges to* $+\infty$ *for each real* $t \neq 0$ *as* $y \to \infty$.

Corollary 4. (Diamond-Erdös [5]) *Suppose that* (1) *holds and that*

$$1 \ll \varphi(p)/p \ll 1 \ .$$

Then φ *is u. d. in* $(0, \infty)$ *if and only if*

$$\sum_{p} (p^{-1} - \mathrm{Re}\ \varphi(p)^{-1-it})$$

converges to a finite number for $t = 0$ *and diverges to* $+\infty$ *for each real* $t \neq 0$.

The following theorem characterizes the case in which the values of a multiplicative function are distributed with zero density.

Theorem 4. *Suppose that* (1) *holds and that there exists a decomposition* $\Pi = \Pi_1 + \Pi_2$ *satisfying the following conditions*:

(i) $\Pi_1(x)\uparrow$ *and* $\Pi_1(x) \ll x/\log x$;

(ii) $\int_{1-}^{x} \exp d\Pi_2$ *satisfies the condition* (A);

and

(iii) $\int_{1-}^{\infty} x^{-1} \exp d\Pi_2 (x) \neq 0$.

Then the values of φ *have zero density in* $(0, \infty)$ *if and only if*

$$F(\sigma):=\sum_{n\geqslant 1} \varphi(n)^{-\sigma}=\mathrm{o}\left(\frac{1}{\sigma-1}\right) \tag{6}$$

holds as $\sigma \to 1+$.

Corollary 5. (Diamond-Erdős [5]) *Suppose that*

$$\sum_{\varphi(p)\leqslant x} 1 \ll x/\log x.$$

Then the values of φ *have zero density in* $(0\ \infty)$ *if and only if* (6) *holds as* $\sigma \to 1+$.

3. Method of Proofs

Our method of proofs follows the general idea of Diamond and Erdős in [5] but with the general idea in [11] as complement. Instead of using perturbation in [5], we make frequent use of convolution techniques.

Given φ satisfying the condition (1). We define a positive valued completely multiplicative function φ_0 by setting

$$\varphi_0(p)=\begin{cases}\varphi(p), & \text{if } \varphi(p)>1.\\ p, & \text{if } \varphi(p)\leqslant 1.\end{cases}$$

Then we have

$$\pi(x)=\pi(x;\varphi)=\pi(x;\varphi_0)=\#\{p:\varphi_0(p)\leqslant x\}.$$

Let

$$N_0(x):=\sum_{\varphi_0(n)\leqslant x} 1.$$

Then

$$dN=dN_0 * dH \tag{7}$$

where

$$H(x) = \sum_{\substack{n, m \\ \varphi_0(n)\varphi(m) \leqslant x}} \mu(n).$$

The condition (1) and the convergence of $\sum_{n\geqslant 1} \varphi(n)^{-\sigma}$ for $\sigma>1$ guarantee the convergence of $\int_{\eta-}^{\infty} x^{-1} \mid dH \mid (x)$ and non-vanishing of $\int_{\eta^-}^{\infty} x^{-1}\, dH(x)$ where η is a positive number such that $H(x)$ has support in $[\eta, \infty)$. The latter two facts imply, from (7), that φ_0 inherits from φ the property (2) in Theorem 1 or the property (4) in Theorem 2, etc., and, by an integral version of Axer's theorem [9], that if

$$N_0(x) = c_1 x - o(x) \tag{8}$$

with a nonnegative constant c_1 holds as $x \to \infty$ then

$$N(x) = cx + o(x)$$

with $c = c_1 \int_{\eta-}^{\infty} x^{-1}\, dH(x)$ holds as $x \to \infty$.

To show (8), we consider

$$dN_0 = dN_1 * \exp d\Pi_2$$

where $dN_1 = \exp d\Pi_1$. Again, the conditions (ii) and (iii) guarantee that dN_1 inherits from dN_0 the property (2) in Theorem 1 or the property (4) in Theorem 2, etc., and that if

$$N_1(x) = c_2 x + o(x) \tag{9}$$

with a nonnegative constant c_2 holds as $x \to \infty$ then (8) with $c_1 = c_2 \int_{1-}^{\infty} x^{-1} \exp d\Pi_2(x)$ holds as $x \to \infty$.

To prove (9), we follow the general idea of Diamond and Erdös in [5]. In this process, we have to translate their arguments into integral version and make frequently minor modifications since dN_1 must be treated in integration. Then the Corollaries can be deduced as a special case of the corresponding Theorems with particular choice of $\Pi_1(x)$ *and* $\Pi_2(x)$ in the decomposition $\Pi = \Pi_1 + \Pi_2$.

4. Examples

Example 1. We recast an example of Diamond [3]. Define $\tau(x)$ by setting

$$\tau(x) = \int_1^x \{1 - \cos(\log t)\} \; (\log t)^{-1}\, dt$$

for $x \geqslant 1$ and $\tau(x)=0$ for $x<1$. Let P_n be the n-th rational prime. We define a completely multiplicative function φ by setting $\varphi(p_n)=\tau^{-1}(n)$ (*i.e.*, $\varphi(P_n)$ is the n-th g-prime of Diamond). Now Theorem 1 applies to this example but Corollaries 1 and 2 do not.

Example 2. Define a sequence $a_k=e^{k^3}$, $k=0,1,2,\dots$ Denote $l_k=k\pi_0(a_k)$ and p_l the l-th rational prime. Then we define a completely multiplicative function φ by taking

$$\varphi(p_l) \in (a_k, a_k+l] \text{ if } l_0+\dots+l_{k-1}<l \leqslant l_0+\dots+l_{k-1}+l_k$$

such that $\varphi(P_l)$ are all distinct and different from rational primes. It is easy to show that Theorem 4 covers this example but Corollary 5 does not.

Example 3. Let φ be the completely multiplicative function defined in Example 2. Let p_n be the n-th rational prime and set

$$f(x)=\pi_0(x)+\sum_{\varphi(p)\leqslant x} 1.$$

We define a completely multiplicative function ψ by setting

$$\psi(p_n)=\min\{x: f(x) \geqslant n\}.$$

Then Theorems 2 and 3 apply to this example but Corollaries 3 and 4 do not.

References

[1] P. T. Bateman, The distribution of values of the Euler function, *Acta Arith.* **21**(1972), 329—345.

[2] H. G. Diamond, Asymptotic distribution of Beurling's generalized integers, *Illinois J. Math.* 14, No. 1 (1970), 12—28, MR 40 #5555.

[3] H. G. Diamond, A set of generalized numbers showing Beurling's theorem to be sharp, *Illinois J. Math.* 14, No. 1 (1970), 29—34.

[4] H. G. Diamond, When do Beurling generalized integers have a density? *J. Reine Angew. Math.* **295**(1977), 22—39, MR 56 #8518.

[5] H. G. Diamond and P. Erdős, Multiplicative functions whose values are uniformly distributed in (0, ∞), *Proc. of the Queen's Number Theory Conference*, 1979 (Kingston, Ont., 1979), 329—378. Queen's papers in Pure and Appl. Math., 54, Queen's Univ., Kingston, Ont., 1980 MR 83e #10063.

[6] R. Dressler, A density which counts multiplicity, *Pacific J. Math.* **34**(1970), 371—378.

[7] P. Erdős, Some remarks on Euler's φ function and some related problems, *Bull. Amer. Math. Soc.* **51** (1945), 540—544.

[8] G. Halász, Über die Mittewerte multiplikativer zahlentheoretischer Funktionen, *Acta Math. Acad. Sci. Hung.*, Vol. **19**(1968), pp. 365—403.

[9] G. Hardy, *Divergent Series*, Oxford Univ. Press (Clarendon), London/New York, 1963.

[10] K. Wooldridge, Mean value theorems for arithmetic functions similar to Euler's phi-function, *Proc. Amer. Math. Soc.* **58**(1976), 73—78.

[11] W.-B. Zhang, Density and O-density of Beurling generalized integers, *J. Number Theory*, (1988).

L. K. Hua, Y. Wang, Beijing

Applications of Number Theory to Numerical Analysis

1981. IX, 241 pp. Hardcover DM 128,- ISBN 3-540-10382-1

Contents: Algebraic Number Fields and Rational Approximation. - Recurrence Relations and Rational Approximation. - Uniform Distribution. - Estimation of Discrepancy. - Uniform Distribution and Numerical Integration. - Periodic Functions. - Numerical Integration of Periodic Functions. - Numerical Error for Quadrature Formula. - Interpolation. - Approximate Solution of Integral Equations and Differential Equations. - Appendix: Tables. - Bibliography.

The publication of this volume marks the beginning of a cooperative venture between Springer-Verlag and the Chinese publisher Science Press to make important results of Chinese mathematicians available to the international mathematical community. Number theoretic methods are used in numerical analysis to construct a series of uniformly distributed sets in the s-dimensional unit cube $G(s \geq 2)$; these sets are then used to calculate an approximation of a definite integral over G_s with the best possible order of error, significantly improving existing methods of approximation. The methods can also be used to construct an approximating polynomial for a periodic function of s variables and in the numerical solution of some integral equations and PDEs.
Many important methods and results in number theory, especially those concerning the estimation of trigonometrical sums and simultaneous Diophantine approximations as well as those of classical algebraic number theory may be used to construct the uniformly distributed sequence in G_s. This monograph, by authors who have contributed significantly to the field, describes methods using a set of independent units of the cyclotomic field and by using the recurrence formula defined by a Pisot-Vijayaraghavan number. Error estimates and applications to numerical analysis are given; the appendix contains a table of *glp* (good lattice point) sets.
The volume is accessible to readers with a knowledge of elementary number theory.

Jointly published by Springer-Verlag and Science Press, Beijing

Springer-Verlag
Berlin
Heidelberg
New York
London
Paris
Tokyo
Hong Kong
Barcelona

L. K. Hua, Beijing

Introduction to Number Theory

Translated from the Chinese by P. Shiu

1982. XVIII, 572 pp. 14 figs. Hardcover DM 138,–
ISBN 3-540-10818-1

Contents: The Factorization of Integers. – Congruences. – Quadratic Residues. – Properties of Polynomials. – The Distribution of Prime Numbers. – Arithmetic Functions. – Trigonometric Sums and Characters. – On Several Arithmetic Problems Associated with the Elliptic Modular Function. – The Prime Number Theorem. – Continued Fractions and Approximation Methods. – Indeterminate Equations. – Binary Quadratic Forms. – Unimodular Transformations. – Integer Matrices and Their Applications. – *p*-adic Numbers. – Introduction to Algebraic Number Theory. – Waring's Problem and the Problem of Prouhet and Tarry. – Schnirelmann Density. – The Geometry of Numbers. – Bibliography. – Index.

This is the English edition of the well-known Chinese original first published in 1957. Apart from giving a broad introduction to number theory and some of its fundamental principles the author emphasises the close relationship between number theory and mathematics as a whole. The book will soon prove itself a worthy successor of the classical book by G. H. Hardy and E. M. Wright "An Introduction to the Theory of Numbers". Various recent results in number theory are presented in such a form as to make this a textbook suitable for teaching purposes.

Springer-Verlag
Berlin
Heidelberg
New York
London
Paris
Tokyo
Hong Kong
Barcelona